Arbind Prasad, Manoj Kumar Sinha, J. Paulo Davim (Eds.)

High-Performance Metallic Biomaterials

Advanced Mechanical Engineering

Edited by
J. Paulo Davim

Volume 15

High-Performance Metallic Biomaterials

Surface Modification and Coating of Implants

Edited by
Arbind Prasad, Manoj Kumar Sinha and J. Paulo Davim

DE GRUYTER

Editors
Dr. Arbind Prasad
Department of Science, Technology and Technical Education
Katihar Engineering College
Government of Bihar
Hirday Gunj, Katihar 854109
Bihar
India
arbind.iitg@gmail.com

Dr. Manoj Kumar Sinha
Department of Mechanical Engineering
National Institute of Technology
136119 Kurukshetra
Haryana
India
mksinha@nitkkr.ac.in

Prof. Dr. J. Paulo Davim
Department of Mechanical Engineering
University of Aveiro
Campus Santiago
3810-193 Aveiro
Portugal
pdavim@ua.pt

ISBN 978-3-11-157137-9
e-ISBN (PDF) 978-3-11-157142-3
e-ISBN (EPUB) 978-3-11-157212-3
ISSN 2367-3796

Library of Congress Control Number: 2025935607

Bibliographic information published by the Deutsche Nationalbibliothek
The Deutsche Nationalbibliothek lists this publication in the Deutsche Nationalbibliografie; detailed bibliographic data are available on the internet at http://dnb.dnb.de.

Cover image: Sky_blue/iStock/thinkstock
Typesetting: Integra Software Services Pvt. Ltd.
Printing and binding: CPI books GmbH, Leck

www.degruyterbrill.com
Questions about General Product Safety Regulation:
productsafety@degruyterbrill.com

MIX
Papier | Fördert gute Waldnutzung
FSC® C083411

Preface

Metallic biomaterials are mainly used for replacing failed hard tissue. The main metallic biomaterials are stainless steels, Co-based alloys, and titanium and its alloys. Recently, titanium alloys have been getting much attention for biomaterials. There are several important challenges associated with the use of implantable metals. For instance, the control over biodegradation kinetics is critical for resorbable metals, such as Mg or Fe, where untimely degradation may lead to premature loss of mechanical strength before the tissue or function has been fully restored. On the other hand, the long-term presence of metals such as steel, cobalt-chromium alloys, or Titanium alloys in the body is associated with an increased risk of development of cutaneous and systemic hypersensitivity reactions, whereas a relatively high modulus of these metals compared to natural bone tissue leads to stress shielding and consequent osteopenia. High-performance metallic materials play a vital role in this scenario to avoid conventional metallic materials problems. This book will also deal with Surface modification of the implants and its impact, laser surface modifications, electrochemical investigation, corrosion stability, and invitro studies of the high-performance metallic biomaterials. This book will consist of 15 chapters. Chapter 1 deals with the introduction phase, processing, and characterization of the metallic materials related to high performance. This chapter also deals with the classification of high-performance metallic materials. Chapter 2 discuss the Chapter 2 introduces design strategy of metallic biomaterials for biomedical applications. Chapter 3 discusses about the stainless implants used in orthopedic fracture fixations. Chapter 4, chapter 5 and 6 investigates regarding Mg based alloys ,Titanium alloys, and its coating towards biomedical applications. As surface topographies and nanocoating have proven enhanced biocompatibility. The impact of multifunctional coating on the metallic materials also leads to increase the efficacy of the metallic biomaterials, which has been discussed in chapter 7 to 9. The chapter 10 comprises of emerging applications, laser employment towards surface modification. the techniques involved as emerging technology such as additive manufacturing of medical implants. Customized implants can be fabricated by this method. Additive manufacturing in the manufacturing sector plays significant role. It not only reduce manpower but the cost as well time. The chapter 11 discuss the laser surface modification of the implants fabricated through metals. The chapter 12 investigate related to electrochemical aspect and corrosion behaviors of the metallic implants. Chapters 13 to 15 deal with invitro studies, design strategies, and current challenges faced by stakeholders and opportunities in the coming days. It will provide in-depth knowledge to readers for easier, faster, efficient, and reliable ways to understand orthopedic implants, surface modification, topographies, presently implemented in research labs and industries. This book has the potential to reach a wide readership, including academics, practitioners, Machinists, mechanical Engineers, postgraduate students, and researchers working in the

https://doi.org/10.1515/9783111571423-202

high-performance metallic biomedical fields. Authors believe that precise and exiting contents of the chapters of the book will open up some of bright minds and will encourage them to take up some of the field for their future research.

Editors
Dr. Arbind Prasad
Dr. Manoj Kumar Sinha
Prof. (Dr.) J. Paulo Davim

Acknowledgement

We express our gratitude to **De Gruyter publishing** and the editorial team for their suggestions and support during completion of this book. We are grateful to all contributors and reviewers for their illuminating views on each book chapter presented in this book **"High-Performance Metallic Biomaterials: Surface Modification and Coating of Implants".**

"This book is dedicated to all engineers, doctors researchers, health professionals and academicians. . ."

https://doi.org/10.1515/9783111571423-203

Contents

Preface —— V

Acknowledgement —— VII

Editors short bio —— XI

List of contributing authors —— XIII

Kamal Kishore, Vivek Singh, Manoj Kumar Sinha and Arbind Prasad
Chapter 1
Introduction to high-performance metallic biomaterials —— 1

Priyanka Dash, Ranjit Barua, Dileep Pathote, Sachin Latiyan,
and Sudipto Datta
Chapter 2
Design strategy of metallic biomaterials for biomedical applications —— 21

Sachin Latiyan, Ranjit Barua, Dileep Pathote, Vilay Vannaladsaysy,
and Sudipto Datta
Chapter 3
Stainless steel-based biomaterial for orthopedic fixations —— 41

Uzwalkiran Rokkala, Mohammad Faseeulla Khan, Ashwin C. Gowda, and Santhosh
Chapter 4
Magnesium alloys for biomedical applications —— 61

Vivek Singh, Sunil Kumar, Kamal Kishore, Rajesh Kumar Sharma, Rakesh Sehgal,
and Sanjay Kumar
Chapter 5
Titanium and titanium alloys in medical applications —— 79

Ramu P. and Deepak Patil
Chapter 6
Titanium dioxide coating for biomedical devices —— 101

Harish Singh Dhami, Srinivasan Chandrasekar
Chapter 7
Surface topographies in the manufacturing of biomedical implants —— 133

Abhinay Thakur, Valentine C. Anadebe, Abdelkader Zarrouk, Omar Dagdag, Ashish Kumar, and Virat Khanna
Chapter 8
Nanocoating for medical devices — 149

Dileep Pathote, Ranjit Barua, Sachin Latiyann, Priyanka Dash, and Sudipto Datta
Chapter 9
Surface modification of bone screws, reconstruction surgeries — 185

Ranjit Barua, Arbind Prasad, and Siddharth Mohan Bhasney
Chapter 10
Emerging application of modern additively manufactured medical implants — 205

Tuhin Kar and Arjyajyoti Goswami
Chapter 11
Laser surface modification of metallic implant materials — 229

Ricardo Luiz Perez Teixeira
Chapter 12
Electrochemical investigations and corrosion stability of the metallic implants — 251

Samir Das, Sachin Latiyan, Arbind Prasad, Bidyanand Mahto, and Sudipto Datta
Chapter 13
In vitro studies of metallic biomaterials — 273

Ranjit Barua, Arbind Prasad, and Siddharth Mohan Bhasney
Chapter 14
Surface modification and coating of metallic implants — 285

Nisha Shankhwar, Satyendra Singh, and Arbind Prasad
Chapter 15
Challenges and prospects of metallic biomaterials: innovations, advances, and future directions — 309

Index — 339

Editors short bio

Dr. Arbind Prasad has completed his Ph.D. (Mechanical Engineering) from **the Indian Institute of Technology Guwahati**, Assam, India. He has also earned the **gold medal** in his **M. Tech**. He is currently working as an Assistant Professor (senior) and Head (Mechanical Engineering) in the Department of Science, Technology and Technical Education, Government of Bihar, at Katihar Engineering College, Katihar, Bihar, India. He has **four granted patents** out of his research work. He has 17 international Journal papers, edited 13 books (Wiley, Elsevier, AAP, De Gruyter, and CRC Press (Taylor & Francis) USA), 38 book chapters, and 15 reputed international conference papers to his credit. Dr. Arbind Prasad has obtained various prestigious awards such as the Sponsored Research Industrial Consultancy (SRIC) award from IIT Kanpur, Best Oral Presentation from the American Chemical Society, and Best Paper awards from IIT Guwahati during Research Conclave. He has completed numerous projects funded by the State Government. He is an Associate Editor for the International Journal of Materials, Manufacturing and Sustainable Technologies (IJMMST) and an early career editor for the International Journal of Mathematical, Engineering and Management Sciences (IJMEMS) Indexed in ESCI/Scopus and DOAJ. He is an editorial board member of various international journals. He acts as an active review board member of 10 prestigious (Indexed in SCI/SCIE/Scopus) and guest editor in special issues of international journals with high impact factor i.e. Bioengineering, Journal of Design, Materials, Journal of functional materials, Molecules, etc. His main areas of research include *Resorbable polymers, recycling of biodegradable polymers, waste processing, Implants biomaterials, materials processing, and bio energies generation*. He is also a lifetime member of the Society of Polymer Science India, the Material Research Society of India, the Society of Biomaterials and Artificial Organs of India, the Asian Polymer Association, and the Indian Society of Technical Education.

Dr. Manoj Kumar Sinha is working as an Assistant Professor, Grade I at the Department of Mechanical Engineering, National Institute of Technology Kurukshetra, Haryana, India. Earlier, he was associated with the National Institute of Technology Hamirpur, H.P., India. Dr. Sinha has completed his B.E., M. Tech. and Ph.D. from N. M.U. Jalgaon, Maharashtra, IIT Guwahati and IIT Delhi respectively. He has been selected as GE Foundation Scholar Leader at IIT Guwahati. He has published more than 35 research papers in peer-reviewed international journals. He has guided 02 PhD students and many UG and PG students. His area of research includes sustainable grinding of Difficult-to-Machine Materials, Additive Manufacturing, Synthesis and Characterization of MMCs, biocomposite characterizations, coatings and implant devices. He is an active reviewer of various reputed international journals. A few earlier selections of Dr. Sinha include Assistant Professor at S.C.E. Saharsa, Bihar; Management Trainee in Coal India Ltd. and Management Trainee in Oil and Natural Gas Corporation Ltd.

https://doi.org/10.1515/9783111571423-205

Prof. J. Paulo Davim is a Full Professor at the University of Aveiro, Portugal. He is also distinguished as an honorary professor in several universities/colleges/institutes in China, India and Spain. He received his PhD in Mechanical Engineering, MSc in Mechanical Engineering (materials and manufacturing processes), and Mechanical Engineering degree (5 years), from the University of Porto (FEUP), the Aggregate title (Full Habilitation) from the University of Coimbra and the DSc (Higher Doctorate) from London Metropolitan University. He is a Fellow (FIET) of IET-London and Eur Ing by Engineers Europe FEANI-Brussels. He is also a Senior Chartered Engineer by the Portuguese Institution of Engineers with an MBA and Specialist titles in Engineering and Industrial Management as well as in Metrology. He has more than 35 years of teaching and research experience in Manufacturing, Materials, Mechanical and Industrial Engineering, with special emphasis in Machining, Tribology & Surface Engineering. He has also an interest in Design, Management, Sustainability, Artificial Intelligence, Engineering Education and Higher Education for Sustainability. He has guided large numbers of postdoc, PhD and master's students as well as has coordinated and participated in several financed research projects. He has received several scientific awards and honors. He has worked as an evaluator of projects for ERC-European Research Council and other international research agencies as well as an examiner of PhD thesis for many universities in different countries. He is the editor-in-chief of several international journals, book Series Editor, and Scientific Advisory for many conferences.

List of contributing authors

Priyanka Dash
Department of Materials Engineering
Indian Institute of Science, C.V. Raman Avenue
Bangalore, Karnataka 560012, India

Ranjit Barua
Mechanical Engineering Department
Omdayal Group of Institutions, Howrah
West Bengal-711316, India

Dileep Pathote
Department of Materials Engineering
Indian Institute of Science, C.V. Raman Avenue
Bangalore, Karnataka 560012, India

Sachin Latiyan
Department of Materials Engineering
Indian Institute of Science, C.V. Raman Avenue
Bangalore, Karnataka 560012, India

Sudipto Datta
Department of Materials Engineering
Indian Institute of Science, C.V. Raman Avenue,
Bangalore, Karnataka 560012, India

Arbind Prasad
Mechanical Engineering Department
Katihar Engineering College
(Under Department of Science, Technology and Technical Education, Government of Bihar)
Katihar, Bihar-854109, India

Vilay Vannaladsaysy
Department of Mechanical Engineering
Faculty of Engineering, National University of Laos, Sokpaluang 01005, Vientiane, Laos.

Uzwalkiran Rokkala
Department of Mechanical Engineering
Aditya University, Surampalem, Andhra Pradesh
India

MD.F. Khan
Department of Mechanical Engineering
College of Engineering, King Faisal University
Al-Ahsa, Saudi Arabia

Ashwin C Gowda
Department of Mechanical Engineering,
Visvesvaraya Technology University, India

Santhosh
Department of Mechanical Engineering
MVJ College of Engineering, India

Vivek Singh
Department of Mechanical Engineering
National Institute of Technology, Hamirpur,
Himachal Pradesh, India, 177005

Sunil Kumar
Department of Mechanical and Industrial Engineering, Indian Institute of Technology
Roorkee, Uttarakhand, India, 247667

Kamal Kishore
Department of Mechanical Engineering
National Institute of Technology, Hamirpur
Himachal Pradesh, India, 177005

Rajesh Kumar Sharma
Department of Mechanical Engineering
National Institute of Technology, Hamirpur
Himachal Pradesh, India, 177005

Rakesh Sehgal
Department of Mechanical Engineering
National Institute of Technology, Hamirpur
Himachal Pradesh, India, 177005

Sanjay Kumar
Department of Mechanical Engineering
National Institute of Technology, Srinagar
Jammu & Kashmir, India, 190006

Ramu P
Department of Production Engineering
National Institute of Technology Tiruchirappalli
Tamil Nadu – 620015

Deepak Patil
Department of Production Engineering, National Institute of Technology Tiruchirappalli, Tamil Nadu – 620015

https://doi.org/10.1515/9783111571423-206

Abhinay Thakur
Division of Research and Development
Lovely Professional University, Phagwara
Punjab, 144411, India

Valentine C. Anadebe
Department of Chemical Engineering
Alex Ekwueme Federal University Ndufu Alike
P. M. B. 1010, Abakaliki, Ebonyi State, Nigeria

Abdelkader Zarrouk
Research Centre, Manchester Salt & Catalysis
unit C, 88-90 Chorlton Rd, M15 4AN, Manchester
United Kingdom
Laboratory of Materials, Nanotechnology and
Environment, Faculty of Sciences
Mohammed V University in Rabat, Av. Ibn
Battouta, P.O. Box. 1014 Agdal-Rabat, Morocco

Omar Dagdag
Department of Mechanical Engineering
Gachon University, Seongnam 13120
Republic of Korea

Ashish Kumar
Nalanda College of Engineering
Bihar Engineering University
Department of Science
Technology and Technical Education
Government of Bihar, 803108, Bihar, India

Virat Khanna
University Centre for Research & Development
Chandigarh University, Mohali, Punjab, 140413
India University Centre for Research &
Development, Chandigarh University, Mohali
Punjab, 140413, India

Tuhin Kar
Department of Mechanical Engineering
National Institute of Technology
Durgapur, 713209, India

Arjyajyoti Goswami
Department of Mechanical Engineering
National Institute of Technology
Durgapur, 713209, India

Ricardo Luiz Perez Teixeira
Federal University of Itajubá Campus
Theodomiro Carneiro Santiago
200 Irmã Ivone Drumond Street
Industrial District II, Itabira, MG, Brazil
Postal Code 35903-087

Samir Das
School of Medical Science and Technology
Indian Institute of Technology Kharagpur –
721302, West Bengal, India

Siddharth Mohan Bhasney
Machino Polymers Limited
plot no.2, Sector 33, Delhi – Jaipur Expy
Phase 2, Info Technology Park, Sector 33
Gurugram, Haryana 122001

Kamal Kishore
Project Research, Scientist -I
Department of Mechanical Engineering
IIT Delhi

Vivek Singh
Department of Mechanical Engineering
NIT Hamirpur

Harish Singh Dhami
Department of Industrial Engineering
Purdue University, West Lafayette, USA

Manoj Kumar Sinha
Department of Mechanical Engineering
NIT Kurukshetra, Haryana-136119, India.

Nisha Shankhwar
Special Centre for Nanoscience
Jawaharlal Nehru University, New Delhi
India, 110067

Satyendra Singh
Special Centre for Nanoscience
Jawaharlal Nehru University
New Delhi, India, 110067

Kamal Kishore, Vivek Singh, Manoj Kumar Sinha* and Arbind Prasad

Chapter 1
Introduction to high-performance metallic biomaterials

Abstract: Over the years, high-performance metallic biomaterials have made great strides in overcoming issues regarding biocompatibility, mechanical performance, and long-term stability. These materials are selected based on several important properties, including strength, fatigue resistance, corrosion resistance, and biodegradation behavior. The ability to tailor microstructure-property relationships has facilitated the design of materials targeted for biomedical applications. Surface modification can improve bioactivity and osseointegration by means of different techniques such as plasma spraying, anodization, and physical vapor deposition. Emerging materials, such as high-entropy alloys and biodegradable metals, present advantages in mechanical properties, corrosion resistance, and regulated biodegradation, diminishing the requirement for secondary surgical procedures. Future studies will work on enhancing the biocompatibility, corrosion, and wear resistance, and developing novel engineering advanced materials with innovative functionalities. Cutting-edge manufacturing and interdisciplinary collaboration will be key to transforming these advancements into life-saving medical devices, revolutionizing modern biomedical implants and improving patient outcomes.

Keywords: High-entropy alloys, metallic biomaterials, implants, cobalt alloys, stainless steel

1.1 Introduction

Materials are the base of any innovation. They are considered one of the pillars that helps the humankind to explore the possibility of certainness. Apart from the various materials, biomaterials are changing the human life rapidly and becoming the key el-

***Corresponding author: Manoj Kumar Sinha**, Department of Mechanical Engineering, National Institute of Technology Kurukshetra, Kurukshetra 136119, Haryana, India, e-mail: mksinha@nitkkr.ac.in
Kamal Kishore, Department of Mechanical Engineering, Indian Institute of Technology Delhi, Delhi 110016, India
Vivek Singh, Department of Mechanical Engineering, National Institute of Technology Hamirpur, Himachal Pradesh 177005, India
Arbind Prasad, Mechanical Engineering Department, Katihar Engineering College (Under Department of Science, Technology and Technical Education, Government of Bihar), Katihar 854109, Bihar, India

https://doi.org/10.1515/9783111571423-001

ements for the human well-being [1]. The history of biomaterials extends across millennia, with indications of metal utilization in dental implants dating back to AD 200 [2]. Nonetheless, substantial growth and advancement in biomaterials occurred until after World War II. Biomaterials can fulfill both transient and enduring functions, augmenting, replacing, or entirely substituting organs, tissues, or other body components to protect or enhance human health [3]. The increasing population and deterioration in lifestyle have resulted in an elevated need for artificial devices constructed from biomaterials to substitute defective bodily components. The selection of biomaterial is contingent upon several criteria, including its composition, mechanical qualities, degradation rate (for temporary applications), capacity to facilitate cell proliferation, and characteristics such as osteoinduction and osteoconduction [4]. A vital element entails reducing implant-associated infections via infection-resistant biomaterials, often accomplished by the application of specialized coatings. The important characteristics of the high-performance metallic biomaterials are shown in Figure 1.1.

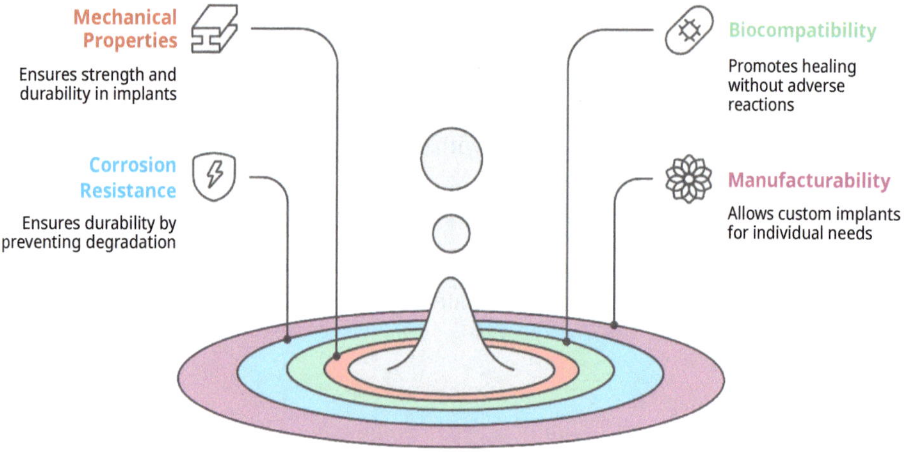

Figure 1.1: Important characteristics of the high-performance metallic biomaterials.

Metallic biomaterials exhibit an excellent reliability in vivo due to their outstanding mechanical properties and hence they are commonly used in implants. They are especially appropriate for load-bearing uses in orthopedic as well as dental operations, where they help to bear the weight of the body on bones and joints and enable restoration of their functions [5, 6]. Previous apprehensions about the enduring characteristics of metallic biomaterials have been alleviated by advancements in regenerative medicine, which demonstrated that the distinctive mechanical and physical properties of metallic biomaterials are both advantageous and essential for biomedical devices. Because of their special properties, metals find great application in biomaterials. A variety of metals are selected based on their compatibility with biological systems,

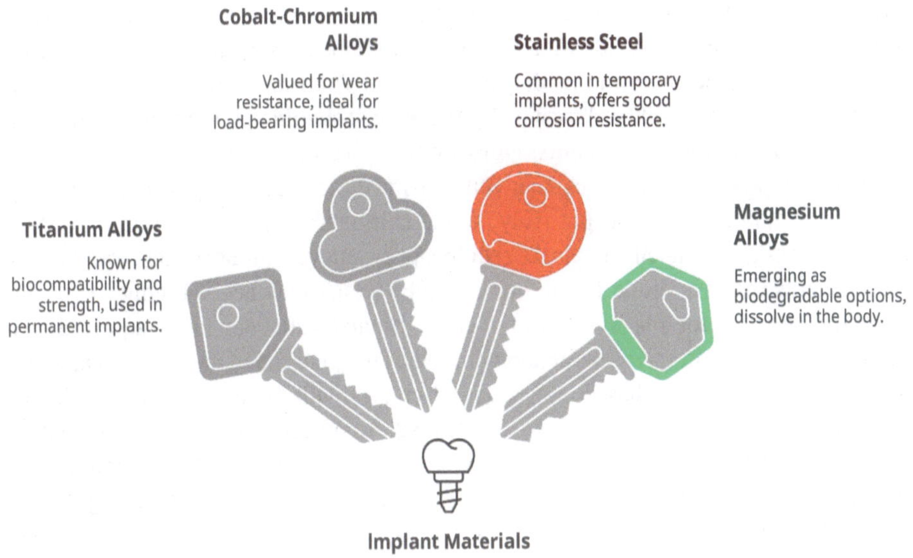

Figure 1.2: Important biometallic materials used for implants.

including titanium (Ti) and its alloys, cobalt and its alloys, steel, tantalum (Ta), nitinol, magnesium (Mg), zinc (Zn), and other alloys [7]. Some of the important biometallic materials used for implants are depicted in Figure 1.2.

1.2 Defining high-performance metallic biomaterials

One of the shared features of biomaterials is their direct interaction with live tissues. A biomaterial is a substance that has been intentionally shaped so that it can be used to control the course of any therapeutic or diagnostic procedure by interacting with parts of living systems. It can be used by itself or as part of a complex system. It is important to note that the prefix "bio" in biomaterials signifies "biocompatible" rather than "biological" or "biomedical," which are common misconceptions [8]. The choice and development of biomaterials are determined by their intended medical use. To ensure long-term functionality without being rejected by human body, a metallic implant must have these following key properties [9]:

(1) Superior biocompatibility by ensuring nontoxicity
(2) Ability to sustain the corrosive environment
(3) Appropriate mechanical strength and durability
(4) Excellent resistance to wear
(5) Effective osseointegration, particularly for bone implants

Since no material remains entirely inert within the human body over an extended period, careful selection is essential. Therefore, it seems reasonable to choose alloying elements that could be found in the human body. The human body is mostly composed of cells having 65–90 wt% water (H_2O). Hence, it should not be surprising that the first two most abundant elements, constituting more than 90% of the mass of the human body, are oxygen and carbon [10]. For ideal biomedical applications, alloying elements should be nontoxic. But truly, no metal is inert or nontoxic in practice. As a result, the design of metallic implants requires using alloys with near-inert elements or the elements that are found in the human body, which may act as highly corrosion-resistant elements for reactive alloys. Metallic biomaterials may be categorized into three groups based on their interaction with cells. First-generation metals, such as stainless steel (SS) and cobalt-based alloys, do not engage with bodily cells and are hence classified as inert materials. The subsequent category comprises second-generation metals (e.g., titanium alloys), including bioactive surfaces to enhance cell regeneration and facilitate integration with adjacent tissues. The third group comprises biodegradable metals (BMs), including magnesium and its alloys [11, 12]. These materials progressively degrade within the human body and are consequently replaced by real human tissues, making them particularly appropriate for applications requiring temporary support. The different generation of metallic biomaterials are represented in Figure 1.3.

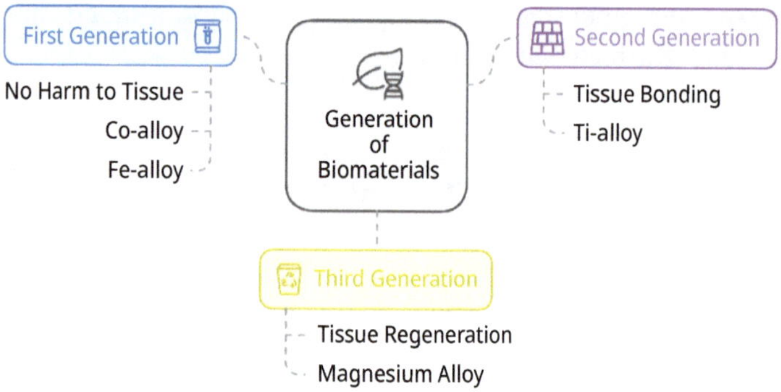

Figure 1.3: Different generation of metallic biomaterials.

1.3 Titanium and its alloy

Commercial pure titanium (Cp-Ti) gained attention in the 1950s, initially in the aerospace industry for its exceptional strength-to-weight ratio and corrosion resistance. Its superior biocompatibility soon made it a preferred material for medical implants [13]. Cp-Ti remains a cornerstone in the biomedical field due to its unique combination of properties that fulfill the rigorous requirements of medical applications. Cp-Ti is now often employed as a biomaterial, due to its superior mechanical and physical characteristics [14]. Moreover, it is well-suited for bone replacement procedures because of its outstanding tensile strength, enhanced stiffness, corrosion resistance, fracture toughness, and biocompatibility [14]. Ti has a modulus of about 112 GPa, which is closer to the modulus of the bone, that is, 20–30 GPa. The surface of Cp-Ti naturally forms a stable titanium dioxide (TiO_2) layer, which prevents corrosion in physiological environments and minimizes adverse reactions. This oxide layer not only offers exceptional corrosion resistance in saline environments but also ensures implant stability and prevents the release of toxic metal ions [15]. Titanium's ability to integrate with the bone tissue is one of its most critical advantages. The surface texture and chemical composition of titanium encourage bone cells to adhere and grow, forming a strong bond between the implant and the surrounding tissue. Pure Ti has limited usage in medical implants because of its vulnerability to abrasive and adhesive wear. The wear debris formation and their entrapment during sliding in load-bearing implants induce bone resorption and affect the implant's long-term stability [16]. Consequently, Cp-Ti is one of the most often utilized metals in medical implant production; however, further attention of researchers is required to improve its wear resistance.

To address the limitations of pure titanium, alloying elements are added to improve mechanical properties, corrosion resistance, and biocompatibility. Titanium alloys must exhibit high strength, low Young's modulus, and excellent corrosion resistance to meet the demands of biomedical applications. Titanium alloys stand out as crucial biomaterials among various options for implants. Their remarkable attributes, including better mechanical and tensile strength, elasticity, biocompatibility, and superior corrosion resistance, make them highly significant biomaterial [15]. Ti-6Al-4V stands as one of the most widely utilized and resilient titanium alloys. Comprising nearly Ti (90%), Al (6%), and V (4%), it finds application in the fabrication of various medical devices. These include artificial pacemakers, bone plates, hip and knee joints, artificial hearts, fracture fixation screws, and cardiac valves [13]. However, aluminum (Al) and vanadium (V) have been associated with allergic reactions and cytotoxicity. This led to the development of second-generation alloys, where Al and V are replaced by safer elements like zirconium (Zr), tantalum (Ta), and molybdenum (Mo). While titanium alloys rank among the most successful biomaterials, their high coefficient of friction and susceptibility to adhesive and abrasive wear remain of significant concerns. Recent studies have also focused on other biocompatible titanium alloys, including Ti-35Nb, Ti-6Al-7Nb, Ti-Nb-Ta-Zr (TNTZ), and Ti-13Nb-13Zr [17].

Titanium alloys can be categorized based on the alloying elements known as stabilizers added to titanium for the specific applications as follows:
- α-Stabilizers: Oxygen (O_2), nitrogen (N_2), carbon (C), copper (Cu), and aluminum (Al)
- β-Stabilizers: Tantalum (Ta), niobium (Nb), vanadium (V), molybdenum (Mo), and others
- Neutral elements: Zirconium (Zr), tin (Sn), and hafnium (Hf)

Some important alloys showing various phases based on stabilizers are:
- α-Phase alloys: Ti-Al, Ti-Cu-Zr, and Ti-Al-Sn
- α + β-Phase alloys: Ti-6Al-7Nb, Ti-6Al-4V, and Ti-Al-Mo
- β-Phase alloys: Ti-Mo, Ti-Nb, and Ti-Ta-Zr

Alloys with a predominant β-phase are preferred in biomedical applications due to their lower modulus of elasticity and improved compatibility with human bone.

Surface modifications, such as coatings, can significantly improve the bioactivity, wear resistance, and osseointegration of titanium and its alloys. An ideal coating should exhibit strong adhesion, superior hardness, exceptional osseointegration, osteoconductive properties, minimal cracking, and a defect-free structure without inclusions. Some vital coating deposition techniques have been mentioned below for biomedical application [18–20]:
- The physical vapor deposition (PVD) method involves vaporizing a material and allowing it to condense onto a metal surface. Among the various PVD techniques, magnetron sputtering is the most widely used for coating deposition. Magnetron sputtering is also known for producing smooth and dense structures of the coatings.
- The chemical vapor deposition method involves introducing a reactant gas into a reactor under high-temperature and high-pressure conditions. The gas reacts with the metal or alloy surface, resulting in the formation of a thin coating.
- The sol-gel process involves immersing a substrate into a colloidal suspension, resulting in the deposition of a gel layer on its surface. The excess liquid is then removed through a drying process.
- Plasma spraying is widely used for rapid deposition with minimal thermal degradation, making it superior to other thermal coating methods. Currently, the only coating approved by the FDA for such applications is plasma-sprayed hydroxyapatite (HAp).
- Micro-arc oxidation (MAO) is a promising technique for creating rough, microporous, and hard coatings on metallic substrates. Its capability to achieve strong adhesion between the coating and the substrate has attracted significant attention. MAO is also known as plasma electrolytic oxidation technique as well as most suitable to produce metal oxide films.

1.4 Cobalt and its alloys

Cobalt-chromium (Co-Cr) alloys are extensively utilized as biometallic materials because of their outstanding mechanical strength and high resistance to corrosion. However, concerns arise due to the release of cobalt ions, which leads to potential toxicity. The researches are ongoing into surface modifications and alternative alloy compositions to enhance the biocompatibility [21]. The popularity of cobalt-based alloys in biomedical applications stems from their advantageous properties. For instance, Co-Cr alloys have high strength, excellent corrosion resistance in physiological environments, and good wear resistance [21]. These properties are crucial for long-term implant performance, especially in load-bearing applications. However, the specific composition of these alloys significantly influences their biocompatibility and mechanical properties. Variations in chromium, molybdenum, and other alloying elements affect corrosion resistance, the release of metal ions, and the overall biological response. The optimization of alloy composition is a critical area of ongoing research, aimed at balancing mechanical strength and biocompatibility [22]. This necessitates a thorough evaluation of the balance between structural effectiveness and the risk of harmful biological impacts.

The mechanical properties of cobalt alloys are crucial for implant design. High tensile strength and yield strength are essential for load-bearing applications such as hip and knee replacements [23]. The elastic modulus, however, needs to be carefully considered. Different types of elastic modulus between the implant and the surrounding bone tissue could lead to stress shielding, where the implant absorbs too much load. This disparity can cause bone loss and maybe implant failure [14]. Research efforts focus on developing alloys with elastic moduli closer to that of bone or employing design modifications to minimize stress shielding [1]. Furthermore, fatigue resistance is critical for long-term implant stability, especially in applications involving cyclical loading [5]. The selection of appropriate alloy compositions and implant designs is critical to ensuring long-term implant success.

The biocompatibility of cobalt alloys, their interaction with cells and tissues, is a complex area of study. While Co-Cr alloys are generally considered biocompatible, concerns remain regarding the release of cobalt ions and their potential toxicity [7]. In vitro studies have investigated cellular responses to Co-Cr surfaces, assessing parameters such as cell adhesion, proliferation, and differentiation [15]. These studies often examine the effect of surface morphology and chemistry on human-cell behavior. In vivo studies provide crucial information on long-term tissue response, including inflammatory reactions and foreign body responses. The potential for adverse cellular responses necessitates research into strategies to improve biocompatibility for the development of biomedical devices using these alloys.

Further, cobalt alloys find extensive use in various biomedical devices. Their exceptional strength, resistance to corrosion, and durability against wear make them ideal for supporting heavy loads. Co-Cr alloys are commonly utilized in orthopedic implants such as knee and hip replacements [15]. Their high strength and wear resistance are essential for load-bearing applications. However, concerns regarding stress shielding and the release of cobalt ions necessitate ongoing research into surface modifications and alloy optimization [23]. The design of orthopedic implants is critical for minimizing stress shielding and promoting osseointegration. Moreover, Co-Cr alloys have also been preferred for dental implants [23]. The need for osseointegration is crucial for dental implant success. Therefore, surface modifications are often used to enhance osseointegration and biocompatibility [21]. Cobalt alloys have also found limited use in cardiovascular applications, particularly in heart valves. Corrosion resistance is also a critical factor for long-term stability in the oral environment. Corrosion resistance and wear resistance are important in these applications. However, the potential for thrombogenicity and the release of cobalt ions are critical considerations. Surface modifications are often employed to reduce thrombogenicity and improve biocompatibility [24]. The selection of appropriate materials and surface modifications is crucial for minimizing complications and ensuring the long-term success of cardiovascular devices.

1.5 Stainless steel

SS's suitability for biomedical applications stems from a combination of inherent properties. Its significant mechanical strength, coupled with its appreciable corrosion resistance and relatively low cost compared to other biomaterials like titanium alloys, makes it a desirable choice for numerous medical devices [25]. But given the wide spectrum of SS grades, each with different chemical compositions, a careful study of how these compositional changes affect the qualities of the material, including most importantly its biocompatibility, is needed [26]. This section provides an insight into the properties that make SS suitable for biomedical applications, examining the common grades employed and the impact of alloying elements on mechanical characteristics and biological compatibility [27].

The strength and durability of SS play a vital role in assessing its appropriateness for different implant structures. Various grades differ in properties such as yield strength, tensile strength, fatigue resistance, and stiffness [28]. For example, certain SS grades offer high tensile strength, making them beneficial for load-bearing applications like orthopedic implants. However, its stiffness is notably higher than that of bone tissue, which may result in stress shielding. Thus, selecting an appropriate SS grade with an elastic modulus closer to that of bone is essential to mitigate stress shielding and enhance implant performance [29].

The corrosion behavior of SS in the biological environment is a critical factor influencing its biocompatibility. Even highly corrosion-resistant SS grades can release small amounts of metal ions into the surrounding tissues [30]. The release of these ions, including chromium, nickel, and molybdenum, can have significant implications for biocompatibility. The concentration of released ions and their potential toxicity depend on several factors, including the specific SS grade, the pH of the surrounding tissues, the presence of oxygen, and the interaction with proteins and other biological molecules. Different corrosion mechanisms, such as pitting corrosion, crevice corrosion, and general corrosion, can contribute to ion release. Pitting corrosion, for example, can lead to localized material degradation and the release of high concentrations of metal ions. Studies have investigated the influence of various factors on the corrosion rate and ion release from SS, leading to the development of surface modification techniques designed to improve corrosion resistance and minimize ion release [24, 31].

The biocompatibility of SS, its interaction with cells and tissues, is a complex area of research. While SS is generally considered biocompatible, its interactions with the biological environment can significantly affect implant success. In vitro studies have investigated cellular responses to SS surfaces, evaluating parameters such as cell adhesion, proliferation, and differentiation [25]. These studies often assess the influence of surface topography and chemistry on cell behavior. In vivo studies, on the other hand, provide crucial information on the long-term tissue response to SS implants, including the inflammatory response and the potential for foreign body reactions. The inflammatory response is a critical factor influencing implant integration and longevity, and it is often modulated by the release of metal ions and the activation of immune cells. The balance between beneficial and adverse cellular responses to SS is a key consideration in its biomedical applications [4]. Figure 1.4 shows the side effect of SS in human body. However, its inherent limitations in biocompatibility necessitate careful consideration of the application and the implementation of surface modification techniques to optimize performance.

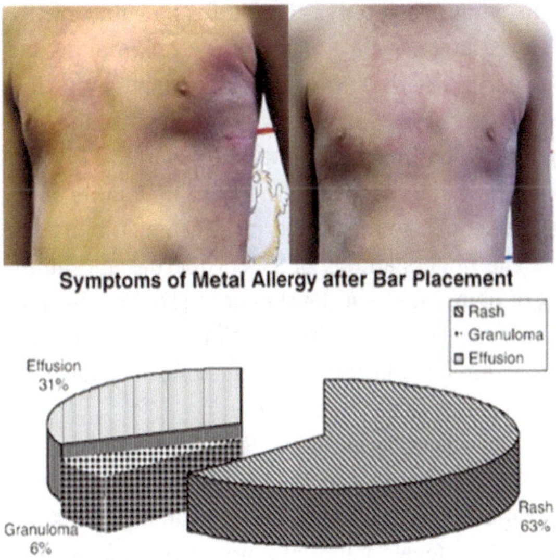

Figure 1.4: Side effect of stainless steel bar in human body [8].

Surface modifications are frequently employed to enhance biocompatibility and osseointegration. The selection of appropriate SS grades and surface treatments is crucial for optimizing implant performance and minimizing complications [8]. The design of orthopedic implants also plays a significant role in their success, and the interaction between material properties, implant design, and surgical techniques is crucial for optimal outcomes. The use of SS in orthopedic implants is well-established, but ongoing research focuses on improving biocompatibility and reducing complications through surface modifications and refined implant designs. The balance between mechanical strength and biocompatibility remains a key consideration in this application [8].

SS has been used in cardiovascular stents, but its inherent thrombogenicity (the tendency to promote blood clot formation) and relatively poor biocompatibility have led to the development of alternative materials. Surface modifications, such as electropolishing, are often employed to reduce thrombogenicity and improve biocompatibility. The challenges of thrombosis, restenosis (re-narrowing of the artery), and late stent failure continue to drive research into improved stent designs and materials. The development of drug-eluting stents, which release medications to inhibit restenosis, has significantly improved the outcomes of cardiovascular interventions. Although SS stents are less prevalent now due to the development of more biocompatible materials, understanding its properties and limitations remains important in the context of stent design and development [7].

SS has been used in dental implants, but its limitations in biocompatibility and corrosion resistance have restricted its widespread use. The need for osseointegration (the direct bonding of the implant to the bone) is paramount in dental implant success. Surface modifications, such as coatings with HAp or other bioactive materials, are often employed to enhance osseointegration and improve biocompatibility. The challenges of corrosion, biofilm formation, and long-term stability continue to drive research into improved dental implant materials and designs. Titanium and its alloys have largely replaced SS in dental implants due to their superior biocompatibility and osseointegration properties. However, the understanding of SS's properties and limitations remains relevant in the context of dental implant development [17]. The development of novel surface modifications and the continued investigation of SS's properties may lead to its future use in specialized dental applications. The need for long-term stability and osseointegration is a major challenge in this application.

This section highlights the significant role of SS in biomedical applications. While its inherent properties offer advantages in terms of mechanical strength and cost-effectiveness, its limitations in biocompatibility and corrosion resistance necessitate the ongoing research and development. The development of innovative surface modification techniques, a deeper understanding of cellular interactions, and rigorous in vivo studies and clinical trials are all vital for enhancing the long-term performance and safety of SS-based implants. Future research should focus on developing novel surface modifications that enhance bioactivity, reduce inflammation, and prevent infection [14]. Advanced characterization techniques are needed to better understand the complex interactions between SS and the biological environment. Finally, comprehensive in vivo studies and clinical trials are essential to validate the efficacy and safety of new SS-based biomaterials and surface modifications before widespread clinical adoption. The continued development and refinement of SS biomaterials will ensure its continued role in the field of biomedical engineering.

1.6 Magnesium and its alloy

Magnesium (Mg) is a natural section of bone structure and shows a crucial role in metabolic processes [32]. Moreover, magnesium-based alloys belong to the third category of implant biomaterials [33]. However, their main drawback is poor corrosion resistance, which causes them to degrade quickly and unpredictably in the body's physiological environment. Passivation occurs on their surface when exposed to air, creating a thin layer of gray magnesium oxide that prevents additional chemical reactions. Research efforts have focused on improving their corrosion resistance to develop Mg-based biodegradable implants suitable for industrial applications [34].

Magnesium has great mechanical strength even if it is totally biodegradable in the body. There are several reasons why these alloys could be used in orthopedic implants. Magnesium is a biocompatible macroelement. Among numerous metallic implants, Mg alloys exhibit a modulus (40–45 GPa) closest to that of cortical bone (10–27 GPa) [35]. In contrast, Ti-based implants and 316L SS have significantly higher moduli of 110 GPa and 193 GPa, respectively. These characteristics make them interesting candidates for uses in cardiovascular interventions as well as bone healing.

Mg-based implants demonstrate excellent mechanical properties for hard tissue applications. However, their high corrosion rate in physiological conditions restricts their use as biodegradable implants. To enhance their viability, the degradation rate of pure Mg must be controlled through surface engineering and alloying [36]. Alloying is a key technique for enhancing the strength, ductility, and corrosion resistance of pure Mg by incorporating different metals in varying concentrations. These improvements are largely attributed to microstructural modifications, particularly grain size refinement. Most research on Mg alloys aims to optimize these properties for commercial applications [37]. As a result, biomedical Mg alloy studies often utilize alloys formerly developed for aerospace, automotive industries, and defense.

Mg-based alloys commonly include Al and rare earth elements (REEs). However, Al is recognized as a neurotoxicant, with its gathering linked to neurological illnesses such as dementia, Alzheimer's, and senile dementia [38]. Similarly, severe hepatotoxic effects have been observed with REEs such as praseodymium, cerium, and yttrium. Consequently, recent research has focused on developing physically nontoxic Mg alloys using safe elements such as Zr, Zn, Mn, and Ca. Other potential alloying elements under investigation include Si, Sr, Li, Bi, Si, Ag, and Cd. Mg-based alloys can be binary, ternary, or more complex with their composition and significantly influencing both mechanical properties and corrosion behavior.

Alloying Mg with Al significantly improves its corrosion resistance by forming a protective oxide layer on the surface. Common Mg-Al-Zn alloys, such as AZ31, AZ61, and AZ91, exhibit moderate corrosion rates. Zn is another widely used alloying element, with studies showing that incorporating 6% Zn in Mg alloys significantly improves corrosion resistance. Calcium (Ca) is also a well-known alloying element, valued for its ability to promote bone growth. The mechanical properties and corrosion behavior of Mg-Ca alloys can be customized by varying the calcium content. Adding Ca to Mg-based alloys improves their corrosion resistance, making them more suitable for various applications. Li et al. [38] studied the biodegradability of binary Mg-Ca alloys with varying Ca content (1–20 wt%) in bone applications. However, alloys with high Ca content, such as Mg-5Ca, Mg-10Ca, and Mg-20Ca, were observed to be extremely brittle. Nonetheless, the mechanical properties and biocorrosion behavior of Mg-Ca alloys can be tailored by adjusting the Ca concentration. As-cast Mg-Ca alloys exhibited a decrease in yield strength, ultimate tensile strength, and elongation with increasing Ca concentration. Cytocompatibility assessments confirmed that Mg-1Ca alloy is nontoxic to cells. Furthermore, both in vitro and in vivo studies demonstrated the formation of a protective surface layer consisting of $Mg(OH)_2$ and HAp during immersion and implantation.

The incorporation of zinc has also been shown to improve the corrosion resistance of magnesium alloys. Huang et al. [39] investigated the in vivo degradation behavior of Mg-Al-Zn alloys by implanting them intramedullary into the femora of rabbits. Over a period of 9 weeks, microscopic analysis revealed the development of a thin calcium phosphate layer around the implants, which plays a crucial role in enhancing bioactivity and corrosion resistance. This protective layer helps regulate the degradation process by acting as a barrier against further corrosion. Additionally, the study found that Mg alloys containing aluminum and zinc, such as AZ31B, exhibited a slower in vivo degradation rate compared to pure magnesium. This suggests that the presence of Al and Zn contributes to improved stability and biocompatibility, making these alloys more suitable for biomedical applications, particularly as temporary implants [38].

Research has demonstrated that strontium (Sr) significantly stimulates osteoblastic activity and accelerates bone formation in vivo. This makes Sr a highly promising biocompatible alloying element for magnesium (Mg) alloys, enhancing their potential for orthopedic and biomedical applications. Li et al. conducted a comprehensive study on the biodegradability of Mg-Zr-Sr alloys through both in vitro and in vivo experiments. Their findings demonstrated that adding more than 2% Sr leads to the formation of rough grain boundaries with a secondary $Mg_{17}Sr_2$ phase, which can induce galvanic effects and accelerate the corrosion of the Mg matrix. However, the investigation also revealed that incorporating less than 2% Sr effectively reduces the corrosion rate of both Mg-Zr-Sr and Mg-Sr alloys, improving their suitability for biomedical applications [40]. Bornapour et al. [41] demonstrated that after immersion in simulated body fluid, a strontium-HAp (Sr-HAp) layer formed on the surface of a binary Mg-Sr alloy. This protective surface layer significantly improved the alloy's corrosion resistance, highlighting its potential for biomedical applications, particularly in implant materials where controlled degradation and biocompatibility are essential [42].

Previous studies have established that the major limitation of pure magnesium and its alloys is their rapid corrosion rate and unpredictable behavior in biological fluids. To better understand their performance in a living system, in vivo testing is crucial. Developing well-controlled and durable material properties requires a systematic and comprehensive approach, ensuring stability and reliability in the complex and dynamic biological environment.

1.7 High-entropy alloys

Conventional metal-based biomaterials such as titanium alloys, cobalt alloy, and SS often exhibit some drawbacks related to their mechanical characteristics and compatibility with biological systems [43, 44]. The inherent bioinertness of some metals, coupled with potential corrosion issues and adverse reactions to long-term implantation, necessitates the development of improved biomaterials. High-entropy alloys (HEAs), distinguished by their unique compositional characteristics and the resulting synergistic properties, offer a compelling alternative to address these limitations [45]. The important features of HEAs are shown in Figure 1.5. This section delves into the current state of research, focusing on the biocompatibility for the development of bio-HEAs. Diverse processing techniques are employed in their fabrication, which enhance their mechanical properties and biocompatibility profiles remarkably and help in promising potential for various clinical applications within biomedical settings.

There are various techniques for the fabrication of bio-HEAs, each possessing its distinct advantages and disadvantages. The selection of technique significantly influences the resulting microstructure, homogeneity, and ultimately, the final properties of bio-HEAs. Traditional methods, such as vacuum arc melting and powder metallurgy, are commonly used for producing bulk bio-HEAs. However, these techniques can sometimes lead to undesirable elemental segregation, compromising the intended homogeneity and potentially affecting the final properties [47]. This segregation arises due to variation in the melting points and solidification behaviors of different elemental compositions. Advanced techniques such as selective laser melting (SLM) and laser powder bed fusion offer significant advantages over these traditional methods [48]. The additive manufacturing techniques allow for precise control over the microstructure and the ability to create complex shapes tailored to specific anatomical requirements. The rapid solidification rates achieved in these processes effectively suppress

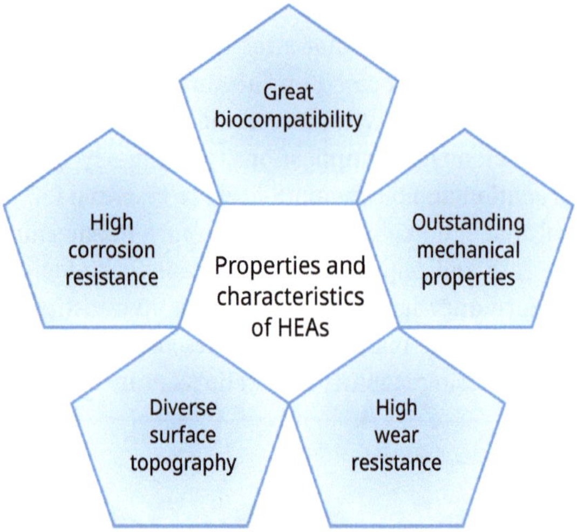

Figure 1.5: The main properties and characteristics of HEAs [46].

elemental segregation, resulting in more homogeneous microstructures [49]. Laser cladding provides another method for producing bio-HEA coatings, offering a cost-effective way to enhance the surface properties of existing biomaterials. This technique allows for the deposition of a thin layer of bio-HEA onto a substrate material, modifying its surface characteristics without the need for complete replacement [50]. Sputtering techniques are also employed to deposit thin films of bio-HEAs, providing further control over surface properties and enabling the creation of functionally graded materials. This approach enables accurate regulation of both the layer's thickness and its composition. Mechanical alloying, followed by sintering, is another effective method for producing homogeneous bio-HEAs by creating a fine mixture of constituent powders that are then consolidated into a dense bulk material [49]. The selection of the appropriate processing method is therefore a crucial step in the development of bio-HEAs with optimized properties.

Bio-HEAs show a wide range of mechanical properties that make them suitable for a variety of biomedical applications. Numerous studies show that HEAs offer significantly greater strength and hardness than traditional biomaterials [51, 52]. This enhanced strength is largely attributed to the high configurational entropy and the solid solution strengthening effect resulting from the multicomponent nature of these alloys. The presence of multiple principal elements leads to a complex interplay of atomic interactions, resulting in a solid solution that is stronger and harder than traditional alloys. However, a major challenge in the development of bio-HEAs is achieving a sufficiently low Young's modulus, similar to that of human bone, to prevent stress shielding [53]. Research efforts are intensively focused on developing strategies to reduce the Young's modulus without compromising the strength and other desirable properties [54]. The introduction of porosity or the engineering of specific microstructural features, such as grain boundaries or precipitates, can significantly contribute to lowering the modulus. The mechanical behavior of an implant under applied forces depends heavily on its tensile properties, which encompass key factors such as yield strength, ultimate tensile strength, and ductility. These characteristics play a vital role in assessing how well the implant can withstand stress and deformation during use. The specific composition of the alloy, along with the chosen processing method, greatly affects these properties, ultimately influencing the implant's overall durability and functionality in biomedical applications. Furthermore, the wear resistance of bio-HEAs is another important consideration. Studies consistently demonstrate superior wear resistance compared to traditional biomaterials making them ideal candidates for applications involving significant frictional forces. Bio-HEAs demonstrate significant promise for diverse biomedical applications, including orthopedic implants, dental restorations, and cardiovascular stents. The excellent mix of strength, biocompatibility, and resistance to corrosion makes them highly suitable for these uses. Adjusting the composition and processing of bio-HEAs allows for customized properties, creating extensive opportunities to develop implants tailored to specific needs [55].

Research is actively exploring the development of bio-HEAs for specific applications, such as minimally invasive implants or shape memory alloys [56]. Shape memory alloys can return to their original form after being deformed, providing distinct benefits for specific implant applications. However, more research is required to gain a comprehensive understanding of the long-term performance of bio-HEAs within the body, particularly their interactions with the immune system and the effects of any degradation by-products [57]. This understanding is critical for ensuring the long-term safety and efficacy of these implants. Additive manufacturing processes, such as SLM and laser powder bed fusion, offer significant opportunities for creating complex geometries and customized implants, tailoring the implant design to the patient's specific anatomy and needs [58]. The development of comprehensive databases and predictive models will significantly accelerate the design and development of new bio-HEAs, allowing for more efficient exploration of the vast compositional space and facilitating the identification of optimal alloy compositions for specific applications.

HEAs represent a substantial advancement in the field of metallic biomaterials. Their unique properties offer significant advantages over traditional materials, but several challenges remain. Further research is necessary to thoroughly understand long-term biocompatibility, degradation mechanisms, and clinical performance. The development of advanced processing techniques and robust predictive modeling tools is crucial for accelerating the transition of bio-HEAs from laboratory research to clinical applications. The ongoing research into bio-HEAs holds immense promise for revolutionizing the field of biomedical implants, potentially leading to improved patient outcomes and expanded treatment options. The ability to tailor the properties of bio-HEAs to meet specific clinical needs presents exciting opportunities for future advancements in this field. Addressing the remaining challenges through continued research and development efforts will pave the way for widespread adoption of bio-HEAs in various biomedical applications.

1.8 Summary

The field of high-performance metallic biomaterials has undergone remarkable advancements, effectively tackling challenges in biocompatibility, mechanical performance, and long-term stability. The selection and optimization of these materials are driven by critical properties such as strength, fatigue resistance, elastic modulus, corrosion resistance, and biodegradation behavior. A deep understanding of the relationship between microstructure and mechanical properties has enabled the development of innovative strategies to tailor these materials for specific biomedical applications.

Surface modification techniques, including plasma spraying, anodization, and PVD, have played a vital role in enhancing bioactivity and osseointegration. Additionally, additive manufacturing has transformed titanium-based implants, allowing for patient-specific designs and optimized structural properties that improve clinical outcomes.

The emergence of novel biomaterials, such as HEAs and BMs, marks an exciting new frontier in biomedical engineering. HEAs exhibit exceptional mechanical properties, superior corrosion resistance, and customizable biocompatibility, positioning them as promising candidates for next-generation implants. Similarly, BMs provide a unique advantage by supporting tissue regeneration while gradually degrading in the body, reducing the need for secondary surgeries.

Looking ahead, the continuous development of high-performance metallic biomaterials will focus on further improving biocompatibility, addressing corrosion and wear issues, and engineering novel materials with advanced functionalities. Cutting-edge manufacturing techniques, including additive manufacturing and surface engineering, will be instrumental in designing next-generation implants tailored to individual patient needs. Moreover, interdisciplinary collaboration among materials scientists, biomedical engineers, and clinicians will be crucial in transforming these innovations into practical, life-saving medical devices. Through sustained research and technological advancements, high-performance metallic biomaterials will continue to push the boundaries of modern medicine, ensuring better patient outcomes and revolutionizing the future of biomedical implants.

References

[1] Festas, A. J., Ramos, A., & Davim, J. P. (2020). Medical devices biomaterials – A review. *Proceedings of the Institution of Mechanical Engineers, Part L: Journal of Materials Design and Applications*, *234*(1), 218–228.

[2] Kishore, K., Varshney, R., Singh, P., & Sinha, M. K. (2022). Materials for 3D Printing in Medicine: Metals, Polymers, Ceramics, Hydrogels. In *Additive Manufacturing with Medical Applications* (pp. 73–95). CRC Press, London, England.

[3] George, E. P., Raabe, D., & Ritchie, R. O. (2019). *Nature Reviews Materials*, *4*, 515–534.

[4] Zhang, E., Zhao, X., Hu, J., Wang, R., Fu, S., & Qin, G. (2021). Antibacterial metals and alloys for potential biomedical implants. *Bioactive Materials*, *6*(8), 2569–2612.

[5] Ghosh, S., Sanghavi, S., & Sancheti, P. (2018). Metallic Biomaterial for Bone Support and Replacement. In *Fundamental Biomaterials: Metals* (pp. 139–165). Edited by Preetha Balakrishnan, Sreekala M S and Sabu Thomas Elsevier.

[6] Davim, J. P. (2008). Machining: Fundamentals and recent advances.

[7] Chen, Q., & Thouas, G. A. (2015). Metallic implant biomaterials. *Materials Science and Engineering: R: Reports*, *87*, 1–57.

[8] Prasad, K., *et al.* (2017). Metallic biomaterials: Current challenges and opportunities. *Materials*, *10*(8), 884.

[9] Eliaz, N. (2019). Corrosion of metallic biomaterials: A review. *Materials*, *12*(3), 407.

[10] Hornberger, H., Virtanen, S., & Boccaccini, A. R. (2012). Biomedical coatings on magnesium alloys – a review. *Acta Biomaterialia*, *8*(7), 2442–2455.

[11] Singh, V., Sharma, R. K., & Sehgal, R. (2024). A comprehensive investigation on nanomechanical, scratch, and tribological characteristics of TaN-Ag nanocomposite coating on Ti6Al7Nb alloy. *Journal of Materials Engineering and Performance*, 1–17.

[12] Singh, V., Sharma, R. K., Sehgal, R., & Wani, M. F. (2024). The effect of RF sputtering power on structural, nanomechanical and tribological properties of single layered TaN coatings. *Surface Topography*, *12*(3), 035003.

[13] Hanawa, T. (2019). Titanium–tissue interface reaction and its control with surface treatment. *Frontiers in Bioengineering and Biotechnology*, *7*, 170.

[14] Veiga, C., Davim, J. P., & Loureiro, A. J. R. (2012). Properties and applications of titanium alloys: A brief review. *Reviews on Advanced Materials Science*, *32*(2), 133–148.

[15] Brunette, D. M., Tengvall, P., Textor, M., & Thomsen, P. (2001). *Titanium in Medicine: Material Science, Surface Science, Engineering, Biological Responses and Medical Applications*. Springer, Heidelberg.

[16] Singh, V., Sharma, R. K., & Sehgal, R. (2024). Impact of deposition pressure on the structural, nanomechanical, and tribological properties of δ-TaN coatings deposited via magnetron sputtering on Ti6Al7Nb alloy. *Proceedings of the Institution of Mechanical Engineers, Part E: Journal of Process Mechanical Engineering*, 09544089241272870.

[17] Shah, F. A., Thomsen, P., & Palmquist, A. (2018). A review of the impact of implant biomaterials on osteocytes. *Journal of Dental Research*, *97*(9), 977–986.

[18] Ibrahim, M. Z., Sarhan, A. A. D., Yusuf, F., & Hamdi, M. (2017). Biomedical materials and techniques to improve the tribological, mechanical and biomedical properties of orthopedic implants–A review article. *Journal of Alloys and Compounds*, *714*, 636–667.

[19] Yetim, T., Turalioğlu, K., Taftali, M., Tekdir, H., Kovaci, H., & Yetim, A. F. (2022). Synthesis and characterization of wear and corrosion resistant Ni-doped Al2O3 nanocomposite ceramic coatings by sol-gel method. *Surface and Coatings Technology*, *444*, 128659.

[20] Chen, J., *et al.* (2023). In vitro degradation, wear property and biocompatibility of nano-Y2O3-containing micro-arc oxidation coating on ZK60 alloy. *Transactions of Nonferrous Metals Society of China*, *33*(5), 1411–1424.
[21] Salaam, A., *et al.* (2024). Bio materials, biocompatibility & its advancements in medical. *E3S Web of Conferences, EDP Sciences*, 01092.
[22] Amirtharaj Mosas, K. K., Chandrasekar, A. R., Dasan, A., Pakseresht, A., & Galusek, D. (2022). Recent advancements in materials and coatings for biomedical implants. *Gels*, *8*(5), 323.
[23] Lin, Y.-C., *et al.* (2024). Hydrogen-treated CoCrMo alloy: A novel approach to enhance biocompatibility and mitigate inflammation in orthopedic implants. *Journal of Materials Chemical B*, 12(32), 7814–7825.
[24] Hauschwitz, P., et al. (2023). Advanced micro- & nanostructuring for enhanced biocompatibility of stainless steel by multi-beam and beamshaping technology. *Biomedical Materials*, *18*(4), 045008.
[25] Hedberg, Y. S., & Odnevall Wallinder, I. (2016). Metal release from stainless steel in biological environments: A review. *Biointerphases*, *11*(1).
[26] Beloti, M. M., De A. Rollo, J. M. D., Filho, A. I., & Rosa, A. L. (2004). In vitro biocompatibility of duplex stainless steel with and without 0.2% niobium. *Journal of Applied Biomaterials & Biomechanics*, *2*(3), 162–168.
[27] Tiwar, S., & Mishra, S. B. (2018). Corrosion of stainless steel and its prevention through surface modification for biomedical application: A review. *Asian Journal of Engineering and Applied Technology*, *7*(S2), 60–66.
[28] Xie, H., Jiang, B., Liu, B., Wang, Q., Xu, J., & Pan, F. (2016, Dec). An investigation on the tribological performances of the SiO2/MoS2 hybrid nanofluids for magnesium alloy-steel contacts. *Nanoscale Research Letters*, *11*(1). doi: 10.1186/s11671-016-1546-y
[29] Ali, S., *et al.* (2022). Microstructure and mechanical properties of modified 316L stainless steel alloy for biomedical applications using powder metallurgy. *Materials*, *15*(8), 2822.
[30] Sidane, D., & Khireddine, H. (2023). Effect of the culture medium on the behavior of the austenitic stainless steel with the specific 316L grade. *Surface and Interface Analysis*, *55*(9), 683–693.
[31] Thangamani, G., Badhai, J., Karthik, S., Sonawane, A., & Palani, I. A. (2023). Biocompatibility and Mechanical Behaviour Studies on Wire-Arc Additive Manufactured Stainless Steel 316L and Shape Memory Alloy (NiTi) Materials for Biomedical Implant. In *Proceedings of the ASME 2023, 18th International Manufacturing Science and Engineering Conference*. ASME, New Jersey, USA.
[32] Saris, N.-E. L., Mervaala, E., Karppanen, H., Khawaja, J. A., & Lewenstam, A. (2000). Magnesium: An update on physiological, clinical and analytical aspects. *Clinica Chimica Acta*, *294*(1–2), 1–26.
[33] Sillekens, W. H. (2013). *Magnesium-based Biodegradable Implants. Thomas Telford Ltd*, Netherlands.
[34] Sezer, N., Evis, Z., Kayhan, S. M., Tahmasebifar, A., & Koç, M. (2018). Review of magnesium-based biomaterials and their applications. *Journal of Magnesium and Alloys*, *6*(1), 23–43.
[35] Zheng, Y. (2015). Magnesium alloys as degradable biomaterials.
[36] Witte, F. (2010). The history of biodegradable magnesium implants: A review. *Acta Biomaterial*, *6*(5), 1680–1692.
[37] Gu, X., Zheng, Y., Cheng, Y., Zhong, S., & Xi, T. (2009). In vitro corrosion and biocompatibility of binary magnesium alloys. *Biomaterials*, *30*(4), 484–498.
[38] Li, Z., Gu, X., Lou, S., & Zheng, Y. (2008). The development of binary Mg–Ca alloys for use as biodegradable materials within bone. *Biomaterials*, *29*(10), 1329–1344.
[39] Huang, J., Ren, Y., Jiang, Y. et al. In vivo study of degradable magnesium and magnesium alloy as bone implant. Front. Mater. Sci. China 1, 405–409 (2007). https://doi.org/10.1007/s11706-007-0074-1
[40] Li, Y., *et al.* (2012). Mg–Zr–Sr alloys as biodegradable implant materials. *Acta Biomaterial*, *8*(8), 3177–3188.
[41] Bornapour, M., Muja, N., Shum-Tim, D., Cerruti, M., & Pekguleryuz, M. (2013). Biocompatibility and biodegradability of Mg–Sr alloys: The formation of Sr-substituted hydroxyapatite. *Acta Biomaterial*, *9*(2), 5319–5330.

[42] Staiger, M. P., Pietak, A. M., Huadmai, J., & Dias, G. (2006). Magnesium and its alloys as orthopedic biomaterials: A review. *Biomaterials, 27*(9), 1728–1734.

[43] Kishore, K., Sinha, M. K., & Setti, D. Grinding and Recent Trends. In *Additive and Subtractive Manufacturing Processes* (pp. 31–50). Edited By Varun Sharma, Pulak Mohan Pandey, 2022 CRC Press.

[44] Kishore, K., Sinha, M. K., Singh, A., Archana, M. K. G., & Korkmaz, M. E. A comprehensive review on the grinding process: Advancements, applications and challenges. *Proceedings of the Institution of Mechanical Engineers, Part C: Journal of Mechanical Engineering Science*, 2022; 236(22), 10923–10952 09544062221110782.

[45] Murty, B. S., Yeh, J.-W., Ranganathan, S., & Bhattacharjee, P. P. (2019). *High-entropy Alloys*. Elsevier, USA.

[46] Ahmady, A. R., Ekhlasi, A., Nouri, A., Nazarpak, M. H., Gong, P., & Solouk, A. (2023). High entropy alloy coatings for biomedical applications: A review. *Smart Materials in Manufacturing, 1*, 100009.

[47] Gao, M. C., Yeh, J.-W., Liaw, P. K., & Zhang, Y. (2016). *High-Entropy Alloys: Fundamentals and Applications*. Springer, Switzerland.

[48] Kishore, K., & Sinha, M. K. (2021). A state-of-the-art review on fused deposition modelling process. *Advances in Manufacturing and Industrial Engineering*, 855–864.

[49] Tsai, M.-H., & Yeh, J.-W. (2014). High-entropy alloys: A critical review. *Materials Research Letters, 2*(3), 107–123.

[50] Farkas, D. (2020). Grain boundary structure in high-entropy alloys. *Journal of Materials Science, 55*(22), 9173–9183.

[51] Yeh, J., *et al.* (2004). Nanostructured high-entropy alloys with multiple principal elements: Novel alloy design concepts and outcomes. *Advanced Engineering Materials, 6*(5), 299–303.

[52] Hori, T., Nagase, T., Todai, M., Matsugaki, A., & Nakano, T. (2019). Development of non-equiatomic Ti-Nb-Ta-Zr-Mo high-entropy alloys for metallic biomaterials. *Scripta Materialia, 172*, 83–87.

[53] Miracle, D. B., & Senkov, O. N. (2017). A critical review of high entropy alloys and related concepts. *Acta Material, 122*, 448–511.

[54] Nagase, T., Iijima, Y., Matsugaki, A., Ameyama, K., & Nakano, T. (2020). Design and fabrication of Ti–Zr-Hf-Cr-Mo and Ti–Zr-Hf-Co-Cr-Mo high-entropy alloys as metallic biomaterials. *Materials Science and Engineering: C, 107*, 110322.

[55] Li, J., Huang, Y., Meng, X., & Xie, Y. (2019). A review on high entropy alloys coatings: Fabrication processes and property assessment. *Advanced Engineering Materials, 21*(8), 1900343.

[56] Gong, P., Li, F., Deng, L., Wang, X., & Jin, J. (2020). Research on nano-scratching behavior of TiZrHfBeCu (Ni) high entropy bulk metallic glasses. *Journal of Alloys and Compounds, 817*, 153240.

[57] Sha, C., Zhou, Z., Xie, Z., & Munroe, P. (2020). High entropy alloy FeMnNiCoCr coatings: Enhanced hardness and damage-tolerance through a dual-phase structure and nanotwins. *Surface and Coatings Technology, 385*, 125435.

[58] Rajaguru, K., Karthikeyan, T., & Vijayan, V. (2019). Additive manufacturing – State of art. *Materials Today Proceedings, No. Xxxx*. doi: 10.1016/j.matpr.2019.06.728.

Priyanka Dash, Ranjit Barua, Dileep Pathote, Sachin Latiyan, and Sudipto Datta*

Chapter 2
Design strategy of metallic biomaterials for biomedical applications

Abstract: The cross talk between science and technology has bestowed the medical industry with innumerable biomaterials for multifarious applications. Among these biomaterials, metallic materials are of great interest to clinical surgeons and research scientists, as they present reliable mechanical and biological attributes. However, their prolonged use is closely associated with various complications, such as poor osteointegration, mechanical instability, inflammation, tissue necrosis, and infection resulting in prolonged patient care, pain, and functional impairment. Tissue integration is crucial when considering metallic implants for any biomedical use. Since an implant functions as a foreign object within the body, the materials used in its construction must be highly compatible with human tissues and bodily fluids. Over the past few decades, various materials, including polymers and metals, have been explored. However, many of these materials have failed to exhibit sufficient compatibility with the human body due to several limitations. Therefore, it is essential to focus on designing and developing metal-based alloys with carefully selected alloying elements to ensure optimal biocompatibility. This chapter discusses major metallic biomaterials, highlighting the existing and emerging key strategies to design and manufacture metallic biodevices. Bulk and surface modification and incorporation of biologics for enhanced biological response are also discussed to present a comprehensive and coherent understanding of the subject and identify future directions.

Keywords: Metallic biomaterials, biodevices, design and development, biomedical applications

***Corresponding author: Sudipto Datta,** Department of Materials Engineering, Indian Institute of Science, C.V. Raman Avenue, Bangalore 560012, Karnataka, India, e-mail: sudiptodatta1990@gmail.com
Priyanka Dash, Dileep Pathote, Sachin Latiyan, Department of Materials Engineering, Indian Institute of Science, C.V. Raman Avenue, Bangalore 560012, Karnataka, India
Ranjit Barua, Mechanical Engineering Department, Omdayal Group of Institutions, Howrah 711316, West Bengal, India

https://doi.org/10.1515/9783111571423-002

2.1 Introduction

Biomaterials, both natural and synthetic, have evolved alongside advancements in living standards and scientific progress, greatly enhancing healthcare outcomes and overall quality of life. The global demand for surgical procedures, including dental implants and bone grafts, continues to rise due to population growth, aging societies, and advancements in medical technology. Metals have been utilized as implants for over a century, beginning in 1895 [1] when Lane introduced an innovative steel plate for bone fracture fixation [2]. In the early stages, metal implants faced challenges such as corrosion and insufficient strength. The introduction of 18-8 stainless steel (SS) in the 1920s, with significantly improved corrosion resistance, quickly gained the attention of clinicians. Since then, metal implants have undergone extensive development and widespread clinical application [3]. The choice of metal for biomedical use depends on the specific implant application. For instance, 316L SS remains the most widely used alloy across various implant fields, from cardiovascular to otorhinolaryngology. However, for implants requiring high wear resistance, such as artificial joints, CoCrMo alloys are preferred. Table 2.1 outlines the types of metals commonly used for different implant categories.

Table 2.1: Types of metals used as implants.

Category	Applications	Materials used
Orthopedic	Artificial joints and bone fixation (plates, screws, and pins)	CoCrMo, Ti6Al4V, Ti6Al7Nb, 316L SS, and Ti
Cardiovascular	Artificial valves and stents	Ti6Al4V, 316L SS, CoCrMo, and Ti
Craniofacial	Plates and screws	316L SS, CoCrMo, Ti, and Ti6Al4V
Otorhinology	Artificial eardrum	316L SS
Dentistry	Fillings and orthodontic wires	AgSn(Cu) amalgam, Au, 316L SS, CoCrMo, TiNi, and TiMo

Metallic biomaterials, including SS, titanium alloys, and cobalt-based alloys, are extensively utilized in medical fields such as dentistry, hip replacements, stents, and orthopedic surgeries. Their widespread use is attributed to their exceptional mechanical strength and efficient heat conductivity. With appropriate processing, these materials demonstrate remarkable durability and resistance to fractures, making them ideal for long-term use in load-bearing applications [4]. Their ease of fabrication and the availability of diverse manufacturing techniques make metals particularly suitable for orthopedics and dentistry, where heavily loaded devices are commonly required [5]. These same advantages have also led to their application in cardiovascular devices and neurovascular implants. People's living standards are rising steadily due to globalization and population increase, which raises the risk of physical illnesses linked to lifestyle choices, such as orthopedic and cardiovascular conditions. Medical professionals worldwide treat numerous cases of human bone or tissue fractures every day, which may be brought on by sports injuries, falls, and auto accidents. However, because of the human body's capacity for self-renewal and regeneration, not all injuries or damaged tissue may necessitate surgical procedures. The steady progress of biomaterials has been attributed in large part to advancements in chemistry, engineering, materials science, and cell and molecular biology. The first generation of biomaterials was created by combining functional characteristics to satisfy tissue needs in an acceptable manner without causing any negative biological environment behavior. Materials like silicone rubber and SS were considered biocompatible because of their inertness. On the other hand, the second-generation biomaterials were created as a result of the first generation's advancements and were intended to create an unbroken response with the surrounding tissue. Drug-eluting endovascular stents and other implants were created based on the bioactivity of these materials. Additionally, throughout that time, biodegradable materials such as polymeric biomaterials emerged. Furthermore, the development of third-generation biomaterials, which aim to promote the regeneration of live or functioning tissues, was made possible by the growing state of the art. Previous studies have shed light on the widespread usage of an untreated cutting tool for these biomaterials' dry, wet, and cryogenic machining processes. However, according to the literature currently in publication, only a small number of recent studies have documented the use of cryo-treated cutting tools during the cryogenic and hybrid machining of these metallic implant biomaterials. There are therefore lots of chances to investigate this field.

Additionally, high electrical conductivity of metals makes them ideal for neuromuscular stimulation devices, such as cardiac pacemakers. The term "biocompatibility" refers to the ability of a metallic implant to integrate with surrounding tissues without causing adverse reactions. Metallic implants are nontoxic, noncarcinogenic, and do not provoke allergic or immunologic responses, making them suitable for medical use [4].

Biomaterials must meet stringent requirements, including biocompatibility, nontoxicity, and appropriate mechanical and physical properties, to serve effectively as replacements or augmentations for biological tissues. One of their primary roles is to replace damaged hard or soft tissues caused by pathological processes, such as infection, fractures, or cancer, which can lead to discomfort, deformity, loss of function, or destruction, as illustrated in Figure 2.1.

A synthetic biomaterial, when designed with specific challenges in mind, may have the potential to repair diseased tissue and restore its function. Currently, metals and polymers are the most commonly used materials for stents. Recent advancements in materials research for implantable metallic devices addressing bone disorders have significantly enhanced the quality of human life, particularly as the global population ages. Improved metals and alloys are widely utilized in biomedical applications due to their exceptional corrosion and fatigue resistance, high strength-to-weight ratio, durability, and overall reliability.

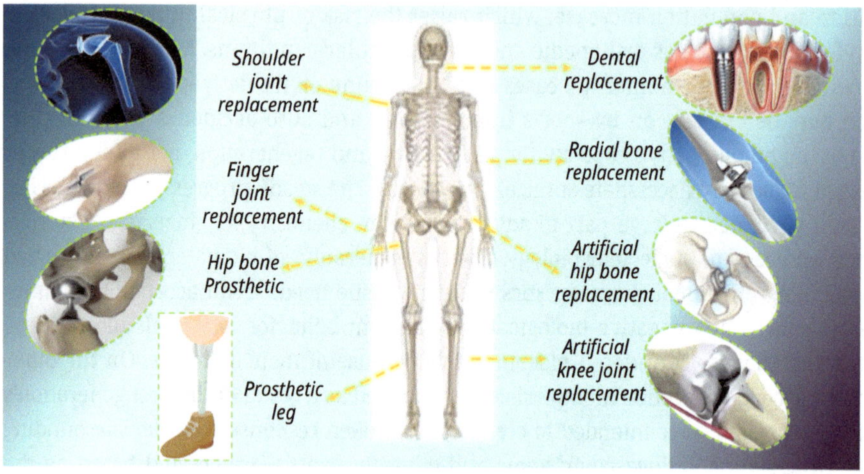

Figure 2.1: Role of biometallic implantations in different parts of the human body [6].

However, these devices face persistent challenges, especially in achieving and maintaining permanent biological fixation at the implant site, which is essential to avoid revision surgeries. In recent years, the concept of biocompatibility has evolved to encompass not only passive implant materials but also active tissue engineering materials. Biocompatibility is guided by two primary principles: biosafety and biofunctionality (or the promotion of body function) [7].

For orthopedic implants, biocompatibility can be further categorized into:

1. **Mechanical biocompatibility**, ensuring proper placement and stability of the implant at the surgical site.
2. **Biological compatibility**, facilitating positive interactions between tissues and the material.
3. **Histocompatibility**, minimizing adverse inflammatory or immune responses.
4. **Blood compatibility**, promoting effective bonding with the host tissue and aiding in the healing of the affected region.

Despite advancements, most metals and alloys used in medical devices were originally developed for aerospace applications, prioritizing mechanical properties rather than biological compatibility. As a result, metallic implants are designed to be inert in physiological environments, enabling nontoxic and biocompatible behavior – critical factors for their effective use (refer to Table 2.2).

Metallic stents offer significantly better strength and stability compared to their polymer-based counterparts, making them the preferred choice in many applications. Popular metallic biomaterials used in stent production include SS (316L SS), nitinol (Ni-Ti), pure iron (Fe), tantalum (Ta), cobalt-chromium (Co-Cr) alloys, platinum-iridium (Pt-Ir) alloys, and magnesium (Mg) alloys. These materials are also extensively

Table 2.2: Mechanical properties of implants – metals and bone.

Metals	Tensile strength (MPa)	Young's modulus (GPa)	Fracture toughness
316 LSS	500–1,000	200	~100
Cp-Ti		110	
CoCrMo	900–1,500	240	~100
Cortical bone	130–150	10–30	2–12

employed in dental implants, orthopedic implants, prosthetic heart valves, and intravascular stents [8]. Their outstanding mechanical properties – such as fatigue strength, fracture toughness, yield strength, and ductility – make metallic biomaterials particularly suitable for load-bearing and high-stress applications.

Recent research has focused on developing novel materials for biomedical use, with surface modification of existing materials emerging as a cost-effective and technologically viable approach. Surface treatments using lasers are gaining traction for modifying metallic biomaterials. From a biomedical engineering perspective, studies have explored degradable and nondegradable metals and alloys, analyzing various surface modification techniques – including treatments with hydrogen peroxide, acids, alkalis, and thermal oxidation – for SSs, titanium, nickel, cobalt-chromium alloys, and others. These chemical treatments enhance the usability of metallic biomaterials in diverse medical applications. Innovation in 3D printing is surpassing the development of new guidelines and standards that target 3D-printed goods in particular. Since its introduction in 2017, the FDA's advice on additive manufacturing (AM) has not been thoroughly updated. Significant research and use of 3D printing in the medical device industry have taken place since then.

2.2 Design innovations in metallic implants

2.2.1 Alloy design for enhanced performance

Advancements in alloy design aim to optimize performance while minimizing risks. For example, reducing nickel content in SSs lowers allergenic potential, while Fe-Cr-Mn-Mo-N alloys improve corrosion resistance and biocompatibility. Nitrogen incorporation enhances strength and stabilizes the austenitic phase [9], as seen in alloys like Ti-13Nb-13Zr, which also reduce elastic modulus (~60 GPa) for better compatibility with bone. The absence of toxic elements such as aluminum and vanadium ensures safer long-term applications. Furthermore, biodegradable magnesium alloys (e.g., Mg-Zn-Ca) are engineered for controlled degradation, offering temporary support in orthopedic applications [10]. Surface coatings and rare-earth elements, like gadolinium, improve corrosion resistance and regulate degradation rates (~0.5 mm/year in physiological conditions).

2.2.2 Surface modification techniques

Innovative surface coatings enhance implant functionality [11, 12]. For instance:
- **Bioactive coatings:** Hydroxyapatite (HA) coatings, applied via plasma spraying, mimic bone mineral matrices, promoting osseointegration with bond strengths exceeding 15 MPa in vivo.
- **Anodization:** Techniques creating titanium oxide nanotubes improve osteoblast adhesion. These surfaces can also deliver localized antibacterial effects, such as vancomycin-loaded nanotubes.
- **Metallic nanoparticles:** Incorporating silver or zinc nanoparticles into implant surfaces reduces bacterial adhesion by over 90%, without compromising biocompatibility.
- **Porous structures:** AM enables precise control over porosity (30–70%) and pore sizes (200–500 µm), optimizing vascularization and bone ingrowth.
- **Functionally graded materials (FGMs):** Gradual compositional changes in FGMs reduce stress at the bone-implant interface, enhancing mechanical compatibility in applications like hip prostheses.

2.2.3 Emerging trends in metallic implants

2.2.3.1 Biodegradable metallic implants

Magnesium alloys, such as Mg-Zn-Ca and Mg-Gd-Dy, are tailored for orthopedic applications, providing osteoconductive properties and uniform degradation. Zinc alloys, like Zn-Mg and Zn-Ca, degrade more uniformly than magnesium, making them suitable for vascular stents and temporary orthopedic devices [13]. Silver and copper nanoparticles embedded in titanium coatings reduce infection risks in high-risk implants. Additionally, implants with polymer coatings can deliver therapeutic agents such as antibiotics, ensuring sustained drug release over extended periods [14].

2.2.3.2 Smart metallic implants

Smart implants integrate sensors to monitor strain, temperature, and performance. Powered by energy harvesters, these implants enable real-time feedback. Shape-memory alloys, like nitinol (NiTi), exhibit superelasticity and shape-memory effects, making them ideal for self-expanding stents and dynamic orthopedic devices [15]. AM techniques like selective laser melting (SLM) facilitate the creation of custom implants tailored to patient anatomy, improving fit, function, and outcomes [16]. Because of its exceptional biocompatibility, titanium and its alloys have a track record of success as biomedical implants. This has to do with the oxide layer that usually forms on the material's surface when it is exposed to an oxidizing environment. Up until recently, the relatively expensive prices and the requirement to prove that the final surfaces had the same level of biocompatibility as implants made using conventional techniques prevented titanium from being processed using the advanced manufacturing technologies of additive layer manufacturing (ALM) and metal injection molding (MIM). This research has shown that titanium alloy structures that are safe and biocompatible for use as medical devices in some applications may be effectively manufactured using advanced and additive manufacturing.

2.3 Design considerations for metallic biomaterials in biomedical applications

Metallic biomaterials are essential in modern medicine due to their unparalleled mechanical properties, corrosion resistance, and adaptability to physiological environments. However, designing metallic implants requires addressing challenges such as biocompatibility, stress shielding, and long-term durability. Biomechanical design, which includes sufficient mechanical qualities like strength, stiffness, and fatigue characteristics required for optimal design, is the first significant difficulty. Second, the biological activation of metal implants and their structural design are also very important. Surface bioactivation aids in implant integration with host bones, especially for bone tissue healing. Bone repair and tissue development are facilitated by the design of porous structures. Third, regenerative tissue engineering is seeing a new trend with metal implants that are biodegradable. Novel metallic bone scaffolds or biodegradable stents require biodegradable qualities. Implants are often thought to be biodegradable after aiding in the healing process and are only required momentarily.

A number of requirements must be met by all biomaterials, including sufficient strength, high corrosion resistance, bioadhesion, biofunctionality, biocompatibility, high wear resistance, and low friction. Nevertheless, not all of the aforementioned conditions are met by the many biomaterials that have been created up to this point. According to reports, the main causes of implant element failure include wear and corrosion. Tribology has several uses in the biomedical field, including the lubrication of artificial heart pumps, pleural surfaces, and the pericardium; wear of heart valves, dentures, screws, and plates in bone fracture repair; and friction between clothing and the person's skin A metallic biomaterial's long-term clinical performance is largely controlled and determined by wear. As the explanation above shows, the biomaterial's ability to withstand wear is crucial to its appropriate performance. A biomaterial's resistance to wear is therefore crucial from a therapeutic standpoint. Numerous investigations have been conducted to examine the tribological characteristics of biomaterials that have been produced. We begin by summarizing the numerous wear test setups that have been employed by different researchers. Next, we describe the various wear characterization approaches and evaluate the various wear and friction data that have been produced by different investigators. Two important characteristics that should be taken into account in the early stages of the development of metallic biomaterials are adequate strength and inertness. Later, inertness gave way to bioactivity in the realm of metallic biomaterials. This concept deviates from established design guidelines for metallic biomaterials. Depending on the needs of the application, more specialized biofunctional designs for metallic biomaterials are suggested. In the meanwhile, additional design guidelines for metallic biomaterials must be carefully taken into account. For clinical uses, metal implants' biologically functioning design is crucial. Innovative metal implants with various biological purposes offer practical methods for human tissue regeneration and repair. The most intriguing problems in biomedical engineering are the design and use of innovative metal implants.

2.3.1 Material selection

Material selection is fundamental to implant design. Factors such as mechanical properties, corrosion resistance, biocompatibility, and manufacturability influence material choice. For example, titanium and its alloys resist corrosion due to their stable titanium dioxide (TiO_2) layer, while Co-Cr alloys benefit from chromium oxide layers. Specific materials are tailored for different applications [17–19]:

- **Orthopedic implants:** Ti-6Al-4V and CoCr are preferred for their strength and wear resistance.
- **Cardiovascular stents:** Nitinol's superelastic properties make it ideal for self-expanding stents.
- **Dental implants:** Pure titanium and Ti alloys support excellent osseointegration.

2.3.2 Balancing strength, ductility, and biocompatibility

Achieving an optimal balance between strength, ductility, and biocompatibility is critical. For instance, alloying titanium with niobium and zirconium ensures biocompatibility while maintaining mechanical strength. Surface engineering techniques such as anodization and bioactive coatings further enhance integration with host tissues [19–21].

2.4 Corrosion and degradation of metallic implants

Corrosion is a significant challenge in using metals and alloys for implants within the human body. To address this issue, a deeper understanding of the underlying principles of corrosion is essential. Metallic implants degrade in the body's aqueous environment through electrochemical reactions, necessitating comprehension of relevant electrochemical mechanisms. The corrosion behavior of implanted alloys resembles the reactions observed in seawater, where metallic components oxidize into ionic forms, and dissolved oxygen reduces to hydroxyl ions. The balance between electron production (oxidation) and consumption (reduction) governs the overall corrosion rate, with the slowest reaction step being the rate-limiting factor [20].

Surgical implants resist corrosion through a protective passive film on their surface. This film minimizes corrosion rates and the release of metal ions [22]. Despite this, all implantable metals experience some level of corrosion due to the complex physiological environment. Common types of corrosion in biomedical alloys include pitting, crevice, galvanic, intergranular, stress-corrosion cracking, corrosion fatigue, and fretting corrosion [23].

2.4.1 Pitting corrosion

Pitting corrosion is a localized and severe form of attack that leads to substantial damage and the release of metal ions. It manifests as small cavities or holes in the material's surface, often beneath an otherwise protective passive film. These pits are formed when aggressive ions compromise weak or defective areas of the film. While some pits are visible, many are microscopic and can initiate stress-corrosion cracking or fatigue cracks, potentially causing implant failure. Maintaining a stable passive state on the material's surface is crucial to mitigating pitting corrosion [24].

2.4.2 Galvanic corrosion

Galvanic corrosion arises when two dissimilar metals come into contact in a conductive fluid medium, such as bodily fluids. This occurs frequently in implants where different metals, like plates and screws, are used together. The metal pairing, ionic and electronic contact areas, and environmental pH influence galvanic corrosion rates. Even in unavoidable scenarios involving dissimilar materials, galvanic corrosion can occur between components such as bone plates and screws, especially where they interface [25, 26].

2.4.3 Fretting corrosion

Fretting corrosion results from repetitive oscillatory motion between contacting surfaces, such as bone plates and prosthetic screws, in a corrosive medium. Even in the absence of corrosion-inducing fluids, fretting can still occur. This phenomenon produces significant corrosion products, potentially causing tissue irritation or initiating cracks that lead to implant failure. Fretting corrosion is influenced by factors such as load, the number of cycles, and material hardness [27].

2.5 Surface engineering and functionalization

Surface engineering plays a pivotal role in enhancing the interaction of biomaterials with biological systems. Key surface properties such as topography, chemistry, and wettability significantly affect biocompatibility and cellular behaviors, including adhesion, proliferation, and differentiation [28]. By engineering these surface characteristics, the integration and performance of implants in biomedical applications can be greatly improved.

Micro- and nanoscale surface features mimic the extracellular matrix, guiding cellular attachment and function. Chemical composition also modulates protein adsorption, which influences cellular responses. Additionally, surface wettability impacts protein adhesion and overall biocompatibility [29].

Modification techniques such as plasma treatments, chemical vapor deposition, and self-assembled monolayers allow for precise control over surface properties. These methods introduce bioactive molecules, enhancing cellular responses [30]. Biofunctionalized surfaces can incorporate growth factors or therapeutic agents to promote tissue regeneration and provide localized treatment, reducing systemic side effects [30, 31].

2.6 Additive manufacturing and customization

AM, or 3D printing, has transformed biomaterial fabrication by enabling complex, patient-specific designs [32]. Techniques such as selective laser sintering and electron beam melting allow precise control over implant properties like porosity and mechanical strength (Figure 2.2). This customization ensures better integration with host tissues and improved implant performance [33–39]. It is evident that processing titanium alloys using sophisticated powder manufacturing techniques such as MIM and ALM, often known as 3D printing, is gaining popularity as a substitute for casting and machining [37]. As numerous of the aforementioned studies have shown, this is due to the accompanying primary advantages, such as increased functionality, lower processing costs, less waste, and design flexibility [40, 41]. Thus, when taken as a whole, the advanced production approaches provide a huge chance to revolutionize the medical device sector.

However, even though ALM provides design freedom, the high cost of machines that can 3D print metals and the expenses related to skill development may prevent its widespread use in implant manufacture. Advanced manufacturing is a fast changing industry that is becoming more dynamic due to the nearly constant appearance of new technology and discoveries. Additionally, it has been demonstrated that ALM can create implants with tailored rough surfaces; however, in certain implant applications, such as joints, a smooth surface is necessary, and at this time, neither ALM nor MIM can create components with extremely smooth finishes to their surfaces. Impressive advancements in the synergistic integration of soft bioelectronics and 3D printing technologies are driving technological advancements in smart medication delivery systems [42–44]. This opens the door to a direct, individualized therapeutic and diagnostic approach. Remote health monitoring, management, and data collecting for outcome assessment are made possible by the integration of electronic devices inside the body [45, 46]. The use of flexible electronics in medicine delivery is becoming more and more significant.

Recent advancements focus on embedding bioactive materials during printing, enabling simultaneous fabrication and functionalization. This innovation ensures uniform bioactive material distribution, enhancing therapeutic benefits. Integrating surface engineering with advanced manufacturing methods makes it possible to create next-generation biomedical implants tailored for personalized medicine [47, 48]. Among AM techniques, powder bed fusion and directed energy deposition are most commonly employed for metallic materials, highlighting their versatility in biomedical applications.

2.6.1 Emerging trends and future directions

The field of biomedical engineering is advancing rapidly with the development of intelligent and biodegradable metallic biomaterials. Metals like magnesium (Mg), zinc (Zn), and iron (Fe), alongside newer materials such as molybdenum (Mo), possess inherent biodegradability. This eliminates the need for secondary procedures to remove implants, making them ideal for applications like cardiovascular stents and orthopedic implants. Their biodegradability, biocompatibility, and customizable mechanical properties – achieved through alloying and surface modifications – highlight their versatility [33, 49]. Direct manufacture of electrical components is another use for 3D printing in smart medicine delivery systems with electronics. This is termed 3D-printed electronics. Within the realm of AM or 3D printing, 3D-printed electronics is a specialized discipline that focuses on creating electrically and electronically functioning components that employ electrical signals to achieve their intended goals [50]. In

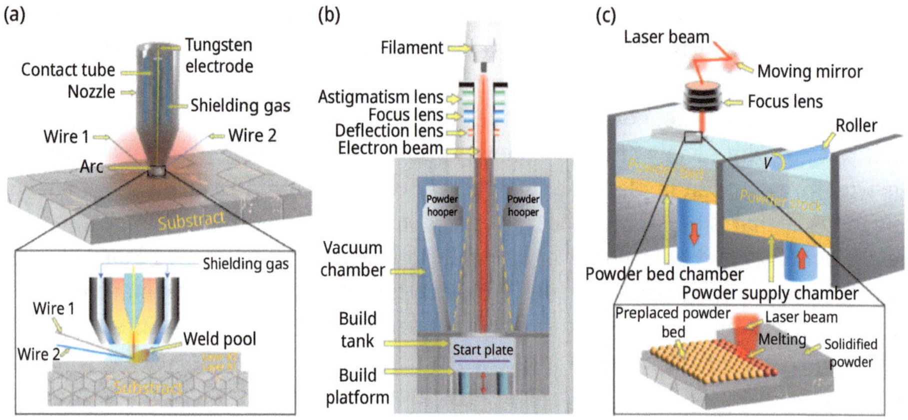

Figure 2.2: Schematic illustrations of 3D printing techniques used to create gradient-structured metals and alloys: (a) directed energy deposition, (b) electron beam melting, and (c) laser powder bed fusion [33].

contrast to traditional 3D printing, 3D electronic printing produces various electronic architectures, from single-material components to multilayer and multiple-material components, by using one or a combination of electrically conducting, semiconducting, and dielectric feedstock materials.

Major sectors including aerospace, automotive, and biomedical device manufacture have seen fast growth in the use of advanced manufacturing techniques such as AM. AM's quick production cycle and design flexibility have greatly benefited the biomedical industry. The primary manufacturing method for metal-based AM implants is powder bed techniques. Alloying metals can improve their qualities, which are rather restricted for pure metals. The majority of metals utilized in engineering applications come in alloy form. Depending on the alloying composition and temperature, the majority of alloys are composed of two or more solid phases that exist as intermetallic compounds or solid solutions. A homogeneous section of a substance with unique chemical and physical properties is referred to as a phase. Every solid solution, intermetallic complex, and pure metal is regarded as a phase. One crucial technique for fortifying alloys and improving their qualities is to alloy metal with finely scattered particles as a second phase. The alloys' overall strength and hardness are increased by the second-phase particles, which act as barriers to dislocation migration.

The focus is shifting toward "smart" materials that go beyond biodegradability by incorporating advanced functionalities such as self-healing, responsiveness to biological stimuli, and controlled drug delivery. Techniques like chemical conversion coatings and micro-arc oxidation are being employed to enhance corrosion resistance, bioactivity, and overall performance. Coupled with innovative alloying methods, these strategies enable precise tuning of mechanical properties and degradation rates to meet both application-specific and patient-specific requirements [34, 35].

Further research is crucial for fully realizing the potential of these materials. Key areas include developing new alloys with superior properties, refining surface modification methods for durability and functionality, and leveraging advanced manufacturing techniques like AM to create complex, personalized implants. Additionally, in silico modeling and comprehensive in vivo studies are essential for predicting and understanding the long-term behavior of these materials within complex biological environments, paving the way for safer, more efficient implantable devices. Long-term implanted devices that are chemically and physiologically inert to prevent damage to human tissues have been the focus of biocompatibility research. However, with recent advances in biotechnology, certain research areas – like tissue engineering, medication, and gene delivery systems, where direct interactions between biomaterials and tissue components are crucial – need a certain amount of biological activity. Biomaterials should have low elastic modulus, shape memory effect or superelasticity, and workability in addition to the advancements in biomedical technology and tissue engineering. They must also completely eradicate any chance of harmful impacts from corrosion, wear, and leaching. Avoiding the usage of nickel while creating metal alloys is one of the issues. A new generation of metallic biomaterials and their innovative processing are the results of this desire.

2.6.2 Nanoengineered surfaces in biomaterials

Nanoengineering is significantly transforming biomaterials science, especially in bone tissue engineering. The nanoscale architecture of bone, comprising collagen fibrils and HA crystals, serves as a model for designing biomaterials that elicit optimal biological responses. By replicating these nanoscale features, researchers are developing surfaces with structures like nanotubes, nanofibers, and nanoparticles, which enhance osseointegration by promoting cell adhesion, proliferation, and differentiation. This biomimetic approach enables the creation of materials that closely resemble the natural bone environment, leading to improved integration and healing.

Beyond mimicking natural structures, nanoengineering introduces novel functionalities. Metallic nanoparticles such as copper, silver, and gold, known for their inherent antibacterial properties, can be integrated into biomaterial surfaces to reduce infection risks in orthopedic and dental implants. Additionally, nanocarriers can be designed to deliver bioactive substances, such as growth factors, in a controlled manner, further enhancing bone regeneration and tissue repair.

Despite these advancements, safety and toxicity remain critical challenges. Comprehensive preclinical evaluations are necessary to address potential issues, including immune responses to nanoparticles and the release of cytotoxic substances. Advanced in vitro and in vivo models, like organ-on-a-chip systems, are essential for ensuring the safety and effectiveness of these materials before clinical application.

2.6.3 Artificial intelligence in biomaterial design

Artificial intelligence (AI), particularly machine learning and deep learning (DL), is revolutionizing biomaterials research and biomedicine. AI is addressing longstanding challenges, such as the slow and resource-intensive nature of traditional clinical trials, by analyzing vast datasets to streamline trial processes, optimize patient matching, and accelerate both fundamental and translational research. Applications in areas like medical imaging, drug development, multiomics analysis, and disease prediction are significantly advancing the field [36].

AI is also transforming biomaterial design by enabling the creation of predictive models that optimize material properties and performance. By integrating biological data from genomics, transcriptomics, proteomics, and other sources, AI systems – utilizing DL architectures such as generative adversarial networks, recurrent neural networks, and convolutional neural networks – can generate virtual libraries of biomaterials. This accelerates the discovery of optimal formulations for various biomedical applications [37]. Furthermore, it is necessary to carry out and encourage thorough research on the use of coated cryo-treated cutting tools in combination with these lubricoolants. Most importantly, there is still a lack of research on the use of promising hybridized combinations of newly developed and existing lubricoolants with coated cryo-treated cutting tools that achieve significant biocompatibility features. This highlights the need for current and aspiring researchers to conduct extensive research in this area.

However, challenges persist. The "black box" nature of many AI models limits their transparency and acceptance in clinical contexts, while data and computational resource limitations remain barriers. Interdisciplinary collaboration among scientists, engineers, and clinicians, as well as advancements in explainable AI, is necessary to overcome these obstacles. The integration of AI into biomaterial design promises to deliver more efficient and effective healthcare solutions [36]. Cryogenic cooling was the only method employed in most previous studies on traditional machining of Mg, NiTi, and Ti-based alloys that produced significant results using a cooling technique other than the typical cutting fluid. Nevertheless, the machining of these biomaterials seldom documented the application of hybrid cooling techniques, which might possibly yield better results, especially the biological reaction of the machined surface. Therefore, to confirm the process capabilities, more biological investigations in vitro and in vivo must be carried out. During conventional machining of biomaterials, it is highly recommended that additional hybrid cooling methods be used in conjunction with cryo-treated cutting tools in future experimental trials. The most recent study's biocompatibility analysis assessed the considerably altered surfaces' short-term in vitro biological response. However, to confirm the long-term biological responses and the ensuing practical use of these changed surfaces, a more thorough in vitro biocompatibility research and an in vivo hemocompatibility inquiry will be carried out. Over the past 10 years, the area of metallic implant biomaterials has grown rapidly, and new alloys and production techniques are being researched to enhance the functionality of biomedical implants inside the human body. Even while biocompatibility, tensile limits or wear, and fatigue resistance have all seen significant advancements, enhancing one of these traits may still need compromising the others or result in low process productivity. However, to increase biocompatibility without sacrificing mechanical and tribological qualities, future trends include porous metallic architectures, functionalized metal surfaces, and novel biodegradable alloys.

2.7 Conclusion

This chapter highlights the design and development of metallic biomaterials for biomedical applications. While metallic biomaterials offer superior mechanical properties and biocompatibility, challenges such as infection and poor osteointegration persist. Strategies to improve biocompatibility include alloy design (e.g., Ti-13Nb-13Zr for a lower elastic modulus or minimizing nickel in SS) and surface modifications like anodization, HA coatings, and antimicrobial nanoparticles. The mechanical performance and biocompatibility of the most widely used metals in the production of medical devices were examined in this study. Selecting an implant material is a complicated challenge that demands numerous distinct factors, including mechanical compatibility, biocompatibility, toxicity of components, and future efficacy and effectiveness. To make use of the complex geometries that AM can produce, the printing process must be improved to produce devices with mechanical performances that are on par with or better than those of traditional manufacturing. This chapter's regulatory section also makes it abundantly evident that certain laws are critical to the technique's future general acceptance and will be a key factor in ensuring its sustained safe and efficient clinical usage. Metal AM has already shown itself to be a fantastic choice for developing novel and inventive medical devices, and this review makes it abundantly evident that there is still more space for this technology to advance and take the lead in patient-centered device production.

The need for turning biomaterials into superior industrially produced human body parts – also referred to as medical implants – has skyrocketed. The well-known types of implants that might be used to save human lives include drug delivery systems, bone plates, screws, cranial, and dental devices. However, a designed implant's precision, surface properties, and capacity to interact with the surrounding tissue or tissues in a given biological context are mostly determined by its processing methods. Furthermore, it is still difficult to manufacture an implant at a high cost while maintaining strict standards.

Emerging trends emphasize biodegradable materials, such as magnesium alloys, and "smart" implants with sensors and energy harvesters for real-time monitoring and personalized responses. AM, particularly techniques like SLM and electron beam melting, plays a crucial role in creating complex, patient-specific implants. Numerous in vitro and in vivo investigations corroborate this finding. To see how cells react to surfaces, the experiments included both human and animal participants in addition to cultured fibroblasts and osteoblasts. It has also been demonstrated that several modern manufacturing processes offer an environment that facilitates the production of implants with customized scaffolds, surfaces, and porosity.

To advance the field, research must focus on intelligent and biodegradable materials, enhanced surface engineering techniques, and AI integration for optimized material discovery and implant design. Comprehensive preclinical evaluations of nanoengineered biomaterials are essential to ensure their safety and efficacy before clinical implementation.

Conflict of interest: No.

References

[1] Lane, W. A. (1895). Some remarks on the treatment of fractures. *British Medical Journal, 1*(1790), 861.

[2] Moghadasi, K., Isa, M. S. M., Ariffin, M. A., Raja, S., Wu, B., Yamani, M., . . . Bin yusoff, N. (2022). A review on biomedical implant materials and the effect of friction stir based techniques on their mechanical and tribological properties. *Journal of Materials Research and Technology, 17*, 1054–1121.

[3] Datta, S., Barua, R., & Das, J. (2020). Importance of alginate bioink for 3D bioprinting in tissue engineering and regenerative medicine. *Alginates – Recent Uses of This Natural Polymer, 10.*

[4] Hermawan, H., Ramdan, D., & Djuansjah, J. R. (2011). Metals for biomedical applications. *Biomedical Engineering-from Theory to Applications, 1*, 411–430.

[5] Barua, Ranjit. "Innovations in Minimally Invasive Surgery: The Rise of Smart Flexible Surgical Robots." In *Emerging Technologies for Health Literacy and Medical Practice*, edited by Manuel B. Garcia and Rui Pedro Pereira de Almeida, 110–131. Hershey, PA: IGI Global, 2024. https://doi.org/10.4018/979-8-3693-1214-8.ch006.

[6] Thanigaivel, S., Priya, A. K., Balakrishnan, D., Dutta, K., Rajendran, S., & Soto-Moscoso, M. (2022). Insight on recent development in metallic biomaterials: Strategies involving synthesis, types and surface modification for advanced therapeutic and biomedical applications. *Biochemical Engineering Journal, 187*, 108522.

[7] Datta, S. & Barua, R. (2024). 3D Printing in Modern Healthcare: An Overview of Materials, Methods, Applications, and Challenges. In M. Garcia & R. de Almeida (Eds.), Emerging Technologies for Health Literacy and Medical Practice (pp. 132–152). IGI Global Scientific Publishing. https://doi.org/10.4018/979-8-3693-1214-8.ch007.

[8] Datta, S., Barua, R., & Das, J. (2020). A review on electro-rheological fluid (ER) and its various technological applications. *Extremophilic Microbes and Metabolites-Diversity, Bioprospecting and Biotechnological Applications.*

[9] Ashby, M. F., Shercliff, H., & Cebon, D. (2018). *Materials: Engineering, Science, Processing and Design.* Butterworth-Heinemann.

[10] Waidi, Y. O., Barua, R., & Datta, S. (2023). Metals, Polymers, Ceramics, Composites Biomaterials Used in Additive Manufacturing for Biomedical Applications. In A. Kumar, P. Kumar, A. Srivastava, & V. Goyat (Eds.), *Modeling, Characterization, and Processing of Smart Materials* (pp. 165–184). IGI Global Scientific Publishing. https://doi.org/10.4018/978-1-6684-9224-6.ch008

[11] National Research Council, Division on Engineering, Physical Sciences, Commission on Physical Sciences, Mathematics, Applications, & Committee on Polymer Science. (1994). *Polymer Science and Engineering: The Shifting Research Frontiers*. National Academies Press.
[12] Barua, R., Datta, S., Datta, P., & Chowdhury, A. R. (2021). Scaffolds and Tissue Engineering Applications by 3D Bio-Printing Process: A New Approach. In I. Management Association (Ed.), *Research Anthology on Emerging Technologies and Ethical Implications in Human Enhancement* (pp. 718–733). IGI Global Scientific Publishing. https://doi.org/10.4018/978-1-7998-8050-9.ch037.
[13] Bandyopadhyay, A., Mitra, I., Goodman, S. B., Kumar, M., & Bose, S. (2023). Improving biocompatibility for next generation of metallic implants. *Progress in Materials Science*, *133*, 101053.
[14] Das, S., Datta, S., Barman, A., & Barua, R. (2023). Smart Biodegradable and Bio-Based Polymeric Biomaterials for Biomedical Applications. In A. Kumar, P. Kumar, A. Srivastava, & V. Goyat (Eds.), *Modeling, Characterization, and Processing of Smart Materials* (pp. 56–82). IGI Global Scientific Publishing. https://doi.org/10.4018/978-1-6684-9224-6.ch003
[15] Datta, S. & Barua, R. (2023). Fluorescent Nanomaterials and Its Application in Biomedical Engineering. In A. Rakha, A. Munawar, V. Khanna, & S. Bansal (Eds.), *Modeling and Simulation of Functional Nanomaterials for Forensic Investigation* (pp. 164–186). IGI Global Scientific Publishing. https://doi.org/10.4018/978-1-6684-8325-1.ch009
[16] Di Schino, A. (2020). Manufacturing and applications of stainless steels. *Metals*, *10*(3), 327.
[17] Hanawa, T. (2012). Research and development of metals for medical devices based on clinical needs. *Science and Technology of Advanced Materials*, *13*(6), 064102.
[18] Long, M., & Rack, H. J. (1998). Titanium alloys in total joint replacement – a materials science perspective. *Biomaterials*, *19*(18), 1621–1639.
[19] Ducheyne, P., Van Raemdonck, W., Heughebaert, J. C., & Heughebaert, M. (1986). Structural analysis of hydroxyapatite coatings on titanium. *Biomaterials*, *7*(2), 97–103.
[20] Besinis, A., De Peralta, T., Tredwin, C. J., & Handy, R. D. (2015). Review of nanomaterials in dentistry: interactions with the oral microenvironment, clinical applications, hazards, and benefits. *ACS nano*, *9*(3), 2255–2289.
[21] Bowen, P. K., Drelich, J., & Goldman, J. (2013). Zinc exhibits ideal physiological corrosion behavior for bioabsorbable stents. Advanced materials (Deerfield Beach, Fla.), 25(18), 2577–2582. https://doi.org/10.1002/adma.201300226
[22] Besinis, A., De Peralta, T., Tredwin, C. J., & Handy, R. D. (2015). Review of nanomaterials in dentistry: interactions with the oral microenvironment, clinical applications, hazards, and benefits. *ACS nano*, *9*(3), 2255–2289.
[23] Chowdhury, S., Jain, N., Waidi, Y. O., Barua, R., Das, S., Prasad, A., & Datta, S. (2024). Biological Smart Biomaterials: Materials for Biomedical Applications. In: Kumar, A., Kumar, A., Kumar, A. (eds) *Applications of Biotribology in Biomedical Systems* (pp. 313–325). Cham: Springer Nature Switzerland.
[24] Wang, J., Chu, J., Song, J., & Li, Z. (2024). The application of implantable sensors in the musculoskeletal system: a review. *Frontiers in Bioengineering and Biotechnology*, *12*, 1270237.
[25] Jain, N., Waidi, Y. O., Chowdhury, S., Barua, R., Das, S., Prasad, A., & Datta, S. (2024). Medical Devices Tribology. In: Kumar, A., Kumar, A., Kumar, A. (eds) *Applications of Biotribology in Biomedical Systems* (pp. 235–250). Cham: Springer Nature Switzerland.
[26] Barua, R., & Datta, S. (2024). Beyond Boundaries: Navigating Future Horizons with Augmented and Virtual Reality in Healthcare and Biomedical Engg. In *Navigating the Augmented and Virtual Frontiers in Engineering* (pp. 199–211). IGI Global.
[27] Rouf, S., Malik, A., Singh, N., Raina, A., Naveed, N., Siddiqui, M. I. H., & Haq, M. I. U. (2022). Additive manufacturing technologies: Industrial and medical applications. *Sustainable Operations and Computers*, *3*, 258–274.

[28] Duan, X., Yang, Y., Zhang, T., Zhu, B., Wei, G., & Li, H. (2024). Research progress of metal biomaterials with potential applications as cardiovascular stents and their surface treatment methods to improve biocompatibility. *Heliyon*, 10(4), e25515. https://doi.org/10.1016/j.heliyon.2024.e25515
[29] Barua, R., Biswas, N., & Das, D. (2024). Emergent Applications of Organ-on-a-chip (OOAC) Technologies with Artificial Vascular Networks in the 21st Century. In *Emerging Technologies for Health Literacy and Medical Practice* (pp. 198–219). IGI Global.
[30] Duerig, T., Pelton, A., & Stöckel, D. J. M. S. (1999). An overview of nitinol medical applications. *Materials Science and Engineering: A*, *273*, 149–160.
[31] Azmat, A., Asrar, S., Channa, I. A., Ashfaq, J., Ali Chandio, I., Chandio, A. D., . . . Devanesan, S. (2023). Comparative study of biocompatible titanium alloys containing non-toxic elements for orthopedic implants. *Crystals*, *13*(3), 467.
[32] Datta, S., Das, S., & Barua, R. (2023). Self-sustained nanobiomaterials: Innovative materials for biomedical applications. *Advanced Materials and Manufacturing Techniques for Biomedical Applications*, 303–323.
[33] Marin, E., & Lanzutti, A. (2023). Biomedical applications of titanium alloys: a comprehensive review. *Materials*, *17*(1), 114.
[34] Ali, W., Mehboob, A., Han, M. G., & Chang, S. H. (2019). Effect of fluoride coating on degradation behaviour of unidirectional Mg/PLA biodegradable composite for load-bearing bone implant application. *Composites Part A: Applied Science and Manufacturing*, *124*, 105464.
[35] Eliaz, N. (2019). Corrosion of metallic biomaterials: a review. *Materials*, *12*(3), 407.
[36] Bijapur, K., Molahalli, V., Shetty, A., Toghan, A., De Padova, P., & Hegde, G. (2023). Recent trends and progress in corrosion inhibitors and electrochemical evaluation. *Applied Sciences*, *13*(18), 10107.
[37] Ernst, P., & Newman, R. C. (2002). Pit growth studies in stainless steel foils. II. Effect of temperature, chloride concentration and sulphate addition. *Corrosion Science*, *44*(5), 943–954.
[38] Peng, H., Wang, W., Jiang, H., Zan, R., Sun, Y., Yu, S., . . . Zhang, X. (2021). Effect of galvanic corrosion on the degradability of biomedical magnesium. *Frontiers in Materials*, *8*, 767179.
[39] Høl, P. J., Mølster, A., & Gjerdet, N. R. (2008). Should the galvanic combination of titanium and stainless steel surgical implants be avoided?. *Injury*, *39*(2), 161–169.
[40] Brown, S. A., & Simpson, J. P. (1981). Crevice and fretting corrosion of stainless-steel plates and screws. *Journal of Biomedical Materials Research*, *15*(6), 867–878.
[41] Long, S., Zhu, J., Jing, Y., He, S., Cheng, L., & Shi, Z. (2023). A comprehensive review of surface modification techniques for enhancing the biocompatibility of 3D-printed titanium implants. *Coatings*, *13*(11), 1917.
[42] Zhu, L., Luo, D., & Liu, Y. (2020). Effect of the nano/microscale structure of biomaterial scaffolds on bone regeneration. *International Journal of Oral Science*, *12*(1), 6.
[43] Kim, H. S., Kumbar, S. G., & Nukavarapu, S. P. (2021). Biomaterial-directed cell behavior for tissue engineering. *Current Opinion in Biomedical Engineering*, *17*, 100260.
[44] Meyers, S. R., & Grinstaff, M. W. (2012). Biocompatible and bioactive surface modifications for prolonged in vivo efficacy. *Chemical Reviews*, *112*(3), 1615–1632.
[45] Nikolova, M. P., & Chavali, M. S. (2019). Recent advances in biomaterials for 3D scaffolds: A review. *Bioactive Materials*, *4*, 271–292.
[46] Ji, W., Zhou, R., Vivegananthan, P., Wu, M. S., Gao, H., & Zhou, K. (2023). Recent progress in gradient-structured metals and alloys. Progress in Materials Science, 140, 101194. https://doi.org/10.1016/j.pmatsci.2023.101194
[47] Khan, A. R., Grewal, N. S., Zhou, C., Yuan, K., Zhang, H. J., & Jun, Z. (2023). Recent advances in biodegradable metals for implant applications: Exploring in vivo and in vitro responses. *Results in Engineering*, *20*, 101526.

[48] Khan, A. R., Grewal, N. S., Zhou, C., Yuan, K., Zhang, H. J., & Jun, Z. (2023). Recent advances in biodegradable metals for implant applications: Exploring in vivo and in vitro responses. *Results in Engineering*, *20*, 101526.
[49] Han, S., & Wu, J. (2024). Artificial intelligence (AI) meets biomaterials and biomedicine. *Smart Materials in Medicine*, *5*(2), 251–255.
[50] Han, S., & Wu, J. (2024). Artificial intelligence (AI) meets biomaterials and biomedicine. *Smart Materials in Medicine*, *5*(2), 251–255.

Sachin Latiyan, Ranjit Barua, Dileep Pathote, Vilay Vannaladsaysy, and Sudipto Datta*

Chapter 3
Stainless steel-based biomaterial for orthopedic fixations

Abstract: Orthopedic implants are vital for treating bone deformities resulting from diseases or injuries, and their application has grown on a large scale. An orthopedic implant's success within the human body depends on its shape, size, and mechanical characteristics. Chemical stability and toxicity are equally crucial for their long-term biocompatibility. Materials widely utilized for orthopedic applications include stainless steel (SS), cobalt-chromium and titanium alloys, ceramics, polyethylene, and polymethyl methacrylate bone cement which are known for their mechanical strength, corrosion resistance, and biocompatibility. However, their prolonged use might result in the release of metal ions, which can cause tissue injury, inflammation, and implant failure. Since the development of SS over a century ago, they have made significant contributions to the medical and manufacturing industry. In this chapter, we have discussed SS, an important biomaterial for orthopedic implants, focusing on its composition, mechanical properties, merits, and current advances in increasing its properties. We also look at future directions in SS-based biomaterials, particularly in terms of corrosion reduction, biocompatibility enhancement, and implant integration with bone tissue.

Keywords: Stainless steel, fracture fixation, orthopedic implants, metallic biomaterials

---

Acknowledgment: Sudipto Datta acknowledges SERB for the National Postdoctoral Fellowship (PDF/2022/000990).

***Corresponding author: Sudipto Datta**, Department of Materials Engineering, Indian Institute of Science, Bengaluru 560012, Karnataka, India, e-mail: sudiptodatta1990@gmail.com
Sachin Latiyan, Dileep Pathote, Department of Materials Engineering, Indian Institute of Science, Bengaluru 560012, Karnataka
Ranjit Barua, Mechanical Engineering Department, Omdayal Group of Institutions, Howrah 711316, West Bengal, India, e-mail: ranjitjgec007@gmail.com
Vilay Vannaladsaysy, Department of Mechanical Engineering, Faculty of Engineering, National University of Laos, Sokpaluang 01005, Vientiane, Laos

https://doi.org/10.1515/9783111571423-003

3.1 Introduction

The development of orthopedic biomaterials was inspired by the requirement for repairing fractures, correcting deformities, and treating degenerative bone diseases with a material that possesses high strength, biocompatibility, and corrosion resistance. Various kinds of metallic implants used within the human body based on the application (deformity or fracture) are shown in Figure 3.1. The use of metals in orthopedic fixation began in the early nineteenth century when surgeons tried different types of materials to immobilize fractures. The first efforts were made with the help of metals like iron, gold, and silver but all of them were associated with certain disadvantages. For example, gold and silver were too soft for use in load-bearing applications, and iron implants had corrosion issues [1–3].

The key change occurred only at the beginning of the twentieth century when the first stainless steel was created, an alloy that combines strength with corrosion resistance. Stainless steel was first used in orthopedic surgery, especially after its success in industrial applications. Stainless steel was adopted as the material of choice for use in bone screws, plates, and intramedullary rods by the mid-twentieth century because of its resistance to corrosion by body fluids. Stainless steel, especially 316L, has been a preferred orthopedic fixation material for many years, despite the advancement of titanium and cobalt-chromium alloys because of its mechanical properties, cost, and ease of procurement [3, 4].

As the population ages, the demand for better implants in orthopedic and dentistry disciplines grows. Over time, the complexity of orthopedic surgeries has grown, with the need for materials that can handle large loads without wearing out the body. Modern implants are comprised of metals and alloys that are strong, corrosion-resistant, and long-lasting. However, obtaining stable, long-term fixation and fostering early bone ingrowth are still prominent challenges. Recent biomaterials research has focused on improving biocompatibility, ensuring that implants promote appropriate healing while avoiding undesirable tissue responses such as inflammation [3, 5, 6].

In this chapter, we discuss why stainless steel remains in practice and evaluate current developments aimed at improving its efficiency. In summary, this chapter illustrates the history of stainless steel as an implant material, the composition of the steel, its mechanical properties and corrosion resistance, and the issues arising from its long-term use. Efforts to improve the performance and utility of stainless steel implants, especially future developments that may help extend the use of stainless steel in orthopedic surgery, are also outlined.

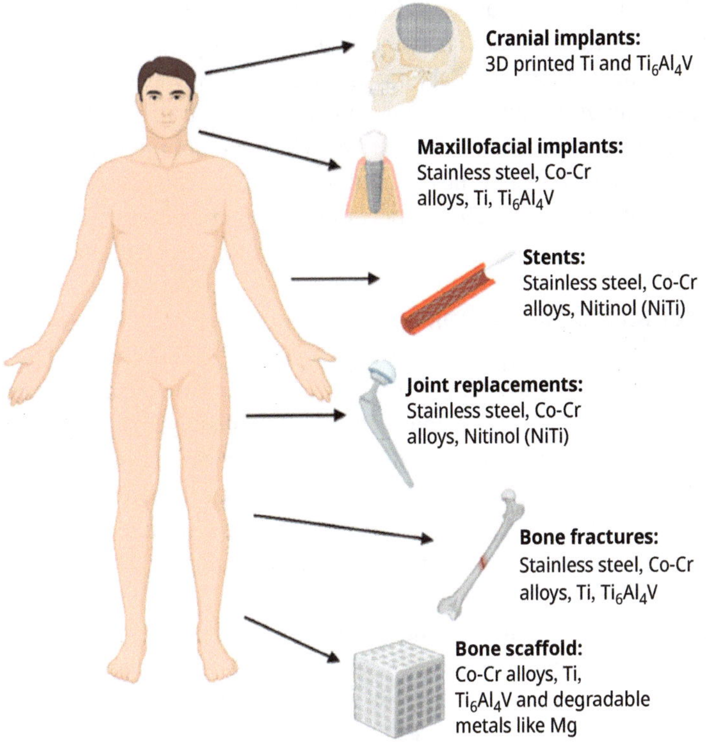

Figure 3.1: Various kinds of metallic implants are used within the human body based on the application (deformity or fracture). Adapted from [1].

3.2 Stainless steel and orthopedics

3.2.1 Stainless steel: a preferred material for implants

Surgical stainless steel was first used in the 1920s, and by the mid-1930s, it had largely replaced many earlier metals because of its greater resistance to corrosion and its superior mechanical characteristics. New elements were introduced into the stainless steel to make better grades, namely chromium formed a passive oxide layer to prevent corrosion, and nickel enhanced ductility and toughness in the material [4].

The development of 316L stainless steel, a low-carbon steel, added to the appropriateness of the material for orthopedic implants. The low carbon concentration decreases the chances of carbide formation when welding and fabricating the material to enhance its corrosion properties. This innovation was a leap forward for implants that are expected to be in the body for a very long time [4, 7].

Stainless steel was widely used in orthopedic fixations all through the second half of the twentieth century. It was suitable for both temporary and permanent applications: from fracture immobilizations to total joint replacements. However, due to the increasing requirements for lightweight and biocompatible materials, alloys such as titanium and cobalt-chromium alloys are being used. While stainless steel is still an important material, the cost of the materials used plays a decisive role; in the case of temporary fixation devices, the material has to perform for a short time only [4].

3.2.2 Other metal alloys for orthopedics

In the contemporary practice of orthopedics, there is a tendency toward the application of alternative materials, including titanium and cobalt-chromium alloys for permanent implants. Titanium alloys are characterized by high specific strength and significantly higher biocompatibility than stainless steel. Also, they have a lower modulus of elasticity, which is nearly equal to bone and thus decreases stress shielding, a factor where an implant with a high modulus of elasticity takes over the load bearing from the bone, leading to resorption of the bone.

On the other hand, cobalt-chromium alloys have excellent corrosion resistance and a high fatigue strength, which allow them to be used in high-stress applications such as joint replacement prostheses. But they are more costly and hard to machine than stainless steel and also prone to corrosion. Stainless steel has not lost its appeal as a material for use in conditions where factors such as cost, material availability, and ease of fabrication are important [8].

Although titanium and cobalt-chromium alloys are used in permanent implants, stainless steel is still used in orthopedic appliances like plates, screws, and rods. Stainless steel implants are preferred particularly in cases where the bone is broken or in circumstances where the implants will be likely to be withdrawn once the bone has healed [4, 8].

3.3 Metallurgical and compositional aspects of stainless steel

3.3.1 Stainless steel: 316L

316L stainless steel, the most popular of all grades used in orthopedic implants, comprises mainly iron, chromium, nickel, and molybdenum. The main ingredient of stainless steel is chromium (16–18%): when exposed to air, it spontaneously develops a thin passive chromium oxide film on the material surface, which acts as a protective barrier to further corrosion. Nickel (10–14%) improves the toughness and ductility of the alloy and hence easy to forge and fabricate. The presence of molybdenum (2–3%) increases the material's resistance to chloride attack, an essential property in the chloride-rich environment of the human body. The "L" in 316L means "low carbon" with a carbon content of not more than 0.03% in it. This lower carbon content minimizes the possibility of carbide formation during welding, which if formed can cause intergranular corrosion. By reducing this risk, 316L stainless steel retains the integrity of the material, whether under mechanical stress or more complicated processes of fabrication [4, 6, 9].

Besides these basic constituents, the 316L stainless steel also contains manganese, silicon, and nitrogen in small proportions that have influences on the characteristics of the alloy. The balanced combination of these components is the reason why 316L stainless steel possesses high strength, good corrosion resistance, and manufacturability [4, 6].

3.3.2 Chromium, nickel, and molybdenum in stainless steel

The unique properties of stainless steel for use in orthopedic fixations largely stem from the roles played by its primary alloying elements [4, 6, 10]:

- Chromium: It is the most important constituent when it comes to the corrosion resistance of stainless steel. Chromium forms a passive oxide layer (Cr_2O_3) when the alloy is exposed to oxygen. This layer provides a shield to the underlying metal to be further corroded even in harsh environments such as chloride-rich environment of the body fluids.
- Nickel: The addition of nickel to stainless steel enhances the overall mechanical performance of the material by increasing its toughness and at the same time reducing brittleness. It also improves the stability of the austenitic phase of the steel, and makes it nonmagnetic, thus improving its workability, particularly at lower temperatures.
- Molybdenum: The addition of molybdenum to stainless steel improves the metal's ability to resist pitting and crevice corrosion, especially in saline environments. It becomes essential for orthopedic use as implants usually come into contact with body fluids, which otherwise would corrode the material.

3.3.3 Effect of microstructure and heat treatment on mechanical properties

It was also found that the microstructure of stainless steel can be altered through heat treatment processes with a view of improving its mechanical characteristics. The austenitic structure of stainless steel, 316L, due to the presence of nickel as an alloying element is nonmagnetic and possesses appropriate strength and ductility. Further enhancement of these properties can be achieved through additional heat treatments, including annealing, to reduce internal stress and increase the homogeneity of the material [4, 6, 11].

Stainless steel tempering involves heating it to a certain temperature of about 1,050 °C and quenching it. It enhances the material's hardness and flexibility and enables easy forging into various products such as screws and plates. Such annealing also minimizes crack formation during subsequent machining or implant insertion and therefore increases the implant success rate in orthopedic operations. The stringent mechanical properties of the orthopedic implants can only be achieved by the proper control of the microstructure of 316L stainless steel by appropriate alloying and heat treatment to provide the desirable combination of strength and corrosion resistance [4, 6, 11].

3.4 Mechanical properties of stainless steel as an orthopedic implant

Mechanical properties of stainless steel are among the most critical characteristics that determine its applicability in orthopedic applications. Bolts and rivets, screws, plates, and rods must be capable of withstanding a considerable degree of mechanical loading without failure of the implant. Stainless steel, especially 316L, is preferred for load-bearing orthopedic fixations because of its strength, toughness, and ductility making it the most popular material [2, 3, 5, 12, 13].

3.4.1 Elastic modulus, tensile strength, and yield strength

Another characteristic of stainless steel is the tensile strength, which is the maximum stress level that a material can be subjected to while under tension before it fractures. The tensile strength of 316L stainless steel is normally within 485–620 MPa. The high tensile strength makes it suitable for use, where the implant is exposed to high forces as in load-bearing conditions [4, 6].

Another important parameter is the yield strength, which is the stress level at which the material starts to undergo permanent deformation. 316L stainless steel has the yield strength of around 170–310 MPa. Thus, before the onset of permanent deformation, the material is capable of sustaining a high degree of stress without fracturing. The higher yield strength is desirable for orthopedic screws and plates, as these are required to maintain their shape during bone healing after a fracture [5, 12, 13].

The Poisson's ratio, which measures the ratio of the lateral strain to the axial strain, is 0.3 for the 316L stainless steel, while the elastic modulus, or the measure of stiffness, is approximately 193 GPa. This value is much higher than that of bone (about 17–25GPa), which may cause stress shielding. Stress shielding is when the implant has borne most of the load and the bone has been subjected to minimal mechanical forces, hence leading to bone resorption. This can be a concern; however, the higher stiffness of stainless steel is still desirable in applications where high stiffness is required to avoid bending or distortion of the implant [13, 14].

3.4.2 Fatigue resistance and durability

Regarding orthopedic applications, the implants experience cyclic loading, which is stress resulting from activities such as walking, running, or lifting. The extent to which an implant material is capable of withstanding these repeated loads without failure is normally known as fatigue resistance. Stainless steel also has good fatigue strength whereby the material can withstand cyclic loading for a relatively long time which is desirable in devices such as intramedullary rods and fracture plates that are expected to remain in position for months or even years [3, 5, 11].

But as with any other material, stainless steel also faces fatigue failure. Fatigue failure usually arises when cracks start at regions of high stress concentration, for instance, at the head of screws or corners of plates. This is controlled by design and manufacturing techniques that avoid stress concentrators and by using surface treatments that improve fatigue strength [11].

3.4.3 Load-bearing applications: screws, plates, and rods

The high tensile and yield strength of the material mean that the material will not deform when subjected to the heavy loads that are placed on implants in load-bearing bones such as the femur or tibia. Due to high tensile strength, stainless steel-derived screws, plates, and rods are used in the fixation of fractured or distorted bones, as shown in Figure 3.2.

Stainless steel screws are applied to secure fractured bone fragments and may be accompanied by plates. This is because the threads of stainless steel screws remain intact during bone healing and do not deform when stressed. The high yield strength of stainless steel ensures that screws remain intact and are not distorted during the healing period [1, 13].

Plates are applied to bridge defects and give mechanical support, especially in long bones. Stainless steel plates should be sufficiently resistant to bending and twisting motions and at the same time comparatively thin to avoid bulk. Their durability under load is one of the ways through which it is possible to achieve the proper healing of the bone [2, 3, 5].

On the other hand, intramedullary rods are used in fractures of long bones; these are fixed in the bone marrow cavity and support much of the body's weight. Stainless

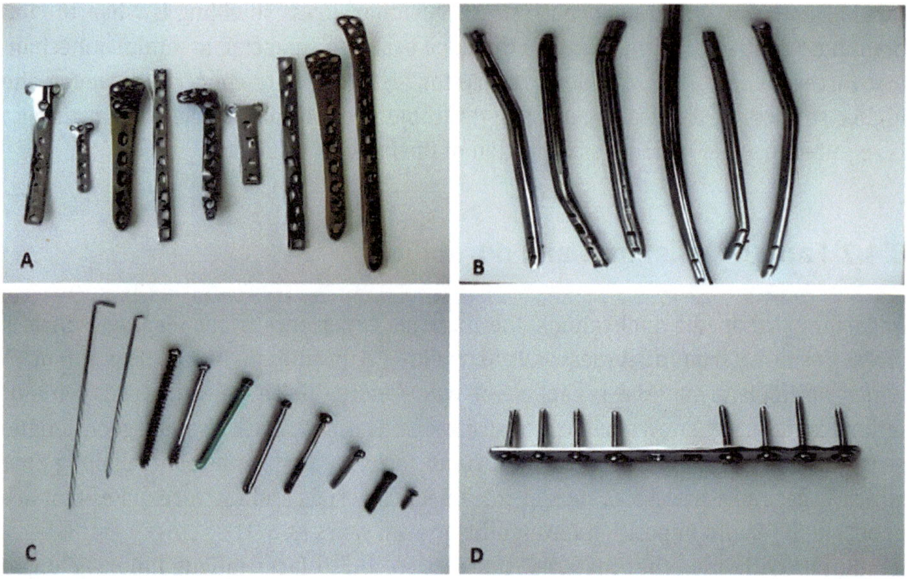

Figure 3.2: Various types of metallic implants: (**A**) stainless steel and titanium plates; (**B**) intramedullary nails made from stainless steel; (**C**) screws and pins made from stainless steel and titanium; and (**D**) stainless steel screws within a plate. Adapted from [15].

steel rods give the support and stability that are necessary to maintain spine alignment during bone healing.

In general, the mechanical properties of stainless steel make it suitable to be used in load-bearing orthopedic devices. Nonetheless, stainless steel has high stiffness, which becomes a problem at some times and can cause stress shielding. However, it continues to be an important element in most orthopedic operations [2, 3, 13].

3.5 Corrosion behavior of stainless steel in the human body

The corrosion resistance makes stainless steel the most widely biomaterial because it does not corrode and therefore lasts longer. But when this steel is used in the environment of the human body, where the steel could be exposed to different electrolytes and chloride-rich fluids, corrosion is still a problem [2, 3, 8].

A study of 50 failed 316L stainless steel implants retrieved from patients in India over 4 years found that implant failures were most common in the femoral area (74%), with lower rates in the knee, tibia, humerus, and radius/ulna. More than 90% of the failures were caused by localized corrosion, mainly pitting and crevice corrosion, which worsened over time. The study found pit-induced fatigue failure in compression bone plates and stress corrosion cracking in intramedullary nails, highlighting the importance of corrosion in implant failure. Thus, knowledge of the processes of corrosion in implants made of stainless steel is necessary to be able to predict the service life of these implants in orthopedic applications [8, 16].

3.5.1 Chloride environment: pitting and crevice corrosion

Corrosion of stainless steel spinal implants can occur mechanically or electrochemically. Fretting is a mechanical kind of corrosion that happens at contact points between materials under load when they are subjected to repetitive cyclic loading or vibration. Electrochemical corrosion includes pitting, galvanic, and crevice corrosion. Galvanic corrosion (a type of electrochemical corrosion) occurs when materials with various electrochemical potentials are close together in an electrolyte solution (such as saline). One substance oxidizes (anode), while the other reduces (cathode) [10].

The human body is rich in saline and acidic environments that pose challenges to metallic implants. Blood, synovial, and interstitial liquids that are present in the human body are known to contain chloride ions (Cl^-), which have a high corrosion effect on metals. In the case of stainless steel, it is protected mostly by the passive oxide layer mainly comprising chromium oxide. This layer acts as a shield that would help avoid any reaction of the metal below it with its surroundings. But, in some circumstances, the passive layer may break, especially where chloride ions are present in high concentrations. Pitting corrosion is a kind of corrosion that occurs in a localized form and causes pits to be formed due to interference with the passive layer as shown in Figure 3.3. These pits can extend with time, and this adversely affects the mechanical stability of the implant. Pitting is most devastating to stainless steel screws and plates because even minimal corrosion can reduce the strength of the material significantly [8, 14].

Another kind of corrosion affecting orthopedic implants is crevice corrosion, which takes place where there is restricted access to oxygen, for example, between the screw and the bone or beneath the plate. In these areas, the protective Cr_2O_3 layer cannot self-heal, and localized corrosion can take place. Crevices are very dangerous because they are concealed, and the corrosion process may proceed unnoticed until the metal is severely damaged.

In the case of pitting and crevice corrosion, the factors that determine its occurrence include implant design, surface finish, and the environment in the body. Implants that move relative to the surrounding tissues, such as in load-bearing joints, are especially at risk because mechanical wear can destroy the oxide layer, which in turn accelerates corrosion.

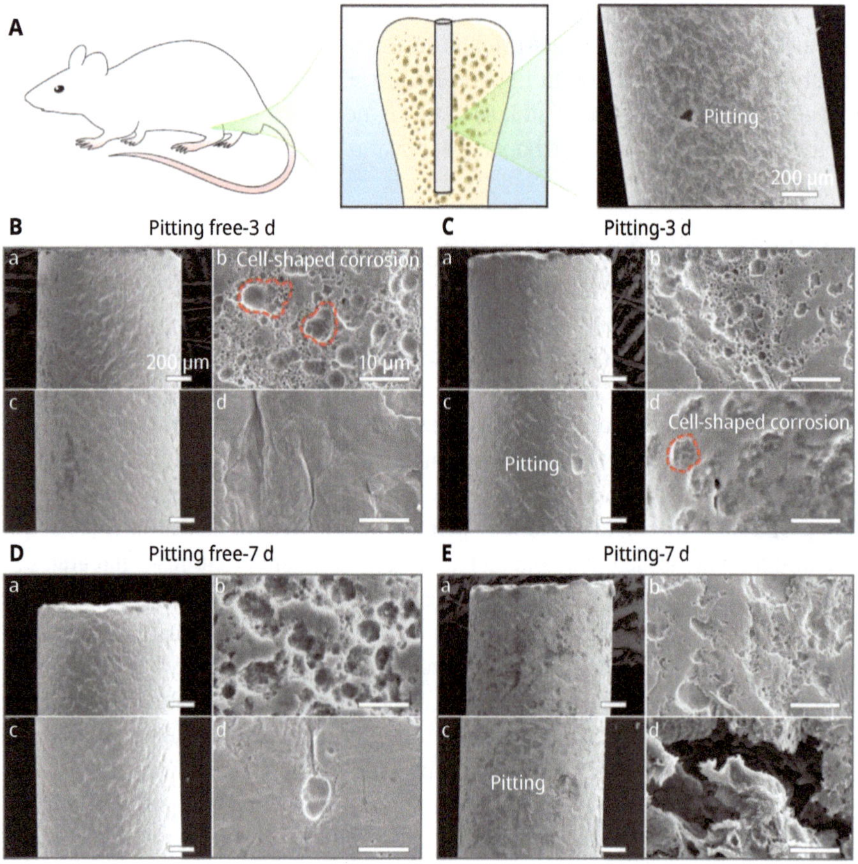

Figure 3.3: In vivo corrosion study of stainless steel (SS) implant. (A) Schematic illustration of in vivo study and morphology of pitting within sample; (B) SEM micrographs of pitting-free 316L SS after 3 days in vivo corrosion study; (C) SEM micrographs of pitting corrosion in 316L SS after 3 days in vivo corrosion study; (D) SEM micrographs of pitting-free 316L SS after 7 days in vivo corrosion study; (E) SEM micrographs of pitting corrosion in 316L SS after 7 days in vivo corrosion study: (a) corrosion morphologies of the end region of samples, (b) zoomed micrographs of (a), (c) corrosion morphologies of the middle region of samples, and (d) zoomed micrographs of (c). Adapted from [17].

3.6 Stainless steel implants and their biocompatibility

Biocompatibility is an important aspect when the given material is a part of biomedical-related products; it means that the material should be capable of performing the intended function, and at the same time, it should not elicit an immune system. For orthopedic fixation, biocompatibility includes not only how the material interfaces with bone and soft tissues but also the ability to minimize other biological responses that are undesired including inflammation, allergy, or infection. 316L-grade stainless steel is perhaps the most popular of all the stainless steel types used in orthopedic implants because of its relatively higher biocompatibility in comparison to other variants. However, the following issues are still contentious, especially about the long-term effects and possible ion release [18].

3.6.1 Stainless steel: bones and soft tissue interactions

Stainless steel after implantation needs to have long-term mechanical stability and at the same time elicit the right biological response within the bone and adjacent soft tissues. The first contact that the implant has with the surrounding tissues includes the generation of a fibrous layer around the implant which is a normal response by the body to injury. In most instances, this fibrous layer is usually thin and does not hinder the performance of the implant. However, the excessive fibrous encapsulation was found to cause a loosening or instability of the implant, especially if it is used in load-bearing applications [18–20].

Previous research has demonstrated that the typical response of bone tissue to 316L stainless steel implants is promising. Particularly, it has been confirmed that osteoblasts (the bone-forming cells) are capable to grow and proliferate on the stainless steel implant surface. However, the material does not have any characteristics that can enhance osseointegration, that is, direct bonding between the bone and the implant surface. The absence of osseointegration may be a drawback in the case of permanent implants, in which bone adhesion is critical. However, since temporary fixture devices involve implants that will be retrieved after bone healing, the biocompatibility of stainless steel is satisfactory [4].

Another important factor is the soft tissue response, especially when the implant runs through tendons, ligaments, or muscles. The stainless steel material usually produces a low-grade inflammatory reaction of tissue and this is usually temporary. Most of the time there are no serious complications when soft tissues grow around the implant. Nevertheless, friction or movement of the implant within soft tissues is undesirable and may cause local irritation, leading to pain or implant failure [7, 18, 20].

3.6.2 Release and toxicity of metal ions

Another general issue that stainless steel has with biocompatibility is corrosion and subsequent release of metal ions into the tissue and blood. Corrosion, especially pitting and crevice corrosion, can cause nickel, chromium, and iron ions to be liberated into the body from the implant. This release may be hastened by mechanical wear that brings out the metal underneath the coating to the biological environment and causes various complications [3, 5].

Nickel is an identified allergen and has been reported to cause both local and systematic reactions in patients from stainless steel implants. Nickel allergy is not rare and its effect may be as mild as skin rash or as severe as inflammation, whereas chromium and iron ions are not as likely to elicit an allergic response as nickel ions. However, these may induce inflammation and, at higher concentrations, may have cytotoxic effects [8].

The degree of ion release and its clinical effects depend on the location of the implant, the patient's immune reactions, and the period of implantation. Stainless steel used in the manufacture of these temporary implants does not normally pose problems since the implants remain in the body for a short time only. However, for permanent implants, metal ion release is a much bigger issue; thus, searching for other materials or modifying the surface of the implants to minimize the release of metals is required [8, 15, 17].

To overcome these issues, implants made from stainless steel are subjected to different treatments, including passivation, coating, or plasma treatments that minimize chances of corrosion and ion release. Similarly, to reduce the risk of an allergic reaction in the body, new low-nickel stainless steel alloys or other biocompatible coatings have been tried [8].

3.6.3 Immune-induced inflammatory responses

Stainless steel implants are usually acceptable by the immune system of the body, and the initial inflammatory response to them is usually moderate and inflammation subsides with healing. Surface characteristics of the implant such as roughness, chemical constituents, and cleanness are critical in determining the immune response at the implantation site. Passivated smooth surfaces provoke a lower grade of inflammation than rough or mechanically distorted surfaces as foreign body response [1, 17, 21, 22].

The foreign body response entails the stimulation of macrophages to surround the implant via a fibrous capsule. This response is characteristic and typical for most of the biomaterials, including stainless steel. However, sometimes the body will initiate a long-term immune response causing inflammation, the loosening of the implant, or pain. The chronic inflammation can be caused by wrong implant positioning, mechanical loads, or infection [20].

Infection is still one of the worst septic complications that can be associated with any orthopedic implant. Stainless steel is not known to encourage bacterial growth but infections may result from bacterial introduction at the time of surgery or development of bacterial growths on the surface of the implant. Implanted stainless steel components are prone to bacterial biofilm formation from *Staphylococcus aureus*, which is a leading cause of orthopedic infections. However, the biofilm is very hard to eliminate without the removal of the implant, once it has formed. As a result, new studies are underway to develop antibacterial coatings on stainless steel implants that could reduce the rate of infection and enhance the quality of patients' lives [21, 22].

3.7 Strategies to mitigate corrosion: coating and surface treatment of stainless steel

Considering the problems of corrosion, wear, and biocompatibility in orthopedic applications, the following surface modification technologies have been employed to improve the performance of stainless steel. These changes are intended to enhance corrosion protection, minimize metal ion leaching, enhance biological integration, and prevent infection [20].

3.7.1 Passivation, alloying, and electropolishing

The use of stainless steel in orthopedic applications has been associated with some challenges, which include corrosion problems. The most used technique to overcome this problem is passivation, which involves the use of chemical solutions to encourage the formation of the passive oxide layer on the surface of stainless steel. Passivation usually involves immersing the exposed surface of the implant into a mild oxidizing agent like nitric acid to initiate the formation of a passive chromium oxide layer [11, 14, 23]. Passivated stainless steel has increased resistance to pitting and crevice corrosion, which is suitable for longer durations of implantation [12, 20].

Another method of enhancing the corrosion resistance of stainless steel is by alloying. The presence of molybdenum and nitrogen in the alloy greatly improves the material's ability to resist pitting and crevice corrosion. Of all the alloying elements, molybdenum enhances the ability of the passive layer to withstand chloride attack, particularly in cases where chloride concentration is high.

Other surface modification techniques that have been used for enhancing the corrosion resistance of stainless steel implants include electropolishing and plasma nitriding. Electropolishing removes small peaks and valleys that act as sites of initiation of corrosion, while plasma nitriding increases the surface hardness and provides better wear and corrosion resistance [2, 3, 11, 23]. Electropolishing is another frequently employed surface treatment that removes the material from the surface of stainless steel and makes it smooth by electrochemical procedures. The electropolished implant possesses a smoother and cleaner surface that is less likely to rust or to develop biofilm. Electropolishing also eliminates micropitting, such as burr and rough surface, which can be stress raisers or starting sites of corrosion. Because electropolished stainless steel implants have smoother surfaces, they are not as suitable for bacterial colonization or stimulation of the inflammation process, and are therefore more biocompatible and long-lasting [12, 20].

Integrating these approaches, stainless steel implants can provide long-term protection against the destructive influences of human fluids, which is why they can be used as orthopedic fixation devices. Nevertheless, more current research still goes on with the aim of improving their performance even under demanding clinical conditions where high loads and exposure to chloride-containing fluids are likely to occur [11, 14, 23].

3.7.2 Plasma nitriding and ion implantation

Plasma nitriding is one of the surface engineering techniques in which nitrogen is incorporated into the surface of stainless steel through plasma. This process results in a thin, but strong layer of nitride material on the surface of the implant and greatly improves the wear characteristics of the material. In addition, plasma nitriding also enhances the corrosion protection of stainless steel, especially in chloride-rich solutions by stabilizing the passive oxide layer [11, 12, 19, 23].

Ion implantation is another process, where stainless steel is subjected to high-energy ions, say nitrogen or titanium, to change its chemical composition. Ion implantation will increase surface hardness, reduce wear rate, and improve corrosion resistance without changing the bulk properties of the material. This makes it especially relevant, where the surface properties of the implant are important for its service life, such as in orthopedic implants [11, 12, 23].

3.7.3 Antibacterial and bioactive coatings

In consideration of the infection threat posed by orthopedic implants, scientists have designed numerous antibacterial coatings for stainless steel. These coatings have been designed to minimize bacterial attachment and biofilm formation that would otherwise lead to implant-related infections. Some of the typical antibacterial coatings are silver, copper, and polymers loaded with antibiotics. It has been observed that silver has antibacterial properties and its coatings on stainless steel implants decrease bacterial adhesion without compromising biocompatibility. Another antimicrobial metal is copper, which is used for coatings, while the use of this metal is not very common because at high concentrations, copper is toxic [21, 22].

Another approach is antibiotic-containing coatings. These coatings are intended to deliver antibiotics slowly and steadily at the implant site; this makes it easier to fight bacterial adhesion. Despite being useful to preventing infections, there are complications associated, such as antibiotic resistance, which has opened new avenues for the exploration of antimicrobial peptides or bioactive molecules [9, 22].

Furthermore, besides antibacterial coatings, other bioactive coatings have emerged that enhance bone-to-implant bonding as well as bone healing. For instance, hydroxyapatite coatings that act like the mineral component of human bone have been used to enhance osseointegration to orthopedic implants. Because hydroxyapatite-coated stainless steel implants have a higher ability to integrate with the adjacent bone, the incidence of implant loosening and the need for revision surgery will be lower in the long term [9, 21, 22].

3.7.4 Trends in surface modifications and future prospects

The current state of surface modification for stainless steel implants is still a relatively new field and hence needs further research to enhance the biological and mechanical properties of orthopedic implants. Nanotechnology plays an important role in surface modification to produce nanostructured surfaces that enhance cell attachment and tissue integration. For instance, the nanotextured surfaces can be designed in a way that the cells prefer to adhere to, like the natural Extra Cellular Matrix. This approach is likely to enhance the osseointegration and thereby reduce the rate of implant failure [12].

Multifunctional coatings are another technique in orthopedic implant technology that has been developed to incorporate various surface characteristics to overcome different problems associated with the use of stainless steel implants. These coatings are intended to enhance mechanical properties and biological properties and provide solutions against wear, corrosion, infection, and poor osseointegration problems [11, 12, 23].

Multifunctional surface coatings are typically a combination of antibacterial and bioactivity to enhance the surface properties. For example, a coating could be composed of a hydroxyapatite layer, which would enhance bone tissue formation, and silver nanoparticles, which would help to prevent bacterial growth and attachment. This dual-function approach not only improves tissue integration of the implant but also prevents postoperative infection which is often a cause of implant failure [14, 23].

Another example of such coatings is that they can be made in the form of drug-releasing coatings, which can deliver drugs such as anti-inflammatory drugs, growth factors, or antibiotics in controlled amounts over time. These coatings are most helpful in the initial days of surgery because they can decrease inflammation, accelerate healing, and minimize infection. New developments in controlled drug release systems make it possible to deliver these agents at the precise concentration and location, avoiding the necessity of a systemic drug intake [14].

Nanotechnology plays a key role in the development of surface-modified implants for better healthcare. The method of developing nanoscale surface topographies to mimic the natural architecture of bone has been used effectively to enhance cellular responses. It is demonstrated that nanostructured coatings improve osteoblast adhesion and increase the rate of bone ingrowth and proper integration. Also, using nanoparticle-based coatings, drugs, or antimicrobial agents can be directly provided to the implant site, which would enhance the patient's treatment tenure [11, 14].

In conclusion, the advancements in surface modification of stainless steel implants can be achieved through the synthesis of multifunctional, responsive, and bioactive coatings. These advancements can address various issues associated with orthopedic implants such as wear rate, corrosion, infection, and tissue integration so that implants could be more durable and biocompatible.

3.8 Stainless steel: applications in orthopedic fixations

Stainless steel, and more specifically the 316L alloy, has been widely employed in many different clinical orthopedic applications, including fracture fixation as well as joint replacement. Newer materials exist but stainless steel remains preferred because of its mechanical characteristics, economic factors, and efficacy in clinical use. This section describes some of the outstanding clinical applications of stainless steel in orthopedic surgery as well as the results connected with these uses [4, 6, 12].

3.8.1 Fracture fixation devices: plates, screws, and pins

The use of stainless steel in orthopedics has been widely seen in fracture fixation more than in any other process. Stainless steel plates, screws, and pins are used to fix bone fractures so that the bones can be set right and heal properly. It has very high tensile strength and corrosion resistance, so it is suitable for the implants to overcome forces exerted by the skeletal system during the healing process [12, 20].

3.8.2 Joint replacement components

Stainless steel has been partially replaced by titanium and other cobalt-chromium alloys in joint replacement arthroplasty but it remains the material of choice for specific components such as in hip and knee prosthesis. Modular components of joint replacements such as screws and pins that hold the implant in place are often made from stainless steel.

In hip replacement, stainless steel is used occasionally for the acetabular cups and femoral stems of joint prostheses. The superior wear resistance and biocompatibility make it suitable for load-bearing applications. Further, it can be combined with other polymers, ceramics, or metals to improve wear resistance such as ultra-high-molecular-weight polyethylene (UHMWPE) for knee replacement surgery [1, 4, 6].

3.8.3 Spinal fixation devices

Stainless steel is also used in spinal fixation and plays an important role during healing. After spine deformities like scoliosis or fracture and dislocation, the spine needs surgical intervention, stainless steel rods, screws, and plates to put the vertebrae back in place. Two of the most critical mechanical properties for this application are tensile strength and fatigue resistance because the implant is subjected to continuous motion and loading.

Moreover, stainless steel rods are applied in spinal fusion surgical procedures to maintain the spinal column's position as the grafted bones fuse and heal. These rods impart the requisite stability to reduce motion at the intervertebral interface, which is essential for healing. The rods are accompanied by stainless steel screws and plates that are used to fix the implant on the spine so that it can be well-aligned and stable [6, 9, 12, 24].

3.9 Merits and demerits of stainless steel in orthopedic fixation

Stainless steel plays a key role in the fabrication of orthopedic implants because of its mechanical properties, relatively cheap, and easily accessible. However, due to some of its disadvantages as highlighted, the search for better material that can be used as an alternative with it has been in progress.

3.9.1 Merits of stainless steel as an orthopedic implant

- High tensile and yield strength: The mechanical properties of stainless steel enable the material to bear the large forces experienced during load-bearing application.
- Corrosion resistance: Stainless steel is prone to pitting and crevice corrosion but it is relatively resistant to corrosion in the saline bodily fluids for short- to medium-term implantation.
- Cost-effective: Stainless steel implants are cheaper compared to those from metals such as titanium or cobalt-chromium alloys, making them more suitable for use where cost is a factor in the clinic.
- Ease of manufacturing: Stainless steel is not very complex to manufacture and its processing techniques are well-defined to ensure performance and quality.
- Established clinical track record: 316L stainless steel has received FDA clearance for orthopedic surgery, and its usage has been documented for various applications [4, 6, 9].

3.9.2 Demerits of stainless steel as an orthopedic implant

- Susceptibility to corrosion: It is prone to pitting and crevice corrosion, which is common in a chloride-rich environment of the body. These demerits restrict the application of stainless steel implants for the long term and lead to implant failure.
- Metal ion release: Nickel, chromium, and iron ions from stainless steel have been reported to have allergic reactions that induce an inflammatory response in patients with associated allergies.
- Stress shielding: Stainless steel, although superior to titanium in terms of mechanical properties, is less similar to bone than the latter. It is stiffer and can cause stress shielding. In the long run, it results in bone resorption as well as deterioration of the bone matrix and structure, which makes the implant loose and eventually fail. This is especially seen in cases where the implant is intended to be a permanent fixture and expected to be stable over time.
- Limited osseointegration: Stainless steel does not facilitate osseointegration, which is critical for the long-term stability of permanent implants. Consequently, there are problems with implant stability and mechanical retention with time, particularly in non-stress-bearing situations when primary osseointegration is crucial.
- Potential for bacterial adhesion: Stainless steel, particularly in the case of non-treated surfaces, is prone to bacterial invasion and may result in biofilm formation. Biofilms that develop on the surface of the implant material can result in recurring infections that are difficult to manage and may require implant removal.
- Allergic reactions: Some patients may exhibit skin allergies to nickel or any other constituents of the stainless steel alloys. This is especially so among patients who have had reactions to metals in the past, which means that surgeons should take into account the various individual characteristics whenever they are choosing this type of implant for use.
- Fatigue and wear resistance: Stainless steel has good fatigue strength yet not as good as titanium or cobalt-chromium materials in service conditions where parts have to last for long under cyclic loading. Wear debris from stainless steel implants can also elicit inflammatory reactions in the adjacent tissues, which may cause implant loosening [4, 6, 9].

3.10 Conclusion

Stainless steel has been widely used in orthopedic fixation devices for many years due to its good mechanical performance, low cost, and wide applications. It is used in various forms such as fracture fixation plates and screws, and spinal stabilization devices; thus, it is an important material in orthopedic surgery. Nonetheless, there are various challenges, including corrosion, metal ion release, stress shielding, and limited osseointegration, which need to be addressed for improving the implant performance and quality of patients' lives.

Recent progress in surface treatment such as passivation, electropolishing, and multifunctional coatings have enhanced the biocompatibility and performance of stainless steel implants. It is expected that future trends in nanotechnology, smart coatings, and drug-eluting systems will have the potential to further increase the functionality of stainless steel in orthopedic applications. Therefore, although stainless steel is still a viable option for most orthopedic applications, a constant search for alternate materials and novel technologies is crucial for the enhancement of patient's quality of life and better orthopedic treatments.

References

[1] Thakur, A., Kumar, A., Kaya, S., Marzouki, R., Zhang, F., & Guo, L. (2022). Recent advancements in surface modification, characterization and functionalization for enhancing the biocompatibility and corrosion resistance of biomedical implants. *Coatings*, *12*(10), 1459. https://doi.org/10.3390/coatings12101459

[2] Hanawa, T. (2010). Overview of Metals and Applications. In *Metals for Biomedical Devices* (pp. 3–24). Elsevier. https://doi.org/10.1533/9781845699246.1.3

[3] Weihong, J., & Chu, P. K. (2019). Orthopedic Implants. In *Encyclopedia of Biomedical Engineering* (pp. 425–439). Elsevier. https://doi.org/10.1016/B978-0-12-801238-3.10999-7

[4] Narahari Prasad, S., & Narayana Rao, M. (2013, September). Stainless steel – A versatile engineering material for critical applications. *Advanced Materials Research*, *794*, 44–49. https://doi.org/10.4028/www.scientific.net/AMR.794.44

[5] Buechel, F. F., & Pappas, M. J. (2015). Properties of Materials Used in Orthopaedic Implant Systems. In *Principles of Human Joint Replacement* (pp. 1–32). Cham: Springer International Publishing. https://doi.org/10.1007/978-3-319-15311-7_1

[6] Nouri, A., & Wen, C. (2021). Stainless Steels in Orthopedics. In *Structural Biomaterials* (pp. 67–101). Elsevier. https://doi.org/10.1016/B978-0-12-818831-6.00008-2

[7] Mosas, A., Kirubaharan, K., Chandrasekar, A. R., Dasan, A., Pakseresht, A., & Galusek, D. (2022). Recent advancements in materials and coatings for biomedical implants. *Gels*, *8*(5), 323. https://doi.org/10.3390/gels8050323

[8] Jacobs, J. J., Gilbert, J. L., & Urban, R M. (1998). Corrosion of metal orthopaedic implants*. *The Journal of Bone and Joint Surgery (American Volume)*, *80*(2), 268–282. https://doi.org/10.2106/00004623-199802000-00015

[9] Szczęsny, G., Kopec, M., Politis, D. J., Kowalewski, Z. L., Łazarski, A., & Szolc, T. (2022). A review on biomaterials for orthopaedic surgery and traumatology: From past to present. *Materials*, *15*(10), 3622. https://doi.org/10.3390/ma15103622

[10] Serhan, H. (2004). Is galvanic corrosion between titanium alloy and stainless steel spinal implants a clinical concern?*1. *The Spine Journal*, *4*(4), 379–387. https://doi.org/10.1016/j.spinee.2003.12.004

[11] Sultana, N., Nishina, Y., & Islam Nizami, M. Z. (2024). Surface modifications of medical grade stainless steel. *Coatings*, *14*(3), 248. https://doi.org/10.3390/coatings14030248

[12] Asri, R. I. M., Harun, W. S. W., Samykano, M., Lah, N. A. C., Ghani, S. A. C., Tarlochan, F., & Raza, M. R. (2017, August). Corrosion and surface modification on biocompatible metals: A review. *Materials Science and Engineering: C, 77*, 1261–1274. https://doi.org/10.1016/j.msec.2017.04.102

[13] Golish, S. R., & Mihalko, W. M. (2011). Principles of biomechanics and biomaterials in orthopaedic surgery. *The Journal of Bone and Joint Surgery-American Volume, 93*(2), 207–212. https://doi.org/10.2106/00004623-201101190-00013

[14] Wang, Q., Eltit, F., & Wang, R. (2019). Corrosion of Orthopedic Implants. In *Encyclopedia of Biomedical Engineering* (pp. 65–85). Elsevier. https://doi.org/10.1016/B978-0-12-801238-3.99863-5

[15] Raman, R. K. S., & Harandi, S. E. (2017). Resistance of magnesium alloys to corrosion fatigue for biodegradable implant applications: Current status and challenges. *Materials*, *10*(11), 1316. https://doi.org/10.3390/ma10111316

[16] Chew, S. A., & Danti, S. (2017). Biomaterial-based implantable devices for cancer therapy. *Advanced Healthcare Materials*, *6*(2). https://doi.org/10.1002/adhm.201600766

[17] Li, M., Wu, J., Geng, W., Gao, P., Yang, Y., Li, X., Xu, K., Liao, Q., & Cai, K. (2024, January). Interaction pathways of implant metal localized corrosion and macrophage inflammatory reactions. *Bioactive Materials*, *31*, 355–67. https://doi.org/10.1016/j.bioactmat.2023.08.017

[18] Williams, D. F. (2008). On the mechanisms of biocompatibility. *Biomaterials*, *29*(20), 2941–2953. https://doi.org/10.1016/j.biomaterials.2008.04.023

[19] Liu, Y., Rath, B., Tingart, M., & Eschweiler, J. (2020). Role of implants surface modification in osseointegration: A systematic review. *Journal of Biomedical Materials Research Part A*, *108*(3), 470–484. https://doi.org/10.1002/jbm.a.36829

[20] Ratner, B. D. (2015). The Biocompatibility of Implant Materials. In *Host Response to Biomaterials* (pp. 37–51). Elsevier. https://doi.org/10.1016/B978-0-12-800196-7.00003-7

[21] Goldmann, W. H. (2021). Biosensitive and antibacterial coatings on metallic material for medical applications. *Cell Biology International*, *45*(8), 1624–1632. https://doi.org/10.1002/cbin.11604

[22] Kumar, M., Kumar, R., & Kumar, S. (2021). Coatings on orthopedic implants to overcome present problems and challenges: A focused review. *Materials Today: Proceedings*, *45*, 5269–5276. https://doi.org/10.1016/j.matpr.2021.01.831

[23] Omarov, S., Nauryz, N., Talamona, D., & Perveen, A. (2022). Surface modification techniques for metallic biomedical alloys: A concise review. *Metals*, *13*(1), 82. https://doi.org/10.3390/met13010082

[24] Filip, N., Radu, I., Veliceasa, B., Filip, C., Pertea, M., Clim, A., Pinzariu, A. C., Drochioi, I. C., Hilitanu, R. L., & Serban, I. L. (2022). Biomaterials in orthopedic devices: Current issues and future perspectives. *Coatings*, *12*(10), 1544. https://doi.org/10.3390/coatings12101544

Uzwalkiran Rokkala, Mohammad Faseeulla Khan, Ashwin C. Gowda, and Santhosh

Chapter 4 Magnesium alloys for biomedical applications

Abstract: Magnesium-based alloys have emerged as highly promising materials for biomedical applications due to their excellent biocompatibility, biodegradability, and mechanical properties that closely resemble those of natural bone. These alloys present a viable alternative to traditional permanent implants, potentially eliminating the need for secondary surgeries for implant removal. The degradable nature of magnesium alloys facilitates a gradual load transfer to the healing bone, potentially enhancing the recovery process. However, issues such as rapid corrosion and hydrogen gas release need to be addressed to optimize their performance. Researchers are exploring advancements in alloy composition, surface treatments, and coating technologies to better control degradation rates and enhance biocompatibility. This review examines the current state of magnesium-based alloys in biomedical applications, highlighting recent progress and future research directions for clinical translation.

Moreover, innovative manufacturing techniques, such as additive manufacturing and powder metallurgy, are being investigated to produce complex and patient-specific implant geometries, further expanding the potential applications of magnesium-based alloys in orthopedics and cardiovascular devices. The integration of bioactive elements and compounds into these alloys aims to promote osteogenesis and angiogenesis, accelerating tissue regeneration and implant integration. Additionally, in vivo studies and clinical trials are increasingly demonstrating the safety and efficacy of magnesium-based implants, paving the way for regulatory approvals and widespread clinical use. The development of multifunctional magnesium alloys that combine structural support is controlled.

Keywords: Magnesium alloys, biodegradable implants, biomedical applications, alloy composition, surface treatments

Acknowledgment: This work was supported by the Deanship of Scientific Research, Vice Presidency for Graduate Studies and Scientific Research, King Faisal University, Saudi Arabia (grant no. KFU250870).

Uzwalkiran Rokkala, Department of Mechanical Engineering, Aditya University, Surampalem, Andhra Pradesh, India
Mohammad Faseeulla Khan, Department of Mechanical Engineering, College of Engineering, King Faisal University, Al-Ahsa, Saudi Arabia
Ashwin C. Gowda, Department of Mechanical Engineering, Visvesvaraya Technology University, India
Santhosh, Department of Mechanical Engineering, MVJ College of Engineering, India

https://doi.org/10.1515/9783111571423-004

4.1 Introduction

In the field of biomedical engineering, magnesium (Mg)-based alloys have attracted substantial attention, particularly for cardiovascular and orthopedic applications. The appealing characteristics of Mg-based alloys are derived from their distinctive combination of biodegradability, biocompatibility, and mechanical properties [1]. Mg is a lightweight metal with a density close to that of human bone (1.74 g/cm^3 compared to bone's 1.8–2.1 g/cm^3), making it particularly attractive for load-bearing implants [2]. Additionally, Mg alloys have an elastic modulus similar to natural bone, which helps reduce stress shielding, a common issue with traditional metallic implants such as titanium and stainless steel [3].

One of the most significant advantages of Mg alloys is their biodegradability. Unlike permanent implants, Mg alloys degrade gradually in the physiological environment, eliminating the need for secondary surgeries to remove implants after healing. As Mg degrades, it releases Mg ions, which are essential for many bodily functions, including bone regeneration and cellular metabolism [4]. However, the rapid degradation rate of Mg in the corrosive physiological environment poses a challenge, as it can lead to premature loss of mechanical integrity and potential adverse effects such as hydrogen gas evolution [5, 6]. Recent research has focused on alloying, surface modification, and coating techniques to improve the corrosion resistance, mechanical properties, and biocompatibility of Mg-based alloys. Several strategies have been developed to tailor the degradation rates to match the healing rates of different tissues, addressing one of the major challenges of Mg-based biomaterials (Figure 4.1) [7].

4.1.1 Alloying with rare earth elements and other metals

Alloying Mg with conventional elements such as calcium (Ca), zinc (Zn), manganese (Mn), and rare earth elements (REEs) like yttrium (Y) and gadolinium (Gd) has shown promising results [8–11]. For example, Mg-Zn-Ca alloys have demonstrated improved mechanical properties and corrosion resistance while promoting osteogenesis [12]. The addition of REEs has been particularly effective in refining the microstructure and enhancing the corrosion behavior of Mg alloys. For instance, Mg-Nd-Ag-Zr, Mg-Gd, and Mg-Nd alloys have shown improved biocompatibility and slower degradation rates, making them suitable for longer-term biomedical applications [13–15]. In cardiovascular applications, Mg-based stents have been explored as alternatives to permanent metallic stents [16]. The biodegradable nature of Mg stents helps reduce the long-term complications associated with permanent stents, such as late stent thrombosis and chronic inflammation. Recent developments include the use of Mg-Zn-Ca and Mg-RE (rare earth) alloys for stents, which have shown promising results in preclinical and early clinical trials, demonstrating controlled degradation rates and excellent biocompatibility [17, 18].

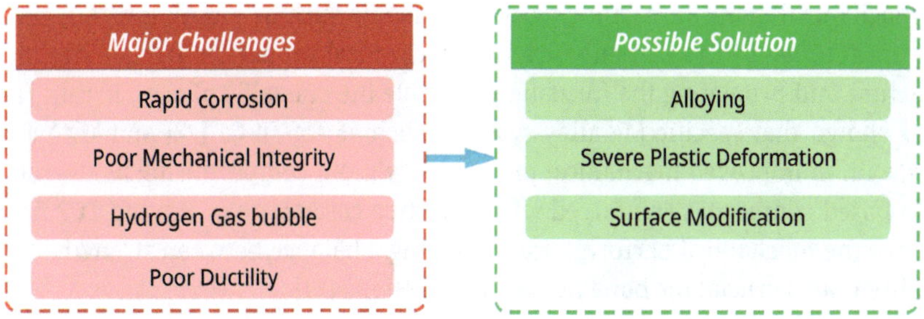

Figure 4.1: Block diagram representing major challenges and possible solutions for Mg alloys.

Surface modification techniques such as micro-arc oxidation, chemical conversion coatings, and ion implantation have been employed to create protective layers that slow down the degradation of Mg-based alloys [19]. For example, MAO creates a ceramic-like coating on the surface, which significantly improves the corrosion resistance of Mg alloys in physiological environments [20]. Recent studies have also explored bioactive coatings like calcium phosphate and hydroxyapatite (HA), which not only enhance corrosion resistance but also improve the bioactivity of the implant, promoting bone integration. Additive manufacturing (AM) techniques, such as 3D printing, have opened new avenues for producing patient-specific Mg implants with complex geometries and controlled porosities. Researchers have also explored Mg-based composites reinforced with bioactive ceramics like tricalcium phosphate (TCP) and bioglass to enhance mechanical properties and control the degradation process. These advancements allow for the customization of implants that better meet the specific needs of patients.

4.2 Properties of Mg alloys relevant to biomedical applications

Mg alloys possess mechanical properties that closely resemble those of human bone, making them ideal for load-bearing orthopedic implants. The elastic modulus of Mg alloys (41–45 GPa) is like that of natural bone (10–30 GPa), which helps reduce stress shielding – an issue often encountered with traditional implant materials like titanium and stainless steel, which have significantly higher elastic moduli. Stress shielding can lead to bone resorption and implant loosening due to the mismatch between the stiffness of the implant and the surrounding bone [21].

In addition, Mg alloys have good tensile strength (90–200 MPa) and compressive strength, which are critical for maintaining the structural integrity of implants under physiological loads. However, the relatively low ductility and yield strength of Mg alloys compared to other metals can be a limitation, particularly in applications requiring high mechanical endurance. Alloying with elements such as aluminum (Al), zinc (Zn), calcium (Ca), and REEs can enhance the mechanical properties by refining the microstructure and promoting the formation of stable intermetallic phases. Recent research has shown that specific Mg alloy systems, such as Mg-Zn-Ca [22] and Mg-Y-RE [23] alloys, exhibit improved mechanical properties without compromising biodegradability. Mg-based composites reinforced with bioactive ceramics like HA and TCP further enhance the mechanical performance, providing a balance between strength and degradability that is crucial for bone healing applications [24].

Biocompatibility is a critical requirement for any biomaterial used in medical implants. Mg alloys are considered biocompatible due to their natural occurrence in the human body and their role in various physiological functions, including bone formation, energy metabolism, and nerve function. Mg ions released during degradation are well tolerated by the body, and moderate levels can even promote bone regeneration and cellular activity, making Mg alloys particularly suitable for orthopedic applications [25]. The biodegradability of Mg alloys is one of their most distinguishing properties, setting them apart from traditional permanent metals like titanium and stainless steel. As Mg degrades in the physiological environment, it gradually dissolves, eliminating the need for secondary surgeries to remove the implant after healing, thus reducing patient morbidity and healthcare costs. This property is particularly advantageous in pediatric orthopedic applications, where the removal of implants can be complex and risky due to the ongoing growth of bones [26]. However, the biodegradability of Mg alloys must be carefully controlled to match the rate of tissue healing. Rapid degradation can lead to premature loss of mechanical support, while slow degradation may cause prolonged inflammation. Alloying and surface modification techniques are used to fine-tune the degradation rate, ensuring that the implant provides sufficient support during the critical healing period and then safely resorbs [27].

The degradation behavior of Mg alloys in physiological environments presents a dual challenge: it facilitates the advantageous biodegradability of these materials while simultaneously introducing substantial difficulties that must be addressed to guarantee their safe and successful employment in medical contexts. Mg alloys have high reactivity and undergo rapid corrosion in the presence of chloride ions present in bodily fluids, resulting in the creation of magnesium hydroxide ($Mg(OH)_2$) and the liberation of hydrogen gas [28]. The accelerated corrosion rate of Mg alloys may jeopardize the structural integrity of implants, leading to their degradation at a pace beyond tissue healing. Moreover, excessive hydrogen evolution might result in gas pocket formation surrounding the implant, potentially hindering bone growth and causing implant failure. Regulating the corrosion rate is consequently a principal objective in the advancement of Mg-based biomaterials.

In order to improve the corrosion resistance of Mg alloys, various strategies have been implemented. These strategies include coating technologies, surface treatments [29], and alloying [30]. The corrosion behavior of an Mg alloy can be improved by alloying with elements such as Ca, Zn, and REEs. This is accomplished by stabilizing the passive layer formed on the surface of the alloy, which in turn reduces the rate of degradation. MAO, anodization, and bioactive coatings are examples of surface changes that can be used to produce protective barriers [31]. These barriers slow down the corrosion process and reduce the amount of hydrogen that becomes released. Bioactive coatings, such as calcium phosphate and HA, not only improve the implant's resistance to corrosion but also boost the implant's bioactivity, which in turn promotes osteointegration and tissue healing. Furthermore, in order to further enhance the effectiveness of Mg implants in clinical settings, novel techniques such as biodegradable polymer coatings that are embedded with anti-inflammatory or antibacterial compounds are being investigated [32].

4.3 Techniques to enhance the properties of Mg alloys

In the following sections, we will explore advanced techniques aimed at enhancing the properties of magnesium (Mg) alloys, particularly for biomedical implant applications. We begin with alloying strategies, where the addition of selective elements improves mechanical strength, corrosion resistance, and biocompatibility. Next, we delve into surface modification techniques, which play a crucial role in tailoring the surface characteristics of Mg alloys to enhance their durability and biofunctionality. We then discuss emerging trends in surface treatments, highlighting innovative approaches that further refine implant performance. Finally, we examine the potential of AM, a cutting-edge fabrication method that enables the development of complex, customized Mg-based implants with superior properties. Together, these advancements pave the way for next-generation biodegradable implants with improved clinical outcomes.

4.3.1 Alloying strategies to enhance Mg properties

Ca is biocompatible and essential to bone metabolism. It is naturally present in the body. Mg alloys containing Ca have better mechanical characteristics and corrosion resistance. Due to their biocompatibility and bone regeneration capability, Mg-Ca alloys are promising orthopedic materials. Mg's yield and tensile strengths improve when Ca refines its grain structure. Ca helps produce stable intermetallic phases like Mg_2Ca, which act as sacrificial anodes and limit the corrosion rate of the alloy [33]. For temporary support like bone scaffolds and fracture repair devices, Mg-Ca alloys' regulated breakdown rate is ideal. However, excessive Ca concentration can cause brittle phases, reducing alloy ductility. Thus, Ca concentration must be optimized to balance strength, corrosion resistance, and ductility.

Furthermore, Zn is a biocompatible alloying element that is frequently utilized in Mg alloys due to the advantageous impact it has on the mechanical strength and corrosion behavior of the alloy. Zn improves the solid-solution strengthening of Mg alloys, enhancing their tensile properties without significantly compromising ductility [13, 34]. Mg-Zn alloys exhibit refined grain structures due to Zn's role in promoting grain boundary pinning, which is critical for achieving high strength. In terms of corrosion resistance, Zn acts as a passivating agent, stabilizing the protective oxide layer formed on the alloy surface, thereby slowing down the degradation rate. Mg-Zn alloys have also demonstrated osteoconductive properties, making them suitable for bone regeneration applications. However, it is essential to control the Zn content, as excessive amounts can accelerate localized corrosion due to galvanic effects.

Manganese is often added to Mg alloys in small amounts to improve corrosion resistance. Mn acts as a scavenger for iron and other impurity elements that can significantly accelerate corrosion. By reducing the harmful effects of impurities, Mn helps to stabilize the microstructure and enhance the overall corrosion resistance of Mg alloys. Additionally, Mn contributes to grain refinement, which further improves the alloy's mechanical properties [35]. Mg-Mn alloys are known for their good balance of strength, ductility, and corrosion resistance, making them suitable for a variety of biomedical applications, including cardiovascular stents and orthopedic devices. The low toxicity of Mn also ensures the biocompatibility of these alloys, which is a critical consideration in implant design.

REEs, such as yttrium (Y), gadolinium (Gd), neodymium (Nd), and cerium (Ce), are increasingly used in Mg alloys due to their remarkable ability to improve mechanical properties, corrosion resistance, and thermal stability [36–38]. REEs refine the microstructure of Mg alloys through grain boundary pinning and precipitation hardening, resulting in enhanced strength and ductility. REEs form fine, stable intermetallic phases within the Mg matrix, such as $Mg_{12}RE$ and Mg_5RE, which act as obstacles to dislocation movement, thereby strengthening the alloy. These intermetallic phases also contribute to grain refinement, improving the alloy's mechanical performance. For example, Mg-Y and Mg-Gd alloys have demonstrated high strength and good ductility, making them suitable for load-bearing applications in orthopedics. The addition of REEs significantly improves the corrosion resistance of Mg alloys by forming a stable, protective surface film that resists degradation in physiological environments. REEs modify the composition and structure of the surface oxide layer, enhancing its protective qualities and reducing the overall corrosion rate. This makes REEs containing Mg alloys particularly attractive for applications where controlled biodegradation is essential. However, the biocompatibility of REEs must be carefully evaluated, as some REEs can be cytotoxic at high concentrations. Research is ongoing to identify optimal REE additions that maximize corrosion resistance and mechanical performance while maintaining biocompatibility.

The microstructure of Mg alloys plays a critical role in determining their mechanical properties, corrosion behavior, and biocompatibility. Alloying elements such as Ca, Zn, Mn, and REEs contribute to microstructural modifications, including grain refinement, phase precipitation, and the formation of protective surface films. Grain refinement is a key strategy to enhance the strength and ductility of Mg alloys. Figure 4.2 shows the grain refinement using friction stir processing (FSP) technique. The as-received base material grain size is 60 ± 5 μm, and after FSP, the grain size significantly reduced to 3 ± 1 μm. Fine grains provide more grain boundaries, which act as barriers to dislocation movement, thereby strengthening the alloy [39].

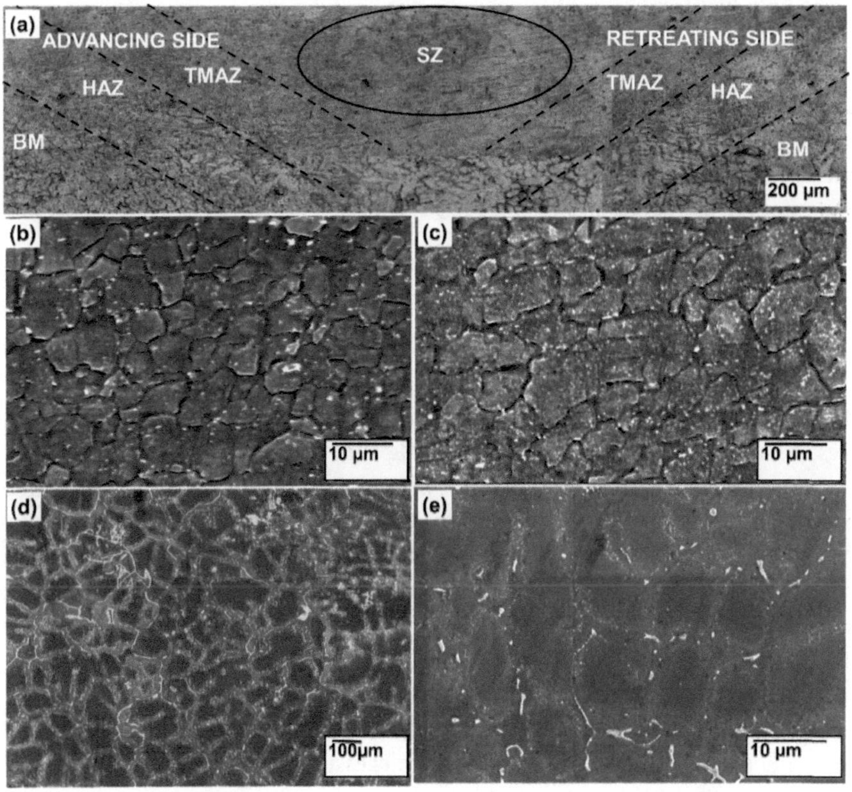

Figure 4.2: Grain refinement using friction stir processing technique: (a) FSP of Mg-Zn-Dy alloy, (b) stir zone (SZ) microstructure, (c) thermal heat-affected zone, (d) heat-affected zone, and (e) base metal [39].

Alloying elements such as Zn, Mn, and REEs are effective grain refiners, promoting the formation of a uniform and fine-grained microstructure that enhances both mechanical properties and corrosion resistance. Precipitation hardening involves the formation of fine intermetallic phases within the Mg matrix, which impede dislocation movement and contribute to increased strength. REEs are particularly effective in promoting precipitation hardening, resulting in alloys with superior mechanical performance compared to unalloyed Mg. The presence of alloying elements can also lead to the formation of stable, protective phases that enhance corrosion resistance. For example, the Mg_2Ca phase in Mg-Ca alloys and the intermetallic phases formed with REEs create protective barriers that slow down the degradation process. Additionally, the surface films modified by REE additions are more stable and protective, reducing the rate of corrosion in physiological environments [40].

4.3.2 Surface modification techniques for enhancing properties of Mg alloys

Surface modification techniques have been developed to enhance the corrosion resistance, biocompatibility, and overall performance of Mg alloys. Key surface modification methods include coating technologies like MAO, chemical conversion coatings, bioactive coatings such as calcium phosphate and HA, and emerging treatments like ion implantation and nanocoatings.

4.3.2.1 Coating technologies

MAO, also known as plasma electrolytic oxidation (PEO), is a widely used surface modification technique for enhancing the corrosion resistance of Mg alloys [41]. MAO involves the application of high voltage to the Mg alloy in an electrolytic solution, generating microdischarges on the surface. This process forms a thick, dense ceramic-like oxide layer, typically composed of MgO, $MgSiO_3$, and other oxides depending on the electrolyte composition. The MAO coating acts as a physical barrier that significantly reduces the corrosion rate of Mg alloys by isolating the alloy from the corrosive environment. Additionally, MAO coatings have a porous structure that can be further modified with bioactive elements such as Ca and phosphorus to enhance bone integration and biocompatibility. The rough surface created by MAO also promotes cell adhesion, which is beneficial for orthopedic implants. However, the coating's porosity can sometimes allow for localized corrosion, necessitating further optimization or sealing of the pores with secondary treatments.

Chemical conversion coatings are another effective method for improving the surface properties of Mg alloys. This technique involves the treatment of Mg alloys in an acidic or alkaline solution, resulting in the formation of a thin, protective film on the surface. Common chemical conversion coatings include phosphate, fluoride, and chromate coatings. Phosphate conversion coatings, such as calcium phosphate, are particularly beneficial due to their bioactivity and ability to form a protective barrier against corrosion. Fluoride coatings are also popular, as they create a stable, compact MgF_2 layer that significantly reduces the corrosion rate. These coatings not only enhance corrosion resistance but also improve the surface bioactivity, promoting better bone-implant integration. However, the thin nature of these coatings makes them susceptible to wear, and they may require combination with other surface treatments for optimal performance.

4.3.2.2 Calcium phosphate coatings

Calcium phosphate coatings, including TCP and HA, are bioactive layers commonly applied to Mg alloys to improve their corrosion resistance and promote osteointegration [42]. Calcium phosphate is a major component of natural bone, making these coatings highly biocompatible and osteoconductive. The application of calcium phosphate coatings can be achieved through various methods, such as dip coating, electrochemical deposition, and biomimetic processes.

These coatings serve as a barrier to corrosion, reducing the degradation rate of Mg alloys while simultaneously enhancing their ability to bond with bone tissue. Additionally, calcium phosphate coatings can be doped with therapeutic ions like strontium and zinc, which further promote bone growth and antibacterial properties [43]. However, the adhesion strength of calcium phosphate coatings can be a concern, particularly under mechanical loading conditions, which necessitates further research into improving their bonding to Mg substrates.

4.3.2.3 Hydroxyapatite (HA) coatings

HA ($Ca_{10}(PO_4)_6(OH)_2$) is the most stable and bioactive form of calcium phosphate used for coating Mg alloys. HA coatings are highly compatible with bone tissue due to their chemical similarity to the mineral component of bone, making them an excellent choice for orthopedic and dental implants [38]. HA coatings are typically applied through plasma spraying, electrophoretic deposition, or sol-gel methods. The HA coating acts as a physical barrier, protecting the underlying Mg alloy from rapid corrosion while also enhancing bone cell attachment and proliferation. The bioactive nature of HA promotes direct bonding with bone, which is crucial for implant stability and long-term success (Figure 4.3) [32]. However, the brittle nature of HA and potential cracking during mechanical stress pose challenges that need to be addressed through improved coating techniques and the development of composite coatings.

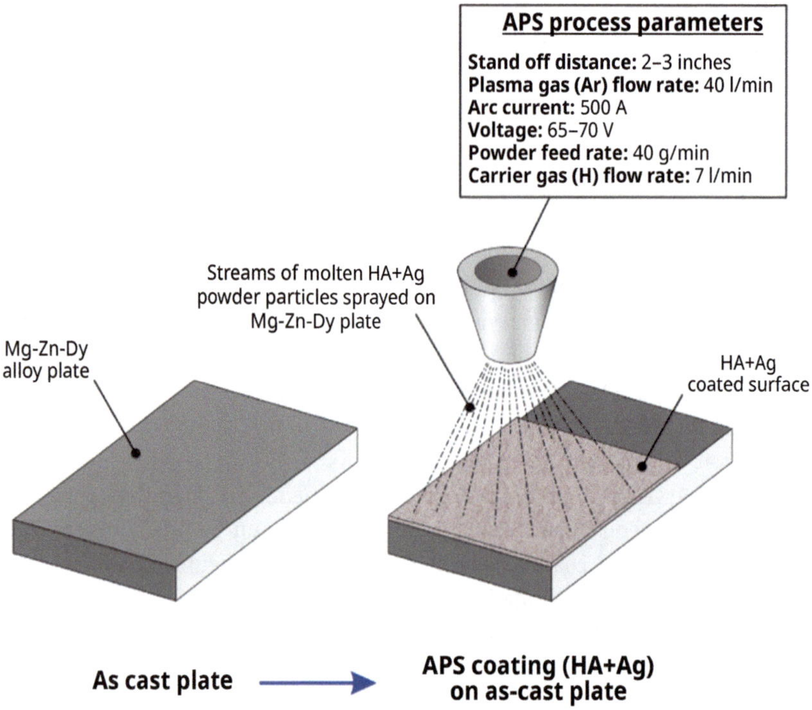

Figure 4.3: Atmospheric plasma spraying (APS) technique used to coat HA/Ag on Mg-Zn-Dy alloy [32].

4.4 Emerging trends in surface treatments

4.4.1 Ion implantation

Ion implantation, which bombards Mg alloys with high-energy nitrogen, carbon, or titanium ions, is a new surface modification method [21]. This method changes alloy surface composition and microstructure without changing bulk properties. Mg alloys are better for biomedical applications after ion implantation because it increases their surface hardness, wear resistance, and corrosion resistance. A thin, nitride-rich surface layer formed by nitrogen ion implantation protects the alloy against corrosion and enhances wear resistance [44]. In applications where implant shape and dimensions must be maintained, ion implantation is ideal since it does not change the surface topography. However, ion implantation equipment is complicated and expensive, which may limit its practical use.

4.4.2 Nanocoatings

Nanocoatings represent a cutting-edge approach to improving the surface properties of Mg alloys. These coatings are typically thin films, often less than 100 nm in thickness, applied using techniques such as atomic layer deposition, sol-gel, or electrochemical deposition [19]. Nanocoatings can be engineered to possess specific functionalities, such as enhanced corrosion resistance, antibacterial properties, or drug delivery capabilities. Nanocoatings, for example, composed of bioactive ceramics, polymers, or metallic oxides, have the ability to considerably slow down the pace of deterioration of Mg alloys while simultaneously promoting cell attachment and subsequent growth [45]. Coatings that are nanostructured offer a vast surface area that can be functionalized with biomolecules, which make them an excellent choice for applications that include targeted medicinal therapies. Additionally, nanocoatings can be designed to release therapeutic agents gradually, enhancing the healing process and reducing the risk of infection. Despite their promise, the long-term stability and biocompatibility of nanocoatings require further investigation to ensure their safety and effectiveness in clinical settings.

In summary, surface modification is essential for biomedical Mg alloy performance. MAO, chemical conversion, bioactive coatings, ion implantation, and nanocoatings improve Mg alloy corrosion resistance, biocompatibility, and mechanical qualities. These surface treatments lengthen Mg-based implants' lifespan and improve their interaction with biological tissues, making them better for medicinal purposes. Continuous research and technology advances in surface modification techniques will expand Mg alloy biomedical applications, delivering safer and more effective implant solutions.

4.5 Additive manufacturing of Mg-based implants

AM, also known as 3D printing, is changing biomedical engineering by making intricate, patient-specific implants possible [46]. For Mg-based alloys, attractive prospects for biodegradable implants, AM offers opportunities to overcome conventional production limitations such poor workability and restricted design freedom. This section highlights Mg alloy 3D printing procedures, specific implant design concerns, and biomedical AM benefits.

4.5.1 3D printing techniques for Mg alloys

Powder bed fusion techniques, which include selective laser melting (SLM) and electron beam melting, are the primary methods utilized in the process of 3D printing composite materials based on Mg [47]. For the purpose of constructing the appropriate structure, these methods involve the selective melting and fusing of Mg powder layer by layer using a high-energy laser or electron beam [48]. These approaches provide fine control over the microstructure, porosity, and surface quality of the printed object, all of which are essential for optimizing the mechanical properties of Mg implants and the rates at which they degrade [49].

SLM, in particular, has been extensively explored for Mg alloys due to its ability to produce dense and complex geometries with high resolution. However, the high reactivity of Mg powders, especially in the presence of oxygen, poses challenges in SLM, such as the risk of oxidation and fire hazards. To mitigate these risks, inert gas atmospheres, optimized process parameters, and specially designed printing setups are employed. Research is also being conducted on the use of novel Mg powders that are less reactive and have improved flowability, which can further enhance the safety and efficiency of the printing process.

4.5.2 Design considerations and customized implants

The design of Mg-based implants using AM is driven by the need for customization, mechanical performance, and controlled biodegradability [50]. Key design considerations include the geometry, porosity, and surface texture of the implant, all of which significantly influence the implant's interaction with surrounding tissues and its degradation rate. For example, porous structures can be designed to match the mechanical properties of bone, reducing stress shielding effects and promoting bone ingrowth. Additionally, the porosity and pore size can be tailored to enhance the implant's bioactivity and facilitate the delivery of nutrients and cells to the healing site. Customized implants are one of the major advantages of AM, allowing for patient-specific designs based on medical imaging data such as CT or MRI scans [51]. This capability enables the creation of implants that precisely match the anatomical features of the patient, improving the fit, stability, and overall performance of the implant. Customization is particularly beneficial in complex orthopedic cases, craniofacial reconstruction, and pediatric implants, where standard off-the-shelf implants may not be suitable.

4.5.3 Advantages of additive manufacturing in biomedical applications

There are numerous benefits to using AM for Mg-based implants in biomedical applications. Initially, it facilitates the creation of intricate and complex structures, such as porous and lattice designs, that are challenging to produce using traditional milling methods. By promoting bone ingrowth, regulating degradation rates, and enhancing the implant's mechanical properties, these intricate structures can improve its functionality. Secondly, AM facilitates iterative design and rapid prototyping, which results in more efficient development cycles and customized solutions [52]. This adaptability is indispensable in the biomedical sector, where implants must be customized to satisfy precise clinical specifications. Additionally, AM is a more sustainable and cost-effective method to produce Mg implants, as it reduces material waste in comparison to subtractive conventional manufacturing methods [53].

Ultimately, the capacity to manipulate the microstructure of Mg alloys during the printing process presents novel opportunities for enhancing material attributes. By modifying the laser parameters, cooling rates, and powder composition, researchers may optimize the mechanical strength and corrosion resistance of the printed implants, thereby improving their performance within the body. The degree of control, along with the customization potential of AM, establishes 3D printing as a revolutionary technology for the future of biodegradable Mg-based implants in biomedical applications. In summary, AM offers a novel framework for the creation of sophisticated Mg-based implants, delivering unparalleled design flexibility, customization, and control over material characteristics. Ongoing research is tackling current issues such as Mg powder reactivity and process optimization, highlighting the significant potential of AM to transform the domain of bioresorbable implants, offering safer, more effective, and personalized solutions in the foreseeable future.

4.6 Future scope and conclusions

The future of Mg-based alloys in biomedical applications holds great promise, with ongoing research aimed at overcoming current limitations. Future developments will likely focus on optimizing the balance between mechanical strength and corrosion rate, further refining alloy compositions, and developing novel surface treatments to enhance biocompatibility and bioactivity. Moreover, the integration of advanced manufacturing techniques, such as 3D printing, will enable the creation of customized implants with tailored properties.

The use of Mg-based alloys in combination with drug delivery systems is another exciting area of research. Coating Mg implants with drug-loaded layers can provide localized, controlled release of anti-inflammatory agents, growth factors, or antibiotics, enhancing the healing process while minimizing systemic side effects. Furthermore, in vivo and clinical studies are crucial to better understand the long-term performance and safety of Mg-based implants in different physiological environments. In conclusion, Mg-based alloys represent a transformative approach to developing next-generation biodegradable implants for orthopedic and cardiovascular applications. While challenges remain, particularly regarding the control of degradation rates and ensuring consistent mechanical performance, recent advancements have significantly expanded the potential of Mg-based materials. With continued research and technological innovations, Mg-based alloys are poised to play an increasingly vital role in the future of biomedical implants, offering patients safer, more effective, and less invasive treatment options.

References

[1] Zheng, Y. F., & Gu FW, X. N. (2014). Biodegradable metals. *Materials Science and Engineering: A, 77*, 1–34.

[2] Zong, Y., Yuan, G., Zhang, X., Mao, L., Niu, J., & Ding, W. (2012). Comparison of biodegradable behaviors of AZ31 and Mg-Nd-Zn-Zr alloys in Hank's physiological solution. *Materials Science and Engineering B. Solid-State Materials for Advanced Technology*, *177*, 395–401. https://doi.org/10.1016/j.mseb.2011.09.042

[3] Li, H., Zheng, Y., & Qin, L. (2014). Progress of biodegradable metals. *Progress in Natural Science: Materials International*, *24*, 414–422. https://doi.org/10.1016/j.pnsc.2014.08.014

[4] Saris, N. E. L., Mervaala, E., Karppanen, H., Khawaja, J. A., & Lewenstam, A. (2000). Magnesium: An update on physiological, clinical and analytical aspects. *Clinica Chimica Acta*, *294*, 1–26. https://doi.org/10.1016/S0009-8981(99)00258-2

[5] Hermawan, H. (2012). Biodegradable metals: State of the art. 13–22. https://doi.org/10.1007/978-3-642-31170-3_2

[6] Tan, L., Yu, X., Wan, P., & Yang, K. (2013). Biodegradable materials for bone repairs: A review. *Journal of Materials Science and Technology*, *29*, 503–513. https://doi.org/10.1016/j.jmst.2013.03.002

[7] Kumar, K., Gill, R. S., & Batra, U. (2018). Challenges and opportunities for biodegradable magnesium alloy implants. *Materials Technology*, *7857*, 1–20. https://doi.org/10.1080/10667857.2017.1377973

[8] Zheng, Y. (2015). Magnesium alloys as degradable biomaterials. https://doi.org/10.1201/b18932

[9] Li, J., Song, Y., Zhang, S., Zhao, C., Zhang, F., Zhang, X., et al. (2010). In vitro responses of human bone marrow stromal cells to a fluoridated hydroxyapatite coated biodegradable Mg-Zn alloy. *Biomaterials*, *31*, 5782–5788. https://doi.org/10.1016/j.biomaterials.2010.04.023

[10] Gu, X. N., Li, S. S., Li, X. M., & Fan, Y. B. (2014). Magnesium based degradable biomaterials: A review. *Frontiers of Materials Science*, *8*, 200–218. https://doi.org/10.1007/s11706-014-0253-9

[11] Rakesh, K. R., Bontha, S., Ramesh, M. R., Arya, S. B., & Das, M. (2018). Laser surface modification of Mg-Zn-Gd alloy: Microstructural, wettability and in vitro degradation aspects. *Materials Research Express*, *5*, 126502.

[12] Bakhsheshi-Rad, H. R., Chen, X., Ismail, A. F., Aziz, M., Abdolahi, E., & Mahmoodiyan, F. (2019). Improved antibacterial properties of an Mg-Zn-Ca alloy coated with chitosan nanofibers incorporating silver sulfadiazine multiwall carbon nanotubes for bone implants. *Polymers for Advanced Technologies*, *30*, 1333–1339. https://doi.org/10.1002/pat.4563

[13] Kottuparambil, R. R., Bontha, S., Rangarasaiah, R. M., Arya, S. B., Jana, A., Das, M., et al. (2018). Effect of zinc and rare-earth element addition on mechanical, corrosion, and biological properties of magnesium. *International Journal of Materials Research*, *33*, 3466–3478. https://doi.org/10.1557/jmr.2018.311

[14] Khan, M. F., & Rokkala, U. (2023). Development of high strength and corrosion resistance Mg-Zn-Dy/HA-Ag composite for temporary implant applications. *Materials Letters*, *347*, 134604. https://doi.org/10.1016/j.matlet.2023.134604

[15] Kumar, A., Fk, M. D., Panigrahi, S. K., & Chaudhari, G. P. (2021). Microstructural evolution and corrosion behaviour of friction stir-processed QE22 magnesium alloy. *39*, 351–360. https://doi.org/doi:10.1515/corrrev-2020-0083

[16] Tie, D., Feyerabend, F., Müller, W.-D., Schade, R., Liefeith, K., Kainer, K., et al. (2013). Antibacterial biodegradable Mg-Ag alloys. *European Cells & Materials Journal*, *25*, 284–298. https://doi.org/10.22203/ecm.v025a20
[17] Fang, H., Wang, C., Zhou, S., Zheng, Z., Lu, T., Li, G., et al. (2020). Enhanced adhesion and anticorrosion of silk fibroin coated biodegradable Mg-Zn-Ca alloy via a two-step plasma activation. *Corrosion Science*, 168. https://doi.org/10.1016/j.corsci.2020.108466
[18] Deng, Y., Yan, H., Li, Q., Chen, J., Xia, W., Su, B., et al. (2023). Enhancing strength and ductility of low RE content Mg-Gd-Y-Zr alloy via a novel thermomechanical treatment based on multi-directional forging. *Journal of Alloys and Compounds*, *958*, 170535. https://doi.org/10.1016/j.jallcom.2023.170535
[19] Yin, Z., Qi, W., Zeng, R., Chen, X., & Gu, C. (2020). Advances in coatings on biodegradable magnesium alloys. *Journal of Magnesium and Alloys*, *8*, 42–65. https://doi.org/10.1016/j.jma.2019.09.008
[20] Lin, X., Wang, X., Tan, L., Wan, P., Yu, X., Li, Q., et al. (2014). Effect of preparation parameters on the properties of hydroxyapatite containing micro-arc oxidation coating on biodegradable ZK60 magnesium alloy. *40*, 10043–10051. https://doi.org/10.1016/j.ceramint.2014.02.104
[21] Hornberger, H., Virtanen, S., & Boccaccini, A. R. (2012). Biomedical coatings on magnesium alloys – A review. *Acta Biomaterialia*, *8*, 2442–2455. https://doi.org/10.1016/j.actbio.2012.04.012
[22] Fang, H., Wang, C., Zhou, S., Li, G., Tian, Y., & Suga, T. (2021). Exploration of the enhanced performances for silk fibroin/sodium alginate composite coatings on biodegradable Mg–Zn–Ca alloy. *Journal of Magnesium and Alloys*, *9*, 1578–1594. https://doi.org/10.1016/j.jma.2020.08.017
[23] Lin, J., Dai, H., Chen, H., & Miao, J. (2024). Microstructure and properties of low-frequency pulsed arc deposition of Mg-Y-RE-Zr alloy under T6 heat treatment. *Materials Letters*, *358*, 135803. https://doi.org/https://doi.org/10.1016/j.matlet.2023.135803
[24] Wang, H., Zhang, D. T., Cao, G. H., & Qiu, C. (2022). Improving room-temperature ductility of a Mg–Zn–Ca alloy through friction stir processing. *Journal of Materials Research and Technology*, *17*, 1176–1186. https://doi.org/10.1016/j.jmrt.2022.01.082
[25] Chen, Y., Hung, F., Lin, Y., & Lin, C. (2020). Biodegradation ZK50 magnesium alloy compression screws : Mechanical properties, biodegradable characteristics and implant test. *Journal of Orthopaedic Science*. https://doi.org/10.1016/j.jos.2020.01.018
[26] Manne, B., Thiruvayapati, H., & Bontha, S. (2018). Surface & coatings technology surface design of Mg-Zn alloy temporary orthopaedic implants : Tailoring wettability and biodegradability using laser surface melting. *Surface and Coatings Technology*, *347*, 337–349. https://doi.org/10.1016/j.surfcoat.2018.05.017
[27] Ralston, K. D., & Birbilis, N. (2010). Effect of grain size on corrosion: A review. Corros J Sci Eng, 66, 1–13. https://doi.org/10.5006/1.3462912
[28] Rong-Chang, Z., Jin, Z., Wei-Jiu, H., Dietzel, W., Kainer, K. U., Blawert, C., et al. (2006). Review of studies on corrosion of magnesium alloys. *Transactions of Nonferrous Metals Society of China*, *16*, 763–771.
[29] Zhang, Y., Zheng, R. N., Liu, Y., Tian, Y. Q., Wei, Y. L., Guo, H. R., et al. (2022). Research progress on surface modification coatings of biomedical magnesium alloys. *SURF Technology*, *51*, 146–155. https://doi.org/10.16490/j.cnki.issn.1001-3660.2022.08.012
[30] Muhammad, H., Tariq, R., Ishtiaq, M., Kang, H., Chaudry, U. M., & Jun, T. (2025). A critical review on the comparative assessment of rare-earth and non-rare-earth alloying in magnesium alloys. Metals,1–21. https://doi.org/10.3390/met15020128
[31] Dong, H., Xu, W., Virtanen, S., & Wang, Y. (2025). Enhanced bioactivity and degradation behavior of zinc via micro-arc anodization for biomedical applications. High Temperature Materials and Processes, 44, 20240065. https://doi.org/10.1515/htmp-2024-0065
[32] Rokkala, U., Jana, A., Bontha, S., Ramesh, M. R., & Balla, V. K. (2021). Comparative investigation of coating and friction stir processing on Mg-Zn-Dy alloy for improving antibacterial, bioactive and

corrosion behaviour. *Surface and Coatings Technology*, *425*, 127708. https://doi.org/10.1016/j.surfcoat.2021.127708

[33] Hikku, G. S., Arthi, C., Jeen Robert, R. B., Jeyasubramanian, K., & Murugesan, R. (2022). Calcium phosphate conversion technique: A versatile route to develop corrosion resistant hydroxyapatite coating over Mg/Mg alloys based implants. *Journal of Magnesium and Alloys*, *10*, 1821–1845. https://doi.org/10.1016/j.jma.2022.06.005

[34] Huan, Z. G., Leeflang, M. A., Zhou, J., Fratila-Apachitei, L. E., & Duszczyk, J. (2010). In vitro degradation behavior and cytocompatibility of Mg-Zn-Zr alloys. *Journal of Materials Science: Materials in Medicine*, *21*, 2623–2635. https://doi.org/10.1007/s10856-010-4111-8

[35] Ahmadkhaniha, D., Ja, A., Zarei-hanzaki, A., Fedel, M., Deflorian, F., & Karjalainen, L. P. (2016). Microstructural modification of pure Mg for improving mechanical and biocorrosion properties. *61*, 360–370. https://doi.org/10.1016/j.jmbbm.2016.04.015

[36] Feyerabend, F., Fischer, J., Holtz, J., Witte, F., Willumeit, R., Drücker, H., et al. (2010). Evaluation of short-term effects of rare earth and other elements used in magnesium alloys on primary cells and cell lines. *Acta Biomaterialia*, *6*, 1834–1842. https://doi.org/10.1016/j.actbio.2009.09.024

[37] Khan, F., & Panigrahi, S. K. (2016). Materials Science & Engineering A Achieving excellent thermal stability and very high activation energy in an ultra fi ne-grained magnesium silver rare earth alloy prepared by friction stir processing. *Materials Science and Engineering: A*, *675*, 338–344. https://doi.org/10.1016/j.msea.2016.08.077

[38] Rokkala, U., Bontha, S., Ramesh, M. R., & Balla, V. K. (2024). Materials for life sciences Multi-step fabrication of bioactive Mg–Zn–Dy–AlO_3/HA composites: Exploring the synergistic effects of plasma spray and friction stir processing. *Journal of Materials Science*. https://doi.org/10.1007/s10853-024-09830-y

[39] Rokkala, U., Bontha, S., Ramesh, M. R., Krishna, V., Srinivasan, A., & Kailas, S. V. (2021). Tailoring surface characteristics of bioabsorbable Mg-Zn-Dy alloy using friction stir processing for improved wettability and degradation behavior. *Journal of Materials Research and Technology*, *12*, 1530–1542. https://doi.org/10.1016/j.jmrt.2021.03.057

[40] Patel, D. K., Sharma, V. K., Kumar, V., Kumar, P., & Kumar, D. (2025). Enhancing hydrophobicity and anti-corrosion properties of Al-6061Aluminum hybrid composites through trace additions of rare earth oxides. *Proceedings of the Institution of Mechanical Engineers, Part E: Journal of Process Mechanical Engineering*, 09544089251317381. https://doi.org/10.1177/09544089251317381

[41] Yan, X., Zhao, M., Yang, Y., Tan, L., Zhao, Y., & Yin, D. (2019). Improvement of biodegradable and antibacterial properties by solution treatment and micro-arc oxidation (MAO) of a magnesium alloy with a trace of copper. *Corrosion Science*, *156*, 125–138. https://doi.org/10.1016/j.corsci.2019.05.015

[42] Saber-Samandari, S., Alamara, K., & Saber-Samandari, S. (2014). Calcium phosphate coatings : Morphology, micro-structure and mechanical properties. *Ceramics International*, *40*, 563–572. https://doi.org/10.1016/j.ceramint.2013.06.038

[43] Amaravathy, P., & Kumar, T. S. S. (2019). Bioactivity enhancement by Sr doped Zn-Ca-P coatings on biomedical magnesium alloy. *Journal of Magnesium and Alloys*, *7*, 584–596. https://doi.org/https://doi.org/10.1016/j.jma.2019.05.014

[44] Luiz, L. A., Kurelo, B., Souza, G. B. D., Andrade, J. D., & Marino, C. E. B. (2021). Effect of nitrogen plasma immersion ion implantation on the corrosion protection mechanisms of different stainless steels. *Materials Today Communications*, *28*, 102655. https://doi.org/https://doi.org/10.1016/j.mtcomm.2021.102655

[45] Rheima, A. M., Abdul-Rasool, A. A., Al-Sharify, Z. T., Zaidan, H. K., Athair, D. M., Mohammed, S. H., et al. (2024). Nano bioceramics: Properties, applications, hydroxyapatite, nanohydroxyapatite and drug delivery. *Case Studies in Chemical and Environmental Engineering*, *10*, 100869. https://doi.org/https://doi.org/10.1016/j.cscee.2024.100869

[46] Jahangir, M. N., Mamun, M. A. H., & Sealy, M. P. (1980). A review of additive manufacturing of magnesium alloys. *AIP Conference Proceedings, 2018*. https://doi.org/10.1063/1.5044305

[47] Karunakaran, R., Ortgies, S., Tamayol, A., Bobaru, F., & Sealy, M. P. (2020). Additive manufacturing of magnesium alloys. *Bioactive Materials*, *5*, 44–54. https://doi.org/10.1016/j.bioactmat.2019.12.004

[48] Ahmadi, M., Tabary, S., Rahmatabadi, D., Ebrahimi, M. S., Abrinia, K., & Hashemi, R. (2022). Review of selective laser melting of magnesium alloys: Advantages, microstructure and mechanical characterizations, defects, challenges, and applications. *Journal of Materials Research and Technology*, *19*, 1537–1562. https://doi.org/https://doi.org/10.1016/j.jmrt.2022.05.102

[49] Ansari, N., Alabtah, F. G., Albakri, M. I., & Khraisheh, M. (2024). Post processing of additive manufactured Mg alloys: Current status, challenges, and opportunities. *Journal of Magnesium and Alloys*, *12*, 1283–1310. https://doi.org/https://doi.org/10.1016/j.jma.2024.04.017

[50] Wang, Y., Fu, P., Wang, N., Peng, L., Kang, B., Zeng, H., et al. (2020). Challenges and solutions for the additive manufacturing of biodegradable magnesium implants. *Engineering*, *6*, 1267–1275. https://doi.org/10.1016/j.eng.2020.02.015

[51] Koppunur, R., Dama, K. K., Rokkala, U., Thirupathi, B., Sagar, N., & Gugulothu, B. (2022). Design and fabrication of patient-specific implant for maxillofacial surgery using additive manufacturing. *Advances in Materials Science and Engineering*, *2022*, 1–7. https://doi.org/10.1155/2022/7145732

[52] Sekar, P., N, S., & Desai, V. (2021). Recent progress in in vivo studies and clinical applications of magnesium based biodegradable implants – A review. *Journal of Magnesium and Alloys*, *9*, 1147–1163. https://doi.org/10.1016/j.jma.2020.11.001

[53] Ni, J., Ling, H., Zhang, S., Wang, Z., Peng, Z., Benyshek, C., et al. (2019). Three-dimensional printing of metals for biomedical applications. *Materials Today Bio*, 3. https://doi.org/10.1016/j.mtbio.2019.100024

Vivek Singh*, Sunil Kumar, Kamal Kishore, Rajesh Kumar Sharma, Rakesh Sehgal, and Sanjay Kumar

Chapter 5
Titanium and titanium alloys in medical applications

Abstract: Biomaterials are becoming increasingly important because they have promising characteristics to dramatically improve the quality and duration of human life. This discipline has become a multimillion-dollar industry as a result of scientific and technological advances. Titanium (Ti) and its alloys are more often employed as biomedical implants than stainless steel and Co-Cr-based alloys because of their lower modulus, great biocompatibility, corrosion resistance, and high strength-to-weight ratio. Implant failure occurs as a result of debris creation caused by low wear resistance and metal ions created during implant operations in the physiological conditions. Superior wear resistance is crucial for medical implants, especially load-bearing orthopedic implants. As a result, increasing the wear resistance of these metals can greatly extend the life of implants in the body. When an implant is inserted into bone during surgery, various biological, physical, and chemical factors influence osseointegration. Ti and its alloys are frequently employed in dental and orthopedic implants due to their superior biocompatibility and strength, though they do not chemically bond with bone tissue. To improve bonding, surface treatments like hydroxyapatite coatings are employed to promote stable bone formation at the implant interface. This chapter includes the tribological, biofunction, and corrosion performance of medical-grade titanium alloys. Further, this chapter discusses numerous surface modification techniques to improve their tribological, biofunction, and corrosion behaviors.

Keywords: Titanium, titanium alloys, surface modifications, coatings

***Corresponding author: Vivek Singh**, Department of Mechanical Engineering, National Institute of Technology, Hamirpur 177005, Himachal Pradesh, India, e-mail: rvivek235@gmail.com
Sunil Kumar, Department of Mechanical and Industrial Engineering, Indian Institute of Technology, Roorkee 247667, Uttarakhand, India
Kamal Kishore, Rajesh Kumar Sharma, Rakesh Sehgal, Department of Mechanical Engineering, National Institute of Technology, Hamirpur 177005, Himachal Pradesh, India
Sanjay Kumar, Department of Mechanical Engineering, National Institute of Technology, Srinagar 190006, Jammu & Kashmir, India

https://doi.org/10.1515/9783111571423-005

5.1 Introduction

Prosthesis and implants for the human body are commonly made using biomedical materials, which can be either natural or synthetic. Their primary purpose is to restore injured or diseased biological tissues, thereby enhancing the quality of human life [1, 2]. As the population ages, more people are experiencing conditions like arthritis and joint pain. This has resulted in a huge surge in demand for restoring malfunctioning hard tissues with artificial implants composed of biocompatible materials. In recent decades, biomedical materials have garnered significant attention due to their growing importance. These biomaterials are employed in numerous body parts of the human, depending on the specific application. Examples include heart valves, intravascular stents, cardiac pacemakers, and replacement implants for knees, hips, shoulders, elbows, and ears, as well as in dental treatments [3].

Implants used for hip and knee joint replacements have incredibly stringent requirements, as conditions like osteoarthritis (joint inflammation), osteoporosis (bone weakening), and trauma often result in pain or loss of function in these areas. As a result, employing biomedical implants to surgically restore the performance of damaged hard tissues is considered the most effective approach for patient treatment. This method offers a reliable solution for improving mobility, function, and quality of life in affected individuals. Data on joint replacement surgeries in the United States show that by 2030, the total number of hip replacement surgeries is projected to rise by 174%, reaching 0.57 million procedures, while knee arthroplasties are anticipated to increase by 673%, totaling 3.48 million procedures from current levels [4]. The increase in surgeries to replace joints is mostly attributable to both original replacement procedures and revision surgeries. Unfortunately, revision surgeries tend to have lower success rates compared to the first implantation and can cause significant pain for patients. The main causes for implant failure include debris generation due to poor wear resistance, response to the physiological environment, release of metal ions, a mismatch in the stiffness (modulus) between the natural bones and the pairing implants, or insufficient strength of the implant to support the load carrying capacity [2]. Therefore, biomaterial implants used in load-bearing applications must possess several key properties to reduce the need for revision surgery. These include appropriate mechanical strength, excellent corrosion and wear resistance, high biocompatibility, the ability to integrate with bone (osseointegration), and non-cytotoxicity, ensuring they do not harm the surrounding tissues [5].

Ceramic-based biomaterials are highly valued for their exceptional hardness and wear resistance characteristics, making them ideal for joint articulating load-bearing surfaces, dental applications, and bone-bonding surfaces in implants. Moreover, polymer-based biomaterials are known for their flexibility and stability, which are frequently employed in low-friction load-bearing surfaces where movement is essential. However, metallic biomaterials are typically chosen for load-bearing applications due to their superior fatigue strength, allowing them to withstand the physical demands of daily activities. Each material is selected based on its ability to meet the specific mechanical and biological requirements of the intended application.

In the field of metallic biomaterials, titanium (Ti) and its alloys are emerging as some of the most promising materials in biomedical engineering. Their exceptional blend of superior strength, low density, excellent corrosion resistance, and biocompatibility makes them ideal for an ample range of medical applications, including orthopedic implants, dental devices, and cardiovascular stents [6]. Additionally, Ti alloys have a lower elastic modulus compared to commonly used materials like 316 stainless steel (SS) and cobalt-chromium alloys, as shown in Figure 5.1. This reduces the risk of modulus mismatch and the stress shielding effect, which occurs when implants are too stiff and cause the surrounding bone to weaken.

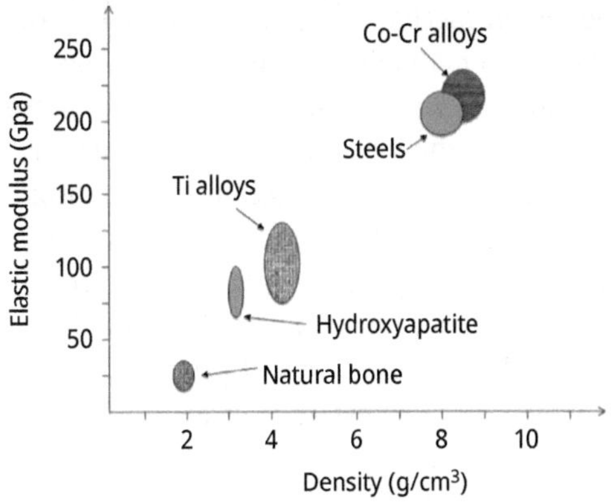

Figure 5.1: Comparison of modulus of different materials to natural bone [65].

This chapter covers the wear, corrosion, and biofunctional characteristics of Ti and its alloys as biomaterials. While Ti alloys offer excellent biocompatibility, their primary limitation is their mediocre tribological performance, characterized by high frictional coefficient and significant wear of the material in oxide formation and adhesion. To address aforementioned challenges, numerous surface modification procedures have been applied to enhance tribological behavior as well as corrosion and osseointegration of Ti alloys, making them more suitable for load-bearing medical implants and other biomedical applications. The chapter includes the most recent advancements in surface treatments for Ti and its alloys, which show significant promise for overcoming these challenges.

5.2 Ti and Ti alloys used in medical applications

Metals and metallic alloys, such as 316 SS, CoCrMo (cobalt-chromium-molybdenum) alloy, and Ti alloys, are extensively utilized in the load-bearing biomedical implants due to their favorable qualities, such as superior strength, low modulus, and toughness. Among these, Ti alloys are particularly favored because they offer superior biocompatibility and mechanical strength. Commercially pure titanium (Cp-Ti) and its alloys, such as Ti6Al4V, Ti6Al7Nb, Ti5Al6Nb, Ti15Mo, and Ti13Nb13Zr, have become essential materials in the biomedical applications due to their stunted density, excellent biocompatibility, and favorable mechanical properties [7]. Initially, Ti and its alloys were developed for aerospace applications. However, in the 1970s, their biocompatibility was discovered, driving demand for their use in biomedical applications. When Cp-Ti is employed instead of its alloys, the loss of mechanical qualities caused by the lack of alloying elements must be compensated to maintain effective performance. Ti alloys have better mechanical and biocompatibility qualities than Cp-Ti. Ti alloys are divided into four separate categories according to their iron and oxygen concentration levels. The most commonly used Ti alloy is Ti6Al4V, which accounts for approximately 50% of all Ti alloys used in bioimplants [8, 9]. Ti6Al4V outperforms other Ti alloy grades in terms of formability, structural stability, biocompatibility, corrosion resistance, and strength-to-weight ratio. These qualities make Ti alloys highly suitable for various biomedical applications, including heart valves, dental prostheses, bone fixators, bone replacements, and artificial joints [10] as shown in Figure 5.2.

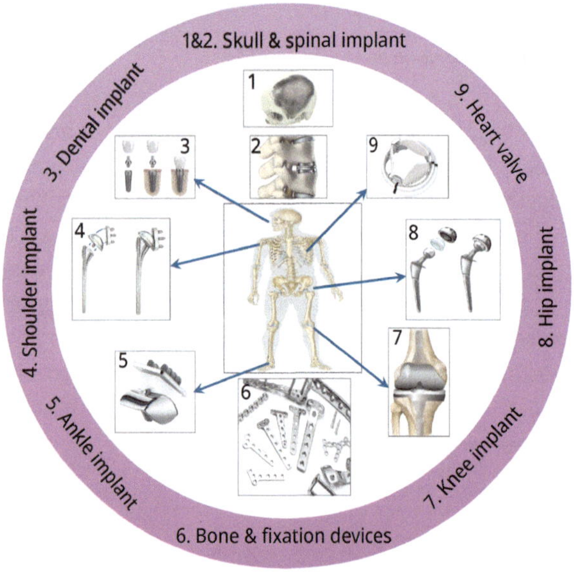

Figure 5.2: Ti and Ti alloys in medical applications.

Biomedical-grade Ti alloys are typically classified into four categories based on their microstructure: alpha (α, such as Ti6Al4V), near-alpha, alpha-beta (α-β), and metastable beta (β, such as Ti6Al7Nb) [11]. These classifications affect their mechanical properties and suitability for various biomedical applications, as listed in Table 5.1. While Ti alloys are commonly utilized in biometallic implants, their high elastic modulus values might create stress shielding effect at the implant-tissue interface. The elastic modulus of Cp-Ti and α-β-type Ti alloys ranges between 100 and 110 GPa (significantly higher than that of natural human bone) and reduces their effectiveness in joint applications. Additionally, the existence of vanadium (V) and aluminum (Al) in certain alloys can steer to the release of toxic ions, such as vanadate, oxovanadium, and aluminum (Al^{3+}), in the physiological environment, potentially causing adverse health effects. As a result, there has been significant interest in developing β-titanium alloys that incorporate elements like zirconium (Zr), tantalum (Ta), niobium (Nb), or molybdenum (Mo) to replace V and Al. These β-alloys offer superior mechanical characteristics, enhanced ductility, improved structural stability, a lower elastic modulus, higher wear resistance, and better corrosion resistance, making them more suitable for biomedical applications [12].

Table 5.1: Implants and medical applications of Ti and Ti alloys [66].

Implants	Use	Ti alloys
Dental	Braces, abutments, fixation devices, bridges, and orthodontics	β-Ti, Cp-Ti, nitinol, and Ti6Al4V
Orthopedic	Joint components, meshes, fixation devices, and bone substitutes	Cp-Ti, Ti6Al4V, Ti6Al7Nb, Ti15Mo, Ti13Nb13Zr, and nitinol
Cardiovascular	Heart valves, guidewires, clips, catheters, ventricular assist devices, stents, and fixation devices	Cp-Ti, Ti6Al4V, Ti6Al7Nb, Ti15Mo, and nitinol
Spinal	Discs, cages, and fixation devices	Cp-Ti, Ti6Al4V
Trauma devices	Nails, plates, rods, and screws	Cp-Ti, Ti6Al4V, and Ti6Al7Nb
Soft tissue	Hernia meshes, breast reconstruction meshes, and fixation devices	Cp-Ti, Ti6Al4V, and Ti6Al7Nb

One of the key drawbacks of Ti alloys is their subpar tribological characteristics, as they tend to exhibit high friction and are prone to abrasive wear [13]. This limits their performance in applications where wear resistance is critical. Additionally, forming a titanium dioxide (TiO_2) layer on the surface of Ti alloys during exposure can protect the material but also hinder the implant-tissue interaction. This passive oxide layer and the accumulation of titanium compounds in the surrounding tissues can lead to implant failure over time [14].

5.2.1 Wear resistance

The main drawback of Ti and its alloys, which hinders broader use, is their inadequate resistance to fretting fatigue and subpar tribological properties stemming from their low hardness. These materials exhibit poor performance in friction-related scenarios, marked by a high friction coefficient, excessive adhesive wear, a higher tendency to seize, and limited abrasion resistance. When Ti contacts either itself or with other surfaces, it experiences rapid wear. This accelerated wear occurs because of the formation of adhesive bonds and the unstable nature of its passive oxide layer, particularly when third-body particles are present. Due to this effect, around 10–20% of total joint replacements involving a Ti head and a polymer cup expect revision within a range of 15–20 years. Moreover, aseptic loosening accounts for over 80% of these modifications [15]. Ti and its alloys have a high friction coefficient, which is the primary cause of implant failure. This causes the release of wear debris into the circulation, which causes inflammation in the surrounding tissue and initiates bone resorption, also known as osteolysis. This process ultimately results in the loosening of the implant, necessitating its replacement with a new one [16].

5.2.2 Biofunctionality or osseointegration

When an implant is inserted through surgery in bone, a variety of biological, physical, chemical, thermal, and other elements come into play, all of which determine whether or not biofunctional or osseointegration occurs well.

Ti and its alloys are extensively employed in dental and orthopedic implants, especially in load-bearing applications, due to their excellent biocompatibility, strength, and fracture toughness. While these materials have been shown to integrate directly with bone, they do not establish a true chemical bond with the bone tissue. In recent years, several new coatings have been developed to enable Ti and its alloys to naturally bond with the living bone. Plasma-sprayed hydroxyapatite (HAp) coatings are frequently employed in cementless articulating joint replacements due to their favorable biocompatibility. However, HAp coatings have certain limitations, including the potential for delamination from the substrate, difficulties in controlling their composition, and degradation that can release debris, contributing to third-body wear [17].

Promoting the generation of compatible bone tissue at the interface of bone-implant can help establish a robust and long-lasting link between the bone and the implant. This can be accomplished through appropriate surface treatments of the implant, such as dipping, electrochemical deposition, and physical vapor deposition (PVD) methods. These techniques augment the surface properties of the implant, encouraging better integration with the surrounding bone tissue [18].

5.2.3 Corrosion resistance

Metals and alloys are susceptible to corrosion when they are exposed to the physiological conditions (body fluids), as the body's environment is particularly harsh due to the presence of proteins and chloride ions. Numerous chemical reactions happen on the subjected surface of implanted alloys. An oxidation reaction occurs on metallic elements, leading to the production of hydroxide ions from dissolved oxygen. Metals and alloys that demonstrate high corrosion resistance typically achieve this by forming a passive layer. In this condition, the metal develops a thin oxide coating on its surface, such as TiO_2 for titanium, called a passive film. This layer serves as a protective barrier, shielding the metal from its surroundings. These films are generally 3–10 nm thick and are made up of metal oxides that resemble ceramic materials. The natural oxide is typically amorphous and has stoichiometric defects. On titanium, the TiO_2 layer is known to promote favorable osseointegration. The durability of this oxide layer largely depends on its thickness, structure, and composition [19]. An oxide film significantly lowers the dissolution rate of a passive metal at a given potential compared to that of an active metal. This reduction is mainly affected by the characteristics of the passive film and how soluble it is in the surrounding media. Aforementioned, oxide films naturally form on the metal's surface and act as barriers, blocking the movement of metallic ions and free electrons. To be effective, these films need to be compact, completely cover the metal surface, and possess an atomic structure that limits ion and electron migration at the metal oxide-solution interface. Moreover, they should remain stable under mechanical stress or abrasion, as seen in orthopedic devices [20].

The development of surface treatments for Ti and its alloys has been driven by their relatively poor tribological properties and potential corrosion challenges. These treatments aim to strengthen the surface layer, boost hardness, and improve resistance to abrasive wear, which also lowers the coefficient of friction. Furthermore, surface modifications help limit or prevent ion release from the material into surrounding tissues, addressing concerns related to both wear and corrosion.

5.3 Need of surface modification

Surface engineering portrays a vital role in significantly enhancing the performance of Ti-based orthopedic devices, extending their durability well beyond their natural capabilities. The primary goals of surface treatments are to improve the implant's tribological behavior, corrosion resistance, and osseointegration. Various coatings and surface modifications are applied to boost wear and corrosion resistance by increasing the surface characteristics of the material. These modifications can be achieved through broadly classified surface modification techniques such as mechanical, chemical, and physical surface modifications, as shown in Figure 5.3.

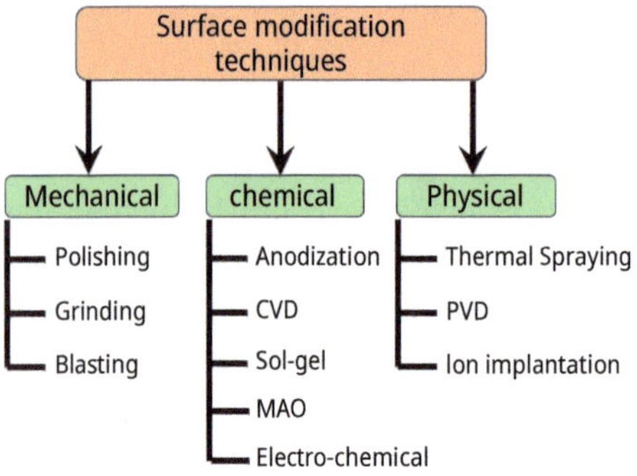

Figure 5.3: Classification of surface modification techniques.

Mechanical methods can directly change a material's surface structure and composition, enhancing the surface morphology of Ti and its alloys for better cell and tissue integration. By changing the surface characteristics, mechanical methods promote a better biological response and support more effective osseointegration. The main mechanical methods for modifying surface morphology are polishing, grinding, and sandblasting. Polishing is a surface finishing technique that uses polishing tools and abrasive particles to reduce surface roughness, achieving a bright, smooth finish. This process is often conducted in multiple steps to ensure a high-quality flat surface. Grinding, another finishing method, utilizes abrasive granules either coated or embedded on a grinding tool to refine the material surface. It involves relative motion between the workpiece and the grinding tool under the controlled pressure, effectively smoothing and shaping the surface. Sandblasting cleans and roughens a substrate by directing a high-speed stream of sand at it, resulting in varying levels of cleanliness and surface roughness. This method enhances the mechanical properties and durability of coatings, increases titanium surface roughness, promotes early cell adhesion, and improves the osseointegration of implants with bone tissue [21].

Physical surface modification alters a material's surface without chemical reactions, using methods such as deposition or coating to enhance biocompatibility and antibacterial properties:

Thermal spraying technology: This process employs various heat sources, including gas flames, electric arcs, plasma arcs, or lasers, to melt metals, ceramics, or other materials. The molten or semi-molten material is then sprayed onto a pretreated surface using high-velocity airflow, creating a strong bond with the substrate. Common thermal spraying methods include flame spraying, plasma spraying, and high-velocity oxygen fuel spraying, providing enhanced wear, corrosion, oxidation, and heat resistance, which are advantageous for biomedical applications.

Physical vapor deposition (PVD): These methods involve evaporating or sputtering a source material, whether solid or liquid, in a vacuum environment. The vaporized or ejected atoms, molecules, or ions are transported through a low-pressure gas or plasma channel and deposited onto the target surface, creating a thin film. Common PVD methods include vacuum vapor deposition, sputter plating, and ion plating. In vacuum vapor deposition, the target material is evaporated in a vacuum using specific heating techniques. The vaporized particles then travel to the surface, forming a cohesive film, which leads to a straightforward process that produces films with high purity and density.

Ion implantation technique: This technique modifies a material by introducing dopant ions into a solid. Ions produced by an ion source are accelerated to high speeds and aimed at the material's surface. The high-energy ions penetrate the surface, embedding themselves and forming a modified layer with improved properties. Ion implantation is generally divided into two categories: plasma immersion ion implantation (PIII) and ion beam ion implantation. PIII is especially effective for modifying biomaterial surfaces due to its thorough and highly reactive process.

Chemical surface modification improves the bioactivity and antimicrobial properties of titanium and its alloys by altering their chemical characteristics. This enhances interactions with cells and bacteria for better performance in biomedical applications. Here are brief descriptions of various chemical surface modification methods:

Anodization: Anodic oxidation forms oxide films on titanium alloys, enhancing surface hardness, wear and corrosion resistance properties, as well as biological activity and antibacterial properties. Anodic oxidation can lead to uneven coating, high costs, and limited adhesion on Ti and Ti alloys. It might also cause surface brittleness and increased complexity in processing and equipment requirements.

Chemical vapor deposition (CVD): This is a method where gaseous or vaporized chemicals undergo a reaction at the interface between gas and solid, resulting in a thin, high-purity film on a material's surface. Unlike PVD, CVD provides better overall coverage of the surface. This technique is also capable of applying coatings to complex surfaces, improving both mechanical and biological properties in titanium and its alloys. However, CVD comes with some drawbacks: it typically has a slow deposition rate, can release harmful by-products, and is less efficient for thin-film applications compared to PVD. Furthermore, CVD requires higher reaction temperatures and is applicable to a more limited range of processes.

Sol-gel coating: In this technique, a liquid solution (sol) is applied to a surface and undergoes chemical reactions to transform into a solid gel. This approach results in thin, uniform coatings that enhance the surface's mechanical and chemical properties. However, the raw materials utilized in this procedure are costly, with some being organic and potentially hazardous. Moreover, the process can be time-consuming, often taking several days or weeks. The gel also contains numerous micropores, which can release gases and organic compounds during drying, leading to shrinkage. Thus, careful condition selection is crucial for obtaining the desired material.

Electrochemical deposition: This method involves applying an electric current to a metal surface that is submerged in an electrolyte solution. Chemical reactions then result in a coating being deposited on titanium and its alloys. These reactions are largely influenced by the composition of the utilized electrolyte and variables of treatment such as current, voltage, and temperature. Electrochemical treatment techniques include electropolishing, electrophoretic deposition, and electrodeposition. Using electrochemical deposition, both metallic and nonmetallic coatings can be applied to Ti and its alloys, improving their hardness, wear resistance, and bioactivity.

Microarc oxidation (MAO): Known as plasma electrolytic oxidation (PEO), this technique creates durable ceramic coatings on metals such as titanium, aluminum, and magnesium. It works by employing arc discharge to intensify and activate reactions at the anode, with adjustments made to the electrolyte composition and electrical settings for optimal results. Evolving from conventional anodic oxidation, this method is widely applied to Ti and its alloys to enhance their mechanical strength, bioactivity, antibacterial properties, and corrosion resistance.

5.4 Important coatings for Ti and Ti alloys for medical applications

Choosing the right coating for titanium alloys is crucial, especially for load-bearing applications like joint prostheses. The selection process depends on factors such as biocompatibility or bioinertness, mechanical strength, and tribological performance. Additionally, coatings produced using different techniques can vary significantly in their properties, including internal stress, morphology, hardness, and toughness, which in turn affect their tribological behavior. Several important coatings used in biomedical applications for wear resistance are highlighted in the following paragraphs. These coatings are recognized for their excellent biocompatibility or bioinertness.

5.4.1 Hydroxyapatite (HAp)

HAp ($Ca_{10}(PO_4)_6(OH)_2$) has emerged as a promising bioactive material for coating metallic biomaterials due to its chemical composition's close resemblance to natural bone tissue and its exceptional biocompatibility. Premphet et al. developed HAp-TiO_2 films with excellent hardness through pulsed DC magnetron sputtering method. These coatings are biocompatible and nontoxic, making them suitable for tissue implants [23]. Due to its weak crystalline structure, the dissolution of HAp coating results in reduced substrate adherence and eventual implant failure. This limitation prompts researchers to explore alternative hard coatings with superior adhesion to substrates and enhanced wear resistance. In their research, Huang et al. [22] employed electrochemical deposition technique to create a hybrid HAp coating on the etched titanium alloy surfaces, with sequential release of BMP-2. Their study revealed that Ti-0.8V, Ti-HAp co-coated BMP workpieces showed the ultimate efficiency in cell proliferation and supported superior cell growth, which contributed to enhanced osseointegration.

5.4.2 Carbon-utilized coatings

The recently synthesized graphite-like carbon (GLC) is a hydrogen-free, amorphous carbon-chromium coating with a hexagonal lattice structure in the lattice layer. Currently, GLC coatings are undergoing testing for use in articulating hip joints, demonstrating no biocompatibility issues. While uncoated hip joints typically last over 10 years, GLC-coated joints with a thickness of 2.50 μm are projected to extend their lifespan to up to 50 years. There is also potential for GLC application in knees and other joints [24].

The widespread interest in implant materials coated with diamond-like carbon (DLC) arises from their favorable biocompatibility [25]. DLC coatings are made up of a combination of sp^2 (graphite-like) and sp^3 (diamond-like) carbon bonds, providing a balance of mechanical strength and tribological performance. DLC possesses crucial characteristics such as corrosion resistance and chemical inertness. These attributes are promising and can serve to safeguard implants effectively [26]. However, various studies revealed poor adhesion, cracking, and delamination to the substrate [27]. Poor adhesion could resist DLC's application in biomedical applications.

5.4.3 High-entropy alloy (HEA) coatings

Currently, researchers are increasingly drawn to high-entropy alloys (HEAs) for their exceptional hardness and wear resistance, both as bulk materials and coatings. HEAs, characterized by their composition of at least five fundamental elements in nearly equal proportions, have garnered significant interest in the biomedical realm. This attention stems from their advantageous properties, including high strength, fracture toughness, hardness, and corrosion resistance. Bio-HEAs typically comprise titanium and other elements from groups IV and V, known for their hypoallergenic and non-toxic properties [28]. Numerous studies indicate that incorporating refractory elements such as Ta, Mo, Cr, Zr, Nb, V, Ti, tungsten (W), and hafnium (Hf) into HEA-based coatings on metal substrates enhances their mechanical properties [29]. However, HEA coating still has certain drawbacks, such as the higher cost than traditional coatings.

5.4.4 Metal oxides

Metal oxides are commonly classified as ceramic coatings and are extensively researched for surface modification of metals or alloys to achieve effective corrosion resistance. Metal oxides have higher chemical stability than pure metals and metal hydroxides. Moreover, oxide coatings predominantly consist of oxides such as Al_2O_3, TiO_2, and ZrO_2 (zirconia), and yttria-stabilized zirconia [30]. Metal oxide coatings can be categorized as either substrate-involving or non-substrate-involving, which depends on whether the substrate is a part of the coating synthesis. Additionally, Al_2O_3 and ZrO_2 exhibit favorable biocompatibility and superior mechanical properties [31]. However, TiO_2 demonstrates strong adhesion to the substrate and superior corrosion resistance. The biomedical application of metal oxide coatings has been primarily hindered by their high modulus and limited ability to interact with physiological conditions [32].

5.4.5 Transition metal nitrides (TMeNs)

Transition metal nitrides (TMeNs) were extensively employed to safeguard metallic components from wear, tear, and corrosion, with the added benefit of high-temperature stability in the past. Titanium nitride (TiN) coatings, in particular, functioned as ornamental coatings. However, in the past decade, TMeN coatings for orthopedic implants have been recommended to protect them from wear. Furthermore, it functions as a diffusion barrier layer, considerably reducing the release of harmful ions from the implanted metal surfaces into the physiological environment [12]. Researchers are increasingly interested in TMeN coatings due to their favorable properties, including better hardness, wear resistance, excellent corrosion resistance, and bioinertness. Coatings such as TiN, TaN, NbN, ZrN, VN, CrN, and TiAlN are well-suited for implants as protective coatings, enhancing their lifespan by improving mechanical properties and reducing wear rate [33–36]. Among aforementioned TMeNs, TaN coatings are attracting researchers due to their complex structural characteristics based on stoichiometry.

5.5 Surface modifications of Ti and Ti alloys for medical applications

Various surface modifications have been done on Ti and its alloys to enhance their wear resistance, biofunction, and corrosion resistance. Some critical investigations have been discussed in the following sections.

5.5.1 Improving wear resistance

Wear resistance is a vital property for biomedical implants, which is need to avoid loosening. Although Ti alloys generally possess superior specific strength, their wear resistance is often poor. Thus, various surface modification methods have been advanced to improve the wear resistance of Ti alloys. Nitriding is a widely used surface modification technique to enhance the wear resistance of Ti alloys by forming a hard TiN layer. Plasma nitriding and PVD techniques are the standard methods for applying the TiN layer on the implants. Samanta et al. [37] studied the effect of plasma nitriding on the Ti6Al4V alloy and found that the wear rate of the plasma-nitrided Ti alloy was about 99% lower than the untreated Ti alloy. It stated that it possesses excellent wear resistance properties. As shown in Figure 5.4, the plasma-nitrided Ti alloy's friction coefficient was 52% lower and the wear track was significantly thinner than the untreated one. Shirazi et al. [38] used a laser nitriding technique to improve wear resistance (hardness, plasticity index, etc.) of Ti6Al4V and Ti35.5Nb7.3Zr5.7Ta alloys.

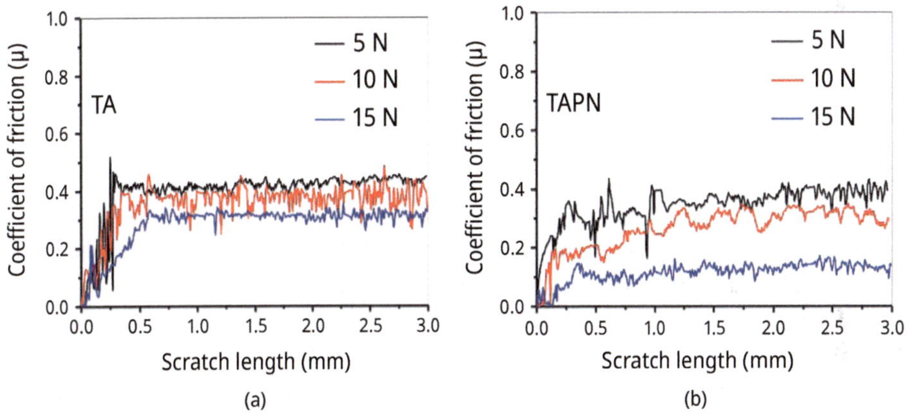

Figure 5.4: The coefficient of friction as a function of scratch length for: (a) Ti6Al4V alloy (TA) and (b) plasma-nitrided Ti6Al4V alloy (TAPN) under different loads [37].

Chan et al. [39] applied fiber laser nitriding to the Ti6Al4V alloy. The nitrided layer obtained, upon optimization at 60% duty cycle, produced much finer wear particles (0.01–0.04 μm) that were less harmful and antimicrobial in nature than larger particles of 1.5 μm obtained from untreated surfaces. This was further enhanced through increasing hardness in the 795 HV to 844 HV range, hence defining a promising approach because of load-bearing applications. Skjöldebrand et al. [40] applied coatings of TiN and TiNbN to joint replacements for surface improvement. The performance of these coatings increases scratch resistance and reduces wear, besides inhibiting ion release, which is rather critical to minimize metal hypersensitivity and tissue reactions. Clinical success was particularly evident for implants coated with TiNbN because their chromium ion release was reduced, and symptoms of allergy were alleviated. Marin et al. [41] significantly improved the wear resistance of titanium alloys by employing a combined nitriding and carburizing process, followed by the deposition of PVD coatings. Both nitriding and carburizing treatments independently contributed to enhanced wear resistance. To evaluate these effects, the researchers conducted wear tests on various treated samples under sliding conditions, using a Ti6Al4V pin counterpart with a tensile load of 50 N, a sliding distance of 2 mm, and a frequency of 1 Hz. The findings of this study are summarized in Table 5.2.

Table 5.2: Volumetric wear loss for various treated samples [41].

Treatment	Volumetric loss (μm^3)
Untreated	$1.6 \times 10^7 \pm 0.3 \times 10^7$
Nitrided	$-2.3 \times 10^5 \pm 0.2 \times 10^5$
Carburized	$6.8 \times 10^5 \pm 0.7 \times 10^5$
TiCN (T1)	$2.1 \times 10^7 \pm 1.1 \times 10^7$
CrN (T2)	$2.9 \times 10^7 \pm 0.9 \times 10^7$
Nitrided + TiCN (N1)	$-2.7 \times 10^5 \pm 0.9 \times 10^5$
Nitrided + CrN (N2)	$-1.7 \times 10^5 \pm 0.6 \times 10^5$
Carburized + TiCN (C1)	$2.87 \times 10^4 \pm 01.5 \times 10^4$
Carburized + CrN (C2)	$1.88 \times 10^5 \pm 0.5 \times 10^5$

DLC coating is another popular choice for wear resistance improvement. DLC coatings are a family of amorphous carbon coatings with diamond-like properties, such as high hardness, low friction, and excellent wear resistance. Choudhury et al. [42] explored the potential of functional DLC-coated Ti6Al4V alloy for prosthetic hip joints. While Ti6Al4V is a popular choice for orthopedic implants, it can experience edge loading and wear issues. They found that DLC-coated Ti6Al4V paired with Si_3N_4 femoral cups offered significant advantages. Its superior hardness, biocompatibility, and reduced wear rates made it a promising candidate.

Moreover, multilayered DLC coatings, especially multilayered 3D, demonstrated enhanced mechanical properties and even less wear than traditional materials. Rodriguez-Rojas et al. [43] discussed the role of submicron DLC coatings on friction and wear behavior of Ti6Al4V alloys in dental implants, focusing on sliding contact in artificial saliva. Indeed, the DLC coating reduced adhesion between surfaces. The coefficient of friction of DLC was about one-quarter that of uncoated Ti6Al4V, while contact stress and wear rate were reduced substantially. The results showed that with DLC film, the wear mode has transformed from mixed abrasion and adhesion to pure abrasion, enhanced by its hardness. HAp is a calcium phosphate compound similar to the mineral composition of bone. It is used to coat titanium implants as it provides a bioactive surface where bone cells can attach and proliferate [44]. Present approaches to deposit HAp include plasma spraying, sol-gel deposition, and electrophoretic deposition. Currently, plasma spraying is the most popular technique for coating HAp. Kumari and Majumdar [45] addressed the wear resistance of plasma spray-deposited HAp-based coatings, namely HAp with 50% TiO_2 and HAp with 10% ZrO_2, onto substrates of the Ti6Al4V alloy. The as-sprayed coatings were found to possess higher porosities and, hence, inferior wear resistance compared to the untreated substrate; heat treatment substantially improved it. The test also showed that the friction coefficients and wear resistance were improved within the two coatings after the heat treatment, with HAp containing 10% ZrO_2 showing the most significant improvement. Jing et al. [46] developed HAp biofunctional gradient coating (BFGC) on Ti6Al4V alloy using laser cladding. With a 700-W laser power, a smooth, dense, and crack-free coating is created, improving the wear resistance considerably. The BFGC helped HAp formation in simulated body fluid and supported osteoblast proliferation without cytotoxicity. Wu et al. [46] developed a photocatalytic TiO_2 nanoceramic coating for titanium implants through a PEO process in order to enhance photocatalytic cleaning under visible light. It showed excellent wear resistance, high hardness, and effective degradation of organic contaminants, maintaining biocompatibility and osteoconductivity. Such features propose a promising approach for deep cleaning of dental implants from peri-implantitis. Consequently, surface treatments applied to Ti and its alloys have significantly enhanced their wear resistance, making them highly suitable for biomedical implant applications.

5.5.2 Improving biofunction

The interaction of biomedical titanium and its alloys with the biological medium, that is, the fluid containing proteins and cells, upon implantation greatly influences implants' long-term behavior. Therefore, tests of biocompatibility have to be performed on these materials. Relatively inert at human body sites, the two commercially used metals, CP-Ti and Ti6Al4V, spontaneously form a thin, protective oxide layer of TiO_2 on the surface. This passive film rapidly regenerates even if it is damaged, which contributes to them possessing excellent biocompatibility. The biocompatibility issues of titanium and titanium alloys have been actively researched since 1976 [47]. For instance, their inert nature can lead to the formation of fibrous tissue capsules around the implants, which can potentially provoke a foreign body response. Certain elements within specific titanium alloys can be toxic when released in elemental form, for example, V within Ti6Al4V and Ni within shape-memory alloys like NiTi. In addition, other properties of implants that are generally desirable include osseointegration and antibacterial character. With the complexity of interactions at the implant-tissue interface, it is evident that the surface composition and topography of such implants are of high importance in their interaction with bone tissue. So far, two strategies have been mainly considered to enhance the biocompatibility of titanium and its alloys: minimizing the use of toxic elements and modifying the surface. β-Type titanium alloys are made from nontoxic elements, and Span Internet revived higher cell viability and growth rates than Ti6Al4V [48]. This has driven a wave of rapid development concerning β-type titanium alloys in recent years. Yet, studies on β-type titanium alloys are still at an early stage and more detailed research should be carried out. Moreover, the surface modifications significantly enhance osseointegration and antibacterial properties of both titanium and its alloys. Hanawa [49] applied the electrodeposition of biofunctional molecules, such as polyethylene glycol, to enhance biocompatibility by inhibiting protein adsorption and promoting cell adhesion. The plasma spraying of HAp, nanoscale TiO_2, and $CaTiO_3$ on dental implants has been discussed [50]. Trino et al. [51] explored the biofunctionalization of titanium surfaces using osteogenic peptides to enhance implant biocompatibility, osseointegration, and corrosion resistance. Titanium surfaces were modified with TiO_2, and different spacers such as 3-(4-aminophenyl)propionic acid and 3-mercaptopropionic acid were used to immobilize dentin matrix protein 1 peptide on the titanium surface. Huang et al. [52] investigated low-temperature plasma polymerization, successfully enhancing the hydrophilic properties of the Ti surface and increasing the concentration of functional amino groups; these modifications are critical for promoting cell adhesion and tissue integration. Zakaria et al. [53] applied PEO technique on Ti6Al4V titanium alloy implants with the incorporation of fluoridated HAp and FAp (fluorapatite) by using an electrolyte composition of trisodium orthophosphate, potassium hydroxide, and calcium fluoride. The layered coating of TiO_2, tricalcium orthophosphate, HAp, and FAp showed a possible improvement in their bioactivity and antibacterial properties that may bring better success to the implants. Ganjali et al. [54] and Moloodi et al. [55] employed laser cladding and sol-gel techniques to incorporate Sr^{2+} and F^- ions onto Ti6Al4V, resulting in improved antibacterial properties, cell adhesion, and diffusivity, and can stimulate the proliferation and differentiation of osteoblasts. Bhattacharjee et al. [56] and Wu et al. [57] used plasma spraying to incorporate Zn^{2+} and F^- ions and achieved enhanced antibacterial properties and improved in vitro biocompatibility of Ti6Al4V. Qiao et al. [58] applied TiO_2 nanotubular and SSAgHAp layers on the Ti substrate by electrochemical deposition, and the 3D interconnected porous structure of the coating is achieved with a uniform distribution of Ag, Sr, and Si ions. The Sr and Si dopants improved the coating's osteogenic potential while mitigating the cytotoxicity of Ag ions. Ke et al. [59] explored the effects of Si^{4+} and Zn^{2+} doping on HAp coatings applied to titanium (Ti) implants. They achieved better osteoblast differentiation and reduced osteoclast activity, indicating their potential to improve implant performance in osteoporosis. Grit blasting is one of the popular mechanical methods used to create rough Ti surfaces to enhance biofunction. In this process, the compressed air is utilized to propel the abrasive particles against a surface in order to attain desired texturing. Grit particles should be chemically stable, biocompatible, and should not hamper the osseointegration of Ti implants [60]. Overall, the biocompatibility of Ti and its alloys is very good; however, in biomaterial applications, the enhancement of osseointegration and antibacterial ability of titanium and its alloys is needed. The surface treatments, ionic incorporation, and use of β-type titanium alloys have shown promising methods to enhance longer-term performances of titanium-based implants.

5.5.3 Improving corrosion resistance

Corrosion resistance of Ti and its alloys is provided by a naturally formed stable TiO_2 layer on their surface. However, in harsh biological environments, especially those rich in chloride ions, such as blood and interstitial fluids, further enhancement of corrosion resistance is crucial for the long-term stability of biomedical implants. Various surface modification techniques discussed above for improvement of wear and biofunction also contribute to improved corrosion resistance, like plasma nitriding, PVD coatings, and surface texturing, which offer improved corrosion resistance by forming either protective layers or surface refinement. Trino et al. [61] addressed a critical concern in orthopedics and dental implantology. The release of titanium ions and debris due to its low corrosion resistance is a significant clinical challenge. They discussed ZnO thin films as a chemical modification strategy to improve corrosion resistance on Ti cp4 substrates, and sol-gel-synthesized ZnO produced a nanostructured film with hydroxyl groups that can be readily functionalized with bifunctional organic molecules, leading to enhanced corrosion resistance. Kim et al. [62] investigated the supplementation role of Mn in developing new dental implant grades of the Ti-Mn alloy. X-ray diffraction, SEM, and TEM techniques characterized the phase composition, mechanical, and corrosion properties. Increasing Mn content showed a change in phase morphology from α-Ti to β-Ti. Among such, the 5 wt% Mn alloy possessed optimum hardness, modulus, and better corrosion resistance to employ as the best candidate grade. Han et al. [63] found that the addition of Nb to Ti-*x*Nb alloys increased the corrosion resistance compared to Cp-Ti. The β and ω phases produced with the increase in the percentage of Nb brought an improvement in microstructure and protected against oxidation. Electrochemical experiments presented an advanced resistance of corrosion and reduced differences in Thermal expansion coefficient between alloys and ceramic coatings. Wang et al. [64] found that the Ag-PDA-functionalized micro/nanostructured surfaces showed very good corrosion resistance compared to the untreated microstructured surface. A necessary barrier layer in the form of polydopamine or active metals such as silver greatly enhanced the corrosion resistance of this material in biomedical media. It is important to note that any method used to enhance corrosion resistance should not compromise the other beneficial properties of Ti implants.

5.6 Conclusions

Titanium-based materials have become widely accepted for medical implants, particularly in orthopedics, due to their excellent corrosion, nonmagnetic properties, mechanical characteristics, and biofunction that closely resemble natural bone. Despite these advantages, the bioinert nature and wear resistance of titanium and its alloys remain a challenge and are an ongoing area of investigation in surface engineering. Various approaches have been explored to enhance the performance of titanium implants, including different methods of surface modifications. The surface features of orthopedic implants play a key role, especially when strong bonding with the surrounding tissue is needed for better healing outcomes. When an implant is inserted into bone during surgery, various biological, physical, and chemical factors influence osseointegration. Ti and its alloys are commonly used in dental and orthopedic implants due to their biocompatibility and strength, although they do not chemically bond with the bone tissue. To improve bonding, surface treatments like HAp coatings are employed to promote stable bone formation at the implant interface. However, hard coatings like transition metal nitrides, oxides, and carbon-based coatings demonstrated superior wear resistance. Therefore, optimizing surface properties is critical to encouraging overall better performance for the typical load-bearing applications. Each available technique to modify the properties of titanium implants comes with its own benefits and limitations. Factors such as implant location, design, surface modification needs, the thickness of the modified layer, and changes in surface composition must all be considered when choosing the appropriate method. Techniques that involve mechanical processing and surface treatment typically leave the surface composition unchanged, so hybrid methods that incorporate surface coatings tend to produce better results. However, challenges related to performance of coating, film thickness, and associated costs remain. Future research should focus on developing hybrid techniques that combine different strategies to further improve the performance of titanium and its alloys for implants.

References

[1] Geetha, M., Singh, A. K., Asokamani, R., & Gogia, A. K. (2009). Ti based biomaterials, the ultimate choice for orthopaedic implants – A review. *Progress in Materials Science*, *54*(3), 397–425, doi: 10.1016/j.pmatsci.2008.06.004.

[2] Abdel-Hady Gepreel, M., & Niinomi, M. (2013). Biocompatibility of Ti-alloys for long-term implantation. *Journal of the Mechanical Behavior of Biomedical Materials*, *20*, 407–415, https://doi.org/10.1016/j.jmbbm.2012.11.014.

[3] Zhang, L. C., & Chen, L. Y. (2019). A review on biomedical titanium alloys: Recent progress and prospect. *Advanced Engineering Materials*, *21*(4), 1–29, doi: 10.1002/adem.201801215.

[4] Kurtz, S., Ong, K., Lau, E., Mowat, F., & Halpern, M. (2007, Apr). Projections of primary and revision hip and knee arthroplasty in the United States from 2005 to 2030. *The Journal of Bone and Joint Surgery*, *89*(4), 780–785, doi: 10.2106/JBJS.F.00222.

[5] Kurtz, S. M., et al. (2007, Oct). Future clinical and economic impact of revision total hip and knee arthroplasty. *The Journal of Bone and Joint Surgery*, *89*(Suppl 3), 144–151. doi: 10.2106/JBJS.G.00587.

[6] Kaur, S., Ghadirinejad, K., & Oskouei, R. H. (2019). An overview on the tribological performance of titanium alloys with surface modifications for biomedical applications. *Lubricants*, *7*(8), doi: 10.3390/lubricants7080065.

[7] Sarraf, M., Rezvani Ghomi, E., Alipour, S., Ramakrishna, S., & Liana Sukiman, N. (2022). A state-of-the-art review of the fabrication and characteristics of titanium and its alloys for biomedical applications. *Bio-Design and Manufacturing*, *5*(2), 371–395, doi: 10.1007/s42242-021-00170-3.

[8] Farooq, M. U., Mughal, M. P., Ahmed, N., Mufti, N. A., Al-Ahmari, A. M., & He, Y. (2020). On the investigation of surface integrity of Ti6Al4V ELI using Si-mixed electric discharge machining. *Materials (Basel)*, *13*(7), doi: 10.3390/ma13071549.

[9] Singh, N., Hameed, P., Ummethala, R., Manivasagam, G., Prashanth, K. G., & Eckert, J. (2020). Selective laser manufacturing of Ti-based alloys and composites: Impact of process parameters, application trends, and future prospects. *Materials Today Advances*, *8*, 100097, https://doi.org/10.1016/j.mtadv.2020.100097.

[10] Elias, C. N., Lima, J. H. C., Valiev, R., & Meyers, M. A. (2008). Biomedical applications of titanium and its alloys. *JOM*, *60*(3), 46–49, doi: 10.1007/s11837-008-0031-1.

[11] Lampman, S. (2012, Jun. 01). Titanium and Its Alloys for Biomedical Implants. In *Materials for Medical Devices 23*, (pp. 0). ASM International, doi: 10.31399/asm.hb.v23.a0005674.

[12] Mosas, K. K. A., Chandrasekar, A. R., Dasan, A., Pakseresht, A., & Galusek, D. (2022). Recent advancements in materials and coatings for biomedical implants. *Gels*, *8*(5), 1–35, doi: 10.3390/gels8050323.

[13] Fellah, M., et al. Tribological behavior of Ti-6Al-4V and Ti-6Al-7Nb alloys for total hip prosthesis. *Advances in Tribology*, *2014*(1), 451387, (2014, Jan), https://doi.org/10.1155/2014/451387.

[14] Delgado-Ruiz, R., & Romanos, G. (2018, Nov). Potential causes of titanium particle and ion release in implant dentistry: A systematic review. *International Journal of Molecular Sciences*, *19*(11), doi: 10.3390/ijms19113585.

[15] Malchau, H., Herberts, P., & Ahnfelt, L. (Oct. 1993). Prognosis of total hip replacement in Sweden. Follow-up of 92,675 operations performed 1978–1990. *Acta Orthopaedica Scandinavica*, *64*(5), 497–506, doi: 10.3109/17453679308993679

[16] Chandra, A., et al. (2011). Life expectancy of modular Ti6Al4V hip implants: Influence of stress and environment. *Journal of the Mechanical Behavior of Biomedical Materials*, *4*(8), 1990–2001, https://doi.org/10.1016/j.jmbbm.2011.06.018.

[17] Liu, F., Wang, F., Shimizu, T., Igarashi, K., & Zhao, L. (2005). Formation of hydroxyapatite on Ti–6Al–4V alloy by microarc oxidation and hydrothermal treatment. *Surface and Coatings Technology*, *199*(2), 220–224, https://doi.org/10.1016/j.surfcoat.2004.10.146.

[18] Hussain, M., et al. (2021). Recent developments in coatings for orthopedic metallic implants. *Coatings*, *11*(7), 1–22, doi: 10.3390/coatings11070791.

[19] Zhu, X., Chen, J., Scheideler, L., Reichl, R., & Geis-Gerstorfer, J. (2004, Aug). Effects of topography and composition of titanium surface oxides on osteoblast responses. *Biomaterials*, *25*(18), 4087–4103, doi: 10.1016/j.biomaterials.2003.11.011.

[20] Kobayashi, E. (2002). Titanium and its alloys as biomaterials. *Japanese Journal of Political Science*, *41*(8), 553–560.

[21] Han, X., et al. (2023). Surface modification techniques of titanium and titanium alloys for biomedical orthopaedics applications: A review. *Colloids Surfaces B Biointerfaces*, *227*(406), 113339, doi: 10.1016/j.colsurfb.2023.113339.

[22] Huang, H. Y., et al. (2018). Effect of hydroxyapatite formation on titanium surface with bone morphogenetic Protein-2 loading through electrochemical deposition on MG-63 cells. *Materials*, *11*(10), doi: 10.3390/ma11101897.
[23] Premphet, P., Prasoetsri, M., Boonyawan, D., Supruangnet, R., Udomsom, S., & Leksakul, K. (2017). Optimization of DC magnetron sputtering deposition process and surface properties of HA-TiO2 film. *Materials Today: Proceedings*, *4*(5), Part 2 6372–6380, https://doi.org/10.1016/j.matpr.2017.06.141.
[24] Fox, V., Jones, A., Renevier, N. M., & Teer, D. G. (2000). Hard lubricating coatings for cutting and forming tools and mechanical components. *Surface and Coatings Technology*, *125*(1), 347–353, https://doi.org/10.1016/S0257-8972(99)00611-8.
[25] Grill, A. (2003). Diamond-like carbon coatings as biocompatible materials – An overview. *Diamond and Related Materials*, *12*(2), 166–170, https://doi.org/10.1016/S0925-9635(03)00018-9.
[26] Hauert, R. (2003). A review of modified DLC coatings for biological applications. *Diamond and Related Materials*, *12*(3), 583–589, https://doi.org/10.1016/S0925-9635(03)00081-5.
[27] Wang, C. T., et al. (2013). Indentation and scratch testing of DLC-Zr coatings on ultrafine-grained titanium processed by high-pressure torsion. *Wear*, *306*(1), 304–310, https://doi.org/10.1016/j.wear.2012.12.033.
[28] Ma, N., et al. (2020). Research progress of titanium-based high entropy alloy: Methods, properties, and applications. *Frontiers in Bioengineering and Biotechnology*, *8*, [Online]. Available https://www.frontiersin.org/articles/10.3389/fbioe.2020.603522.
[29] Ahmady, A. R., Ekhlasi, A., Nouri, A., Haghbin Nazarpak, M., Gong, P., & Solouk, A. (2023). High entropy alloy coatings for biomedical applications: A review. *Smart Materials in Manufacturing*, *1* (September 2022), 100009, doi: 10.1016/j.smmf.2022.100009.
[30] Yadav, V. S., Sankar, M. R., & Pandey, L. M. (2020). Coating of bioactive glass on magnesium alloys to improve its degradation behavior: Interfacial aspects. *Journal of Magnesium and Alloys*, *8*(4), 999–1015, https://doi.org/10.1016/j.jma.2020.05.005.
[31] Montazerian, M., Hosseinzadeh, F., Migneco, C., Fook, M. V. L., & Baino, F. (2022). Bioceramic coatings on metallic implants: An overview. *Ceramics International*, *48*(7), 8987–9005, doi: 10.1016/j.ceramint.2022.02.055.
[32] Nilawar, S., Uddin, M., & Chatterjee, K. (2021). Surface engineering of biodegradable implants: Emerging trends in bioactive ceramic coatings and mechanical treatments. *Materials Advances*, *2*(24), 7820–7841, doi: 10.1039/D1MA00733E.
[33] Thanka Rajan, S., Subramanian, B., & Arockiarajan, A. (2022). A comprehensive review on biocompatible thin films for biomedical application. *Ceramics International*, *48*(4), 4377–4400, doi: 10.1016/j.ceramint.2021.10.243.
[34] Hovsepian, P. E., Ehiasarian, A. P., Purandare, Y., Sugumaran, A. A., Marriott, T., & Khan, I. (2016). Development of superlattice CrN/NbN coatings for joint replacements deposited by high power impulse magnetron sputtering. *Journal of Materials Science: Materials in Medicine*, *27*(9), 147, doi: 10.1007/s10856-016-5751-0.
[35] Singh, V., Sharma, R. K., Sehgal, R., & Wani, M. F. (2024). The effect of RF sputtering power on structural, Nanomechanical and tribological properties of single layered tan coatings. *Surface Topography-Metrology and Properties*, [Online]. Available http://iopscience.iop.org/article/10.1088/2051-672X/ad5b0b.
[36] Singh, V., Sharma, R. K., & Sehgal, R. (2024, Aug). Impact of deposition pressure on the structural, nanomechanical, and tribological properties of δ-TaN coatings deposited via magnetron sputtering on Ti6Al7Nb alloy. *Proceedings of the Institution of Mechanical Engineers, Part E: Journal of Process Mechanical Engineering*, 09544089241272870, doi: 10.1177/09544089241272870.

[37] Samanta, A., et al. (2018). Nano- and micro-tribological behaviours of plasma nitrided Ti6Al4V alloys. *Journal of the Mechanical Behavior of Biomedical Materials*, *77*, 267–294, doi: 10.1016/j.jmbbm.2017.09.013.

[38] Shirazi, H. A., Chan, C. W., & Lee, S. (2021). Elastic-plastic properties of titanium and its alloys modified by fibre laser surface nitriding for orthopaedic implant applications. *Journal of the Mechanical Behavior of Biomedical Materials*, *124*(August), 104802, doi: 10.1016/j.jmbbm.2021.104802.

[39] Chan, C. W., Quinn, J., Hussain, I., Carson, L., Smith, G. C., & Lee, S. (2021). A promising laser nitriding method for the design of next generation orthopaedic implants: Cytotoxicity and antibacterial performance of titanium nitride (TiN) wear nano-particles, and enhanced wear properties of laser-nitrided Ti6Al4V surfaces. *Surface and Coatings Technology*, *405*(November 2020), 126714, doi: 10.1016/j.surfcoat.2020.126714.

[40] Skjöldebrand, C., Tipper, J. L., Hatto, P., Bryant, M., Hall, R. M., & Persson, C. Current status and future potential of wear-resistant coatings and articulating surfaces for hip and knee implants. (2022). *Materials Today Bio*, *15*(January), doi: 10.1016/j.mtbio.2022.100270.

[41] Marin, E., Offoiach, R., Regis, M., Fusi, S., Lanzutti, A., & Fedrizzi, L. (2016). Diffusive thermal treatments combined with PVD coatings for tribological protection of titanium alloys. *Materials & Design*, *89*, 314–322, doi: 10.1016/j.matdes.2015.10.011.

[42] Choudhury, D., et al. (2016). Improved wear resistance of functional diamond like carbon coated Ti–6Al–4V alloys in an edge loading conditions. *Journal of the Mechanical Behavior of Biomedical Materials*, *59*, 586–595, doi: 10.1016/j.jmbbm.2016.04.004.

[43] Rodríguez-Rojas, F., et al. (2023). Effect of a DLC film on the sliding-wear behaviour of TI6AL4V: Implications for dental implants. *Surface and Coatings Technology*, *460*(January), 0–8, doi: 10.1016/j.surfcoat.2023.129409.

[44] De Viteri, V. S., & Fuentes, E. (2013). J. Gegner, (Ed.). *Titanium and Titanium Alloys as Biomaterials* (pp. 5). Rijeka: IntechOpen. doi: 10.5772/55860.

[45] Kumari, R., & Majumdar, J. D. (2018). Wear behavior of plasma spray deposited and Post Heat-Treated Hydroxyapatite (HA)-based composite coating on Titanium Alloy (Ti-6Al-4V) substrate. *Metallurgical and Materials Transactions A: Physical Metallurgy and Materials Science*, *49*(7), 3122–3132, doi: 10.1007/s11661-018-4626-z.

[46] Jing, Z., Cao, Q., & Jun, H. (2021). Corrosion, wear and biocompatibility of hydroxyapatite bio-functionally graded coating on titanium alloy surface prepared by laser cladding. *Ceramics International*, *47*, 24641–24651, https://doi.org/10.1016/j.ceramint.2021.05.186.

[47] Castleman, L. S., Motzkin, S. M., Alicandri, F. P., Bonawit, V. L., & Johnson, A. A. (1976). Biocompatibility of nitinol alloy as an implant material. *Journal of Biomedical Materials Research*, *10*(5), 695–731.

[48] Anene, F., Aiza, J., Zainol, I., Hanim, A., & Suraya, M. T. (2020). Additively manufactured titanium alloys and effect of hydroxyapatite coating for biomedical applications: A review. *Proceedings of the Institution of Mechanical Engineers, Part L: Journal of Materials: Design and Applications*, *234*(11), 1450–1460, doi: 10.1177/1464420720942560.

[49] Hanawa T. Biological reactions on titanium surface electrodeposited biofunctional molecules BT – Interface Oral Health Science 2009. In: Sasano T, Suzuki O, editors., Tokyo: Springer Japan; 2010, p. 83 -9

[50] Hanawa, T. (2010). Biofunctionalization of titanium for dental implant. *Japanese Dental Science Review*, *46*(2), 93–101, doi: 10.1016/j.jdsr.2009.11.001.

[51] Trino, L. D., Bronze-Uhle, E. S., Ramachandran, A., Lisboa-Filho, P. N., Mathew, M. T., & George, A. (2018). Titanium surface bio-functionalization using osteogenic peptides: Surface chemistry, biocompatibility, corrosion and tribocorrosion aspects. *Journal of the Mechanical Behavior of Biomedical Materials*, *81*, 26–38.

[52] Huang, M. S., et al. (2020). Preparation of a biofunctionalized surface on titanium for biomedical applications: Surface properties, wettability variations, and biocompatibility characteristics. *Applied Sciences*, *10*(4), doi: 10.3390/app10041438.

[53] Zakaria, A., Shukor, H., Todoh, M., & Jusoff, K. (2020). Bio-functional coating on Ti6Al4V surface produced by using plasma electrolytic oxidation. *Metals (Basel)*, *10*(9), 1–16, doi: 10.3390/met10091124.

[54] Ganjali, M., Mousavi, S., Nikzamir, S., Milan, P. B., & Mozafari, M. (2022, Jul). Effect of laser cladded co-doped strontium fluorapatite nanopowder coating on the antibacterial and cell attachment of Ti-6Al-4V implants for bone applications. *Materials Technology*, *37*(8), 829–841, doi: 10.1080/10667857.2021.1898716.

[55] Moloodi, A., Toraby, H., Kahrobaee, S., Razavi, M. K., & Salehi, A. (2021). Evaluation of fluorohydroxyapatite/strontium coating on titanium implants fabricated by hydrothermal treatment. *Progress in Biomaterials*, *10*(3), 185–194.

[56] Bhattacharjee, A., Bandyopadhyay, A., & Bose, S. (2022). Plasma sprayed fluoride and zinc doped hydroxyapatite coated titanium for load-bearing implants. *Surface and Coatings Technology*, *440*, 128464, https://doi.org/10.1016/j.surfcoat.2022.128464.

[57] Wu, J., et al. (2023). Investigation of the antibacterial effect, osteogenic activity, and tracing properties of hydroxyapatite co-doped with Tb3+ and Zn2+. *Nanoscale*, *15*(8), 3940–3951, doi: 10.1039/D2NR06795A.

[58] Qiao, H., et al. (2019). Si, Sr, Ag co-doped hydroxyapatite/TiO2 coating: Enhancement of its antibacterial activity and osteoinductivity. *RSC Advances*, *9*(24), 13348–13364, doi: 10.1039/c9ra01168d.

[59] Ke, D., Banerjee, D., & Bose, S. (2019). In vitro characterizations of Si4+ and Zn2+ doped plasma sprayed hydroxyapatite coatings using osteoblast and osteoclast coculture. *ACS Biomaterials Science & Engineering*, *5*(3), 1302–1310.

[60] Jiang, P., et al. (2023). Advanced surface engineering of titanium materials for biomedical applications: From static modification to dynamic responsive regulation. *Bioactive Materials*, *27* (November 2022), 15–57, doi: 10.1016/j.bioactmat.2023.03.006.

[61] Trino, L. D., et al. (2018). Zinc oxide surface functionalization and related effects on corrosion resistance of titanium implants. *Ceramics International*, *44*(4), 4000–4008, doi: 10.1016/j.ceramint.2017.11.195.

[62] Kim, J. W., Hwang, M. J., Han, M. K., Kim, Y. G., Song, H. J., & Park, Y. J. (2016). Effect of manganese on the microstructure, mechanical properties and corrosion behavior of titanium alloys. *Materials Chemistry and Physics*, *180*, 341–348, doi: 10.1016/j.matchemphys.2016.06.016.

[63] Han, M. K., Kim, J. Y., Hwang, M. J., Song, H. J., & Park, Y. J. (2015). Effect of Nb on the microstructure, mechanical properties, corrosion behavior, and cytotoxicity of Ti-Nb alloys. *Materials (Basel)*, *8*(9), 5986–6003, doi: 10.3390/ma8095287.

[64] Wang, G., Wan, Y., Wang, T., & Liu, Z. (2017). Corrosion behavior of Titanium implant with different surface morphologies. *Procedia Manufacturing*, *10*, 363–370, doi: 10.1016/j.promfg.2017.07.006.

[65] Satyanarayana, C. P., Raju, L. S., Dumpala, R., & Sunil, B. R. (2024). A review on strategies to enhance the performance of the titanium based medical implants. *Materials Today Communications*, *38*, 107985, doi: https://doi.org/10.1016/j.mtcomm.2023.107985.

[66] Marin, E., & Lanzutti, A. (2024). Biomedical applications of titanium alloys: A Comprehensive Review. *Materials (Basel)*, *17*(1), doi: 10.3390/ma17010114.

Ramu P. and Deepak Patil*

Chapter 6
Titanium dioxide coating for biomedical devices

Abstract: The application of titanium dioxide (TiO_2) coatings on medical devices has shown significant promise in improving their performance and compatibility with biological systems. TiO_2's chemical stability, non-toxicity, and bioactive characteristics make it highly effective in promoting osseointegration and inhibiting bacterial colonization on implant surfaces. This review explores various methods of TiO_2 coating application, including sol-gel, sputtering, and electrochemical deposition techniques, highlighting their impact on surface morphology, hydrophilicity, and mechanical properties. Additionally, the antimicrobial properties of TiO_2 under ultraviolet light and its role in promoting cellular adhesion and proliferation are discussed. Recent advancements in doping TiO_2 with other elements, such as silver or nitrogen, further enhance its functionality. The review concludes by addressing the challenges of scaling up TiO_2 coatings for clinical applications and outlining future research directions to optimize these coatings for specific biomedical applications.

Keywords: Nanocoatings, titanium dioxide (TiO_2), photocatalysis, biomedical devices, biomaterials, surface modification

6.1 Introduction to titanium dioxide (TiO_2)

Titanium dioxide (TiO_2) is a crucial pigment, renowned for its ability to reflect visible light and absorb UV radiation. Their characteristics contribute to enhanced whiteness, luminosity, and opacity of plastic materials [1]. Commercially, TiO_2 is available in two principal crystal structures: anatase and rutile. Rutile TiO_2 is generally preferred for industrial applications due to its superior light-scattering efficiency, stability, and durability compared to anatase. The choice of TiO_2 grade significantly influences key characteristics like chalk resistance, gloss, color retention in paints, and resistance to discoloration from factors such as mildew and dirt [2]. Most commercial TiO_2 is not in its pure form; rather, it undergoes surface treatments, either organic or inorganic, applied through various processes like precipitation or mechanical blending. These

*Corresponding author: Deepak Patil, Department of Production Engineering, National Institute of Technology Tiruchirappalli, Tiruchirappalli 620015, Tamil Nadu, India, e-mail: deepakdeelip@nitt.edu
Ramu P, Department of Production Engineering, National Institute of Technology Tiruchirappalli, Tiruchirappalli 620015, Tamil Nadu

https://doi.org/10.1515/9783111571423-006

treatments further enhance the performance of TiO_2 in specific applications. White pigments, including TiO_2, differ from colorful pigments in that they achieve opacity through the scattering of light rather than absorption [3]. The high-temperature transformation of anatase and brookite into rutile further emphasizes the rutile's stability and suitability for various industrial applications [4].Titanium dioxide crystals can exist in multiple forms, each with unique shapes and properties, necessitating different production methods for various use. For example, anatase TiO_2 is typically produced through solution-based processes or low-temperature chemical vapor deposition, whereas rutile TiO_2 is often created using high-temperature deposition techniques or by heating other TiO_2 forms. These methods enable specific applications, such as in paints or coatings, where high-performance TiO_2 is required to ensure durability and efficiency. The functionality of TiO_2 is significantly influenced by the specific crystal structures found in its anatase, rutile, and brookite forms. Rutile has a tetragonal crystal structure and forms a stable truncated octahedron (Figure 6.1a(B)), which contributes to its higher light-scattering ability. Anatase also has a tetragonal shape but with a longer c-axis than the a-axis, presenting as a truncated octahedron with low-energy crystal planes (Figure 6.1a (A)). In contrast, brookite exhibits an orthorhombic crystal structure, which incorporates a larger unit consisting of eight TiO_2 molecules. (Figure 6.1a(C)). With their wide range of applications, titanium dioxide nanomaterials have garnered considerable interest over recent decades. These applications extend beyond pigment use to include areas such as sunscreens, photovoltaic cells, and environmental and biomedical technologies, including pollutant degradation, water purification, biosensing, and drug delivery [5]. The rapid advancement in the understanding and production of TiO_2 nanoparticles has been driven by their diverse and critical applications [6]. With the expansion of industrial growth and the increase in fossil fuel consumption, issues like environmental pollution and the depletion of renewable energy resources have become urgent global challenges. The extensive release of toxic substances and manufacturing by-products underscores the urgency for environmentally responsible and sustainable remedies [7]. One of the most promising solutions lies in the development of clean energy technologies, with solar energy emerging as a key player. Solar energy, one of Earth's most abundant renewable resources, can be easily converted into electricity and chemical energy. For example, the process of solar water splitting can produce clean hydrogen (H_2), a sustainable energy source. Therefore, solar energy technologies must be developed on a large scale to address global environmental and energy issues [8]. Among the various emerging energy technologies, photocatalysis has been identified as one of the most promising. The conversion of light energy into chemical energy through sunlight-driven chemical reactions is known as photocatalysis. This scientific field traces its roots back to experiments conducted in the early twentieth century. In 1901, the Italian chemist Giacomo Camunian, one of the pioneers of modern photochemistry, systematically studied the effects of light on chemical reactions. However, his experiments did not involve photocatalysts. The scientific community first encountered the term "photocatalysis" in 1911. Subsequently, scientists uncovered that zinc oxide (ZnO) possessed photocatalytic properties, facilitating various chemical reactions. These reactions encompassed the bleaching of Prussian blue and the transformation of Ag^+ into Ag. By 1932, it was discovered that TiO_2 and niobium pentoxide (Nb_2O_5) could facilitate the photocatalytic reduction of gold and silver compounds. However, practical applications for photocatalysis were not immediately apparent, and research in this field did not gain momentum until much later. The oil crisis of the early 1970s, coupled with increasing awareness of environmental degradation, pushed scientists to seek alternative and renewable energy sources. This marked the resurgence of interest in photocatalysis. In 1968, Boddy found that TiO_2, when illuminated with ultraviolet (UV) light in an electrolytic cell, could produce oxygen (O_2). Four years later, Fujishima and Honda achieved a breakthrough in photocatalysis by discovering that water splitting could produce both hydrogen (H_2) and oxygen (O_2) using a TiO_2 electrode under UV light. Due to its chemical stability, affordability, and non-toxic nature, TiO_2 has become the most promising candidate for photocatalytic water splitting. In 1977, scientists successfully demonstrated the generation of hydrogen and oxygen in a 2:1 molar ratio through the process of photocatalytic water splitting using TiO_2 powder [9]. A significant advancement was made by Kawai and Sakata in 1980 when they found that incorporating methanol as a sacrificial agent into a photocatalytic solution with water substantially improved the generation of hydrogen. In 1979, researchers discovered that various semiconductors, including TiO_2, could be used for the photocatalytic reduction of carbon dioxide (CO_2). This finding presented a new potential application for TiO_2 in the effort to reduce greenhouse gas emissions [10]. These early studies demonstrated the potential of TiO_2 as a photocatalyst for various environmental and energy-related applications, sparking widespread research in the field. In the past 20 years, titanium dioxide (TiO_2) has become the most popular and effective "golden" catalyst for heterogeneous photocatalysis. Its applications span various fields, including the breakdown of pollutants in both air and water environments, as well as the direct splitting of water using solar energy to generate hydrogen [11]. TiO_2-based photocatalysis has expanded rapidly, but much remains to be understood about the fundamental processes governing these reactions. The process of TiO_2 photocatalysis typically encompasses several crucial stages. These include the generation, segregation, relaxation, capture, and movement of charge carriers, along with reactions that break and form chemical bonds. Understanding these fundamental processes is crucial for improving TiO_2-based photocatalytic technologies. Researchers have utilized sophisticated surface analysis methods, including scanning tunneling microscopy (STM), to investigate these reactions on various TiO_2 single crystal surfaces, encompassing both rutile and anatase forms. STM, along with spectroscopy methods, allows researchers to investigate the adsorption, diffusion, and reactions of molecules on TiO_2 surfaces at the molecular level, providing valuable insights into the mechanisms of TiO_2 photocatalysis.

Researchers have extensively examined various TiO_2 crystal surfaces, including rutile (R)-TiO_2(110), R-TiO_2(011), and anatase (A)-TiO_2(101), to gain insights into their functions in photocatalytic processes. Among these, the rutile R-TiO_2(110) surface has become a primary focus for fundamental photocatalysis research due to its chemical stability and relatively simple preparation method. These studies have provided valu-

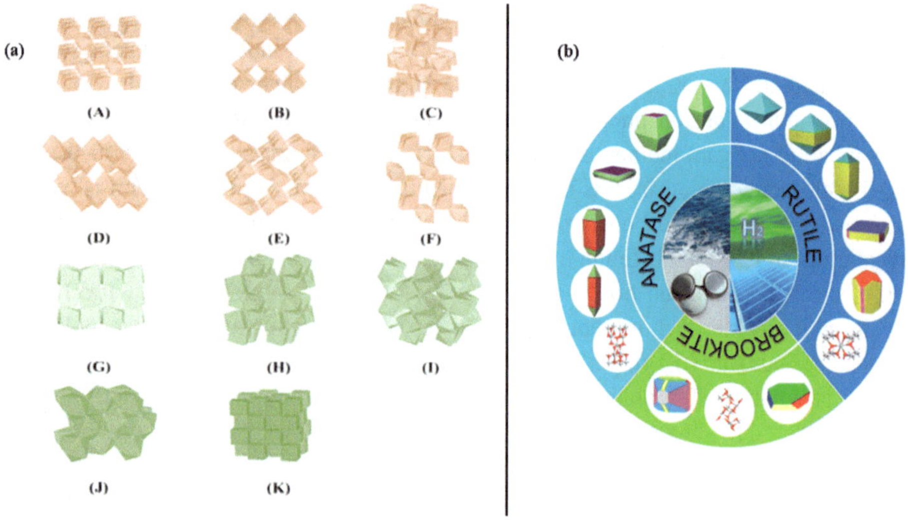

Figure 6.1: (a) Polyhedron models depicting the structures of 11 TiO_2 phases: (A) rutile, (B) anatase, (C) brookite, (D) TiO_2(B), (E) hollandite-like TiO_2(H), (F) ramsdellite-like TiO_2(R), (G) columbite-like TiO_2(II), (H) baddeleyite-like, (I) TiO_2(OI), (J) cotunnite-like TiO_2(OII), and (K) fluorite-like cubic phases. Thin lines outline the unit cells. **(b)** Crystal structures of titanium dioxide in its (a) anatase, (b) rutile, and (c) brookite forms.

able information about adsorbate behavior, diffusion patterns, and interactions with TiO_2 surfaces, thereby enhancing our understanding of TiO_2 photocatalysis in real-world environmental conditions. The field of TiO_2 photocatalysis has witnessed significant advancements over recent decades, with applications spanning from environmental cleanup to sustainable energy generation. Nevertheless, numerous challenges persist, particularly in elucidating the underlying mechanisms involved in photocatalytic reactions. By gaining deeper insights into these processes, researchers can develop more efficient photocatalysts and improve the performance of TiO_2-based technologies. TiO_2 will continue to be a key focus in photocatalytic studies as the need for clean and renewable energy sources increases. This compound is expected to play a vital role in tackling worldwide environmental and energy issues in the coming years.

6.2 Properties of TiO_2

The remarkable characteristics of titanium dioxide nanoparticles have greatly broadened their use in diverse areas of application [4]. As an example, their exceptional ability to convert light has been applied in the development of energy-generating devices [12]. Moreover, these nanoparticles possess chemical durability, exhibit clarity in thin layers, and can be produced cost-effectively, making them beneficial for various environmental restoration techniques. These applications include the purification of wastewater, mitigation of air pollution, and improvement of soil quality [13]. In recent times, titanium dioxide (TiO_2) nanoparticles have been utilized in cancer photothermal therapy (PTT), taking advantage of their capacity to promote non-radiative recombination [14]. The subsequent sections will provide a more comprehensive examination of the distinctive characteristics of TiO_2 nanoparticles. The following sections will offer a more detailed analysis of the unique properties of TiO_2 nanoparticles.

6.2.1 Crystal properties

The formation of nanocrystalline titanium dioxide (TiO_2) can result in three primary polymorphs: rutile, anatase, and brookite. The specific polymorph that emerges depends on the manufacturing process and subsequent heat treatment applied to the material [15]. TiO_2(B), a less common fourth polymorphic form, exists alongside the other three. Researchers have also reported synthesizing two additional high-pressure phases derived from rutile. These include TiO_2(II), which exhibits a structure similar to PbO_2 [17], and TiO_2(H), whose structural characteristics resemble those of hollandite [18]. The tetragonal crystal structures are shared by anatase and rutile phases, despite their classification in different phase groups. In contrast, brookite exhibits an orthorhombic structure, while the rare TiO_2(B) phase displays a monoclinic structure [13–15]. The octahedral structure of the anatase phase exhibits a slightly greater distortion compared to that of rutile [19]. Despite the rutile phase being less thermodynamically favored than anatase at 0 K, the energetic difference between these two phases is relatively minor (approximately 2–10 kJ/mol). For solar cell applications, the anatase phase of TiO_2 is generally favored due to its superior properties, including low density, high-electron mobility, and low-dielectric constant [20]. Additionally, the (101) facets in anatase crystals exhibit lower reactivity compared to the (001) facets [19]. The anatase phase's low density makes it susceptible to a phase transition to rutile at elevated temperatures, typically ranging from 450 to 1,200 °C. This transformation is not solely dependent on temperature but is also affected by various factors such as dopant concentration, initial phase, and particle size [21]. As particle sizes approach a specific threshold, the brookite and anatase forms of TiO_2 tend to convert into the rutile phase. Notably, when particle dimensions exceed 14 nm, rutile becomes more thermodynamically favorable than anatase [22]. Furthermore, the growth rate of the rutile phase surpasses that of anatase once it begins to form [20]. A comprehensive overview of the crystalline characteristics for various TiO_2 polymorphs is provided in Table 6.1.

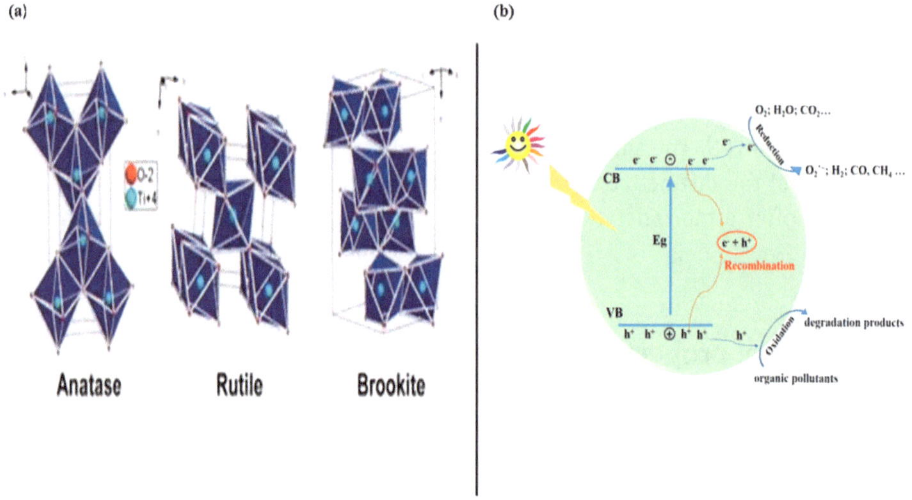

Figure 6.2: **(a)** Anatase, rutile, and brookite comprise the crystal structures of TiO_2 phases. **(b)** A depiction of the overall process of TiO_2 in photocatalysis using solar energy (reproduced with permission from MDPI Publications [18]).

6.2.2 Optical properties

TiO_2 nanoparticles are extensively used in optical devices due to their outstanding properties. These include exceptional resistance to mechanical wear, high levels of transparency within the visible light range, and robust stability when exposed to water-based environments [23]. Research has shown that multiple structural elements affect the optical characteristics of TiO_2 nanoparticles, including their phase composition, band gap, crystal quality, size distribution, shape, porosity, and dimensions [24]. Interestingly, when the size of TiO_2 nanoparticles is decreased from 200 nm to approximately 10 nm or less, their optical properties undergo a transformation. This change causes them to shift from being opaque to transparent in the visible-light spectrum, making them suitable for use as efficient UV light shields [25]. Anatase, one of the pure phases of TiO_2, demonstrates superior catalytic efficiency and electron conductivity in comparison to rutile and brookite phases. This makes anatase particularly well-suited for applications in photovoltaics and photocatalysis [20]. The anatase phase's increased photo reactivity is attributed to its reduced oxygen adsorption, greater hydroxylation, and marginally higher Fermi level [10]. On the other hand, the rutile phase is characterized by its substantial refractive index and optical absorption, making it well-suited for optical communication devices such as modulators, switches, and isolators [25].

Upon exposure to UV light, TiO_2 nanoparticles undergo a process where electrons in the valence band (VB) become energized and move to the conduction band (CB),

Table 6.1: The crystal properties of TIO_2 [26].

Properties	Anatase	Rutile	Brookite
Crystal structure	Tetragonal	Tetragonal	Orthorhombic
Density (g/cm^2)	3.894	4.250	4.120
Space group	$I4_1$/amd	$P4_2$/mnm	Pbca
Molecule (cell)	2	2	4
Lattice constant (Å)	$a = 3.784$ $b = 9.515$	$a = 4.594$ $b = 2.959$	$a = 9.184$ $b = 5.447$ $c = 5.154$
Ti-O bond length (Å)	1.937(4) 1.9659(2)	1.949(4) 1.980(2)	1.87–2.04
O-Ti-O bond angle	77.7° 92.6°	81.2° 90.0°	77.00–105.0°
Volume/molecule (Å^3)	34.061	31.216	32.172

creating holes (h^+) in the VB (Figure 6.2b). The excited electrons (e^-) occupy 3d states, and their transition probability decreases due to differing parity, which results in reduced e^-/h^+ recombination [20]. The anatase phase is considered more photocatalytically active due to its chemical characteristics and ability to produce charge carriers (e^- and h^+). It absorbs UV light in accordance with its band gap [27]. On the other hand, the rutile phase is typically regarded as an ineffective photocatalyst due to the occurrence of bulk h^+ recombination. This process restricts the retention of h^+ on the surface to only that in close proximity, which subsequently undergoes upward band bending and migrate to the surface [28]. The h + that remains interacts with water molecules to create hydroxyl radicals (OH•). These radicals, along with h^+, oxidize organic compounds in close proximity to the particle's surface [27]. Concurrently, the electron (e^-) in the CB engages in a reduction reaction with oxygen molecules in the air, generating superoxide radical anions. It's crucial to understand that in an ideal photocatalyst, the redox potential of the electron-hole (e^-/h^+) pair should be situated within the material's band gap [29]. The energy level at the CB's bottom determines the photoelectrons' reducing capability, while the energy level at the VB's top influences the photogenerated h^+ oxidizing power [20]. These properties make TiO_2 nanoparticles highly desirable as photocatalysts, primarily due to the redox selectivity and potent oxidizing capability of their h^+. Consequently, there is ongoing research focused on developing photocatalysts based on nano-scale TiO_2 [29].

6.2.3 Electrochemical properties

The electronic structure, dimensions, form, surface attributes, and arrangement of nano-TiO_2 affect its chemical and physical characteristics. The electronic features of TiO_2 nanoparticles play a crucial role in determining their particle and crystal size distribution [30]. Pure TiO_2 is classified as a wide band gap n-type semiconductor, with its rutile, anatase, and brookite phases exhibiting indirect energy band gaps of 3.02, 3.2, and 2.96 eV, respectively [17, 26, 27]. Furthermore, studies have shown that rutile's Fermi level is roughly 0.1 eV below that of anatase [20]. Anatase also possesses a lower electron-effective mass than rutile, resulting in improved charge carrier mobility. This characteristic makes anatase particularly suitable for the development of optoelectronic devices. The CB of TiO_2 is composed of titanium's 3d orbitals, while its VB consists of oxygen's 2p orbitals hybridized with titanium's 3d orbitals [20]. The lattice oxygen sites in TiO_2 are particularly important for its super hydrophilic properties, which result from chemical conformational changes on the nanoparticle surfaces [31]. Certain trapped holes (h^+) at oxygen lattice sites may specifically interact with TiO_2, causing a weakening of the bonds between oxygen ions and titanium within the lattice. This interaction enables H_2O molecules to disrupt these weakened bonds, resulting in the creation of new –OH groups. These newly formed –OH groups, which are singly linked, are thermodynamically unstable and possess high surface energy. Consequently, they contribute to the development of a super hydrophilic TiO_2 surface [27, 32]. A significant challenge in utilizing TiO_2 nanoparticles for photoelectrochemical devices is the high density of states present in the band gap. This characteristic leads to electron-hole (e^-/h^+) recombination and facilitates electron transfer reactions at the interface between the electrolyte and oxide [30]. Nevertheless, this characteristic proves beneficial for the attachment of redox-active substances to TiO_2 nanoparticle surfaces, a feature particularly relevant in dental implant applications. For example, certain blood proteins such as fibrinogen are known to undergo chemical adsorption through electron transfer on the surface of TiO_2 nanoparticles [30]. Research has shown that the thrombogenicity of fibrinogen is influenced by the electronic properties of TiO_2 nanoparticles. This is attributed to fibrinogen's semiconducting characteristics, which are essential in the blood clotting process. The adsorption and subsequent breakdown of the protein may be affected by alterations in the Fermi energy level or band size of TiO_2 nanoparticles, as the band structures of both fibrinogen and TiO_2 nanoparticles are in alignment.

6.3 Coating technique use to coat TiO_2

6.3.1 Spin coating

Spin coating is an advanced, economical, and easily manageable technique for evenly applying nanoparticles (NPs) to various substrate surfaces (Figure 6.3). The process can be broken down into three primary phases following the initial preparation of materials and substrates. These phases include: stage i, preparing the substrate and dispensing the solution; stage ii, creating the wet film; and stage iii, evaporating the solvent and forming the final coating (Figure 6.5a).

6.3.1.1 Stage i: substrate preparation and solution dispensing

Initially, the substrate is positioned on the spin coater's platform. A prepared solution is then applied to the upper surface of the glass substrate. This step involves depositing a small amount of the liquid material, which can either be done statically or dynamically, depending on whether the substrate is spinning at low speed during dispensing [33].

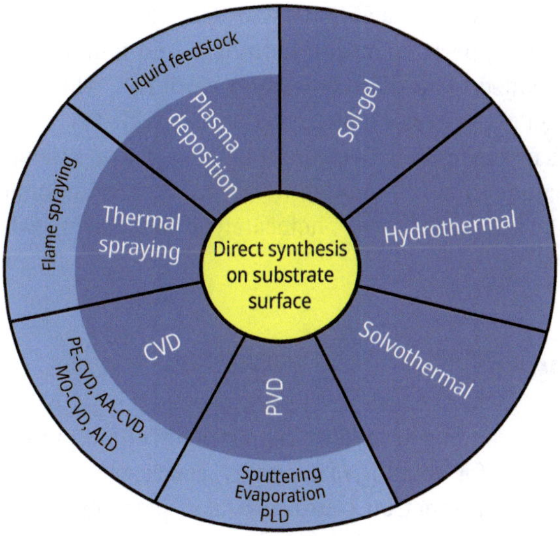

Figure 6.3: Coating technique used to coat TiO_2 (reproduced with permission from ELSEVIER Publications [32]).

6.3.1.2 Stage ii: formation of wet film

During the operation of a spin coater, the substrate develops a wet film due to centrifugal forces. The rotating motion causes the excess fluid to move from the center toward the substrate's edges. Both the rotation speed and time significantly affect the thickness of the final film, with faster speeds typically resulting in thinner coatings [34].

6.3.1.3 Stage iii: solvent evaporation and final coating formation

In the final phase, the coating solidifies into a thin film as the solvent evaporates. For certain applications, additional calcination may be necessary to crystallize the deposited coatings. By adjusting the spinning speed and number of cycles, the thickness of the coating can be precisely controlled, allowing for consistent application across various surfaces [35]. A study by Peeters et al. [36] described the creation of a self-cleaning borosilicate coating by applying a TiO_2 film containing gold (Au) nanoparticles (NPs) onto glass via spin coating. The coated glass underwent calcination at 550 °C for 3 h, with a 1 °C/min heating rate, yielding uniform, smooth, highly transparent, and photoactive Au NPs within anatase TiO_2 thin coatings. The Au NPs, with an average diameter of 16 ± 4 nm, were dispersed at various depths in the TiO_2 coatings. Compared to commercial Pilkington Activ™, the addition of Au NPs substantially improved the photocatalytic self-cleaning performance in both dark and illuminated conditions. This enhancement is credited to the plasmonic effect of Au NPs, which effectively addresses the poor visible light response typically associated with TiO_2 coatings. As a result, self-cleaning glass featuring Au-TiO_2 composite coatings shows significant promise for practical use under natural sunlight. Nevertheless, additional research is required to confirm the efficacy of these photocatalytic coatings in real-world self-cleaning scenarios.

6.3.2 Magnetron sputtering

As photocatalytic TiO_2-based coatings continue to advance rapidly, magnetron sputtering has become a promising technique for creating high-quality, consistent coatings on glass surfaces. The core process of magnetron sputtering takes place in a low-pressure environment, typically between 0.1 and 0.5 Pa. As shown in Figure 6.4a, the combination of electric and magnetic fields ionizes gas atoms, producing gas ions and electrons. The high-energy ions collide with the target (or cathode), causing metal atoms or ions to be expelled from its surface. These particles then accumulate on the substrate, forming a thin film. By introducing reactive gases like oxygen or nitrogen, it's possible to create composite films consisting of oxides or nitrides. Magnetron sputtering can be classified into two types based on the current mode: direct current magnetron sputtering, which is appropriate for conductive targets, and radio frequency (RF) magnetron sputtering, which can be utilized for any target material. Magnetron sputtering techniques can be categorized as either single-target or multi-target sputtering, based on the number of targets employed. This adaptability renders magnetron sputtering an essential method for diverse applications in the fields of materials science and technology (Figure 6.4a).

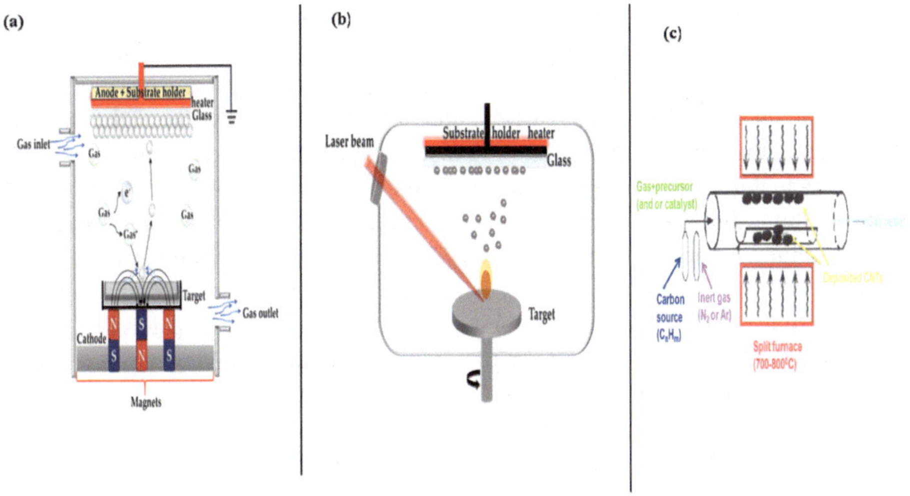

Figure 6.4: (a) Illustration of the magnetron sputtering technique. **(b)** Diagram of the pulsed laser deposition method (reproduced with permission from MDPI Publications [37]). **(c)** Conceptual representation of chemical vapor deposition.

A recent investigation by Mazur et al. [38] involved the application of TiO_2 and SiO_2 thin films onto conventional microscope slide glasses. The researchers employed microwave-assisted reactive magnetron sputtering, utilizing high-purity Ti and Si targets for the deposition process. The resulting antireflection multilayer coatings displayed an amorphous structure, which was likely attributed to the relatively low temperature of the sputtering technique, not surpassing 100 °C. Compared to uncoated glass substrates, the TiO_2/SiO_2-coated glass exhibited significantly improved properties, showcasing exceptional features including anti-reflective qualities, photocatalytic capabilities, enhanced durability, and superior light transmission. These multi-layered anti-reflection coatings show great potential for use in self-cleaning and protective surface applications.

6.3.3 Pulsed laser deposition

Pulsed laser deposition (PLD) is a technique in physical vapor deposition that employs high-energy laser pulses to create plasma plumes by vaporizing a target material. The resulting atoms are subsequently transferred and deposited onto a substrate, often glass, to create a coating with specific optical characteristics [38, 39]. The deposition technique can be executed under various conditions, including vacuum, reactive gas PLD, or liquid PLD environments [40]. As illustrated in Figure 6.4b, PLD enables the adjustment of porosity and density in layered TiO_2 films through precise control of process variables [41]. The resulting unique vertical structures can enhance the movement of photo-generated charge carriers and effectively scatter visible light, making PLD an ideal method for creating photocatalytic coatings with intricate structures and composite materials [41, 42]. PLD is an advanced method for producing thin films, notably titanium dioxide (TiO_2), which plays a crucial role in diverse applications, including photovoltaics and photocatalysis. The technique involves directing a high-powered laser beam at a TiO_2 target, causing the material to evaporate and generate a plasma plume consisting of atoms and molecules. This plume then expands and settles onto a substrate, where the film formation takes place. By modifying various factors such as laser energy, temperature of the substrate, and the composition of surrounding gases (e.g., oxygen or argon) during the deposition process, the properties of the resulting film can be precisely controlled. These adjustments influence the film's morphology, crystallinity, and other physical properties, making PLD particularly advantageous for producing complex stoichiometric materials with precise control over their structure and properties. Consequently, TiO_2 films produced through PLD demonstrate advantageous characteristics for use in photovoltaic cells and optical instruments, highlighting the technique's adaptability and efficiency in creating thin films [43].

6.3.4 Chemical vapor deposition (CVD)

Chemical vapor deposition (CVD) serves as an efficient method for applying titanium dioxide (TiO_2) coatings to substrates, with widespread use in fields such as photovoltaics, photocatalysis, and optical devices. The CVD technique involves introducing gaseous precursors, typically titanium tetrahalide (e.g., $TiCl_4$), along with oxygen or other oxidizing agents into a reaction chamber. The substrate is heated to promote the chemical reactions that result in TiO_2 deposition on its surface [44]. During the CVD process, the gaseous precursors undergo reactions at high temperatures, leading to the formation of high-purity TiO_2 films. The adaptability of CVD allows for precise control over the deposited films' characteristics by modifying parameters like substrate temperature, pressure, and precursor gas flow rates. This parameter control enables manipulation of film thickness, morphology, and crystallinity, making CVD particularly beneficial for producing complex stoichiometric materials. TiO_2 films generated through CVD demonstrate desirable properties for use in solar cells and environmental remediation, showcasing the method's versatility and efficacy in thin-film production [45]. Research has demonstrated that utilizing plasma-enhanced CVD techniques can enhance deposition rates and film quality by improving precursor fragmentation and increasing reaction efficiency at lower temperatures, which is advantageous for heat-sensitive substrates. This approach is especially useful when depositing TiO_2 films on substrates that are unable to withstand high temperatures during the CVD process (Figure 6.4c) [46].

6.3.5 Electrodeposition

The use of titanium dioxide (TiO_2) in processes has become increasingly prevalent due to its diverse characteristics and broad spectrum of applications. Notably, TiO_2 exhibits significant potential as an electrode component for lithium-ion capacitors, owing to its substantial ability to store electrical charge. To overcome the challenges of TiO_2's poor electrical conductivity and limited reactivity with Li^+ ions, it is necessary to integrate it with carbon at the nanoscale level. One effective method for accomplishing this is through the use of vacuum liquid pulse CVD techniques. These processes enable the straightforward production of TiO_2/porous–carbon nanocomposites. This technique involves the uniform distribution of TiO_2 nanoparticles, roughly 4 nm in size, within the cavities of meso- or micro-porous carbon materials. The porous-carbon framework provides efficient conductive pathways for the TiO_2 nanoparticles, allowing them to achieve a high discharge capacity of approximately 200 mA h/g-TiO_2 (calculated based on the weight of TiO_2) [47]. The TiO_2/carbon composites demonstrate exceptional rate capabilities and cycling stability, rendering them ideal for high-rate energy storage applications in lithium-ion capacitors. Furthermore, electrodeposited TiO_2 thin films are crucial components in photovoltaic devices, particularly in the fabrication of flexible dye-sensitized solar cells (DSSCs). Electrodeposition methods can generate highly porous, stable crystalline mesoscopic anatase TiO_2 films, which are crucial for effective light absorption and charge transfer within solar cells [48]. This improves the overall efficiency of DSSCs. Additionally, TiO_2 films created through electrodeposition are significant in photocatalytic applications, especially for purifying air and water. When exposed to UV light, TiO_2's photocatalytic characteristics allow it to break down organic contaminants, making it particularly effective for self-cleaning surfaces and systems that treat wastewater. In conclusion, the adaptability of titanium dioxide in electrodeposition techniques opens up its applications in various fields, such as energy storage, photovoltaics, and photocatalysis. Tailoring the characteristics of TiO_2 films through precise deposition methods improves their effectiveness across the applications shown in Figure 6.5b [49].

6.3.6 Spray coating

Compared to dip coating and spin coating, the spray coating process typically results in a rougher surface, which can increase surface area and enhance photocatalytic activity [50, 51]. The classic spray coating technique is cost-effective and uses simple equipment, making it efficient for achieving high-purity coatings [51, 52]. This technique involves several steps. Initially, a specialized suspension is created and the substrate undergoes pretreatment. Subsequently, the prepared suspension is applied to the glass substrate using a spray nozzle. Following application, the coating undergoes a drying process and is then calcined under carefully regulated time and temperature conditions (Figure 6.5c). Zong et al. [53] applied fluorosilicone/SiO_2-TiO_2 (FSi/SiO_2-TiO_2) coatings to glass surfaces using spray coating techniques. The resulting coatings exhibited remarkable self-cleaning capabilities and photocatalytic properties. When exposed to UV light, these coatings effectively prevented the adhesion of various natural contaminants to outdoor glass surfaces that had been treated with the coating. Nevertheless, the fluorosilane used for modifying surfaces to achieve low surface energy is expensive and potentially harmful to the environment. Consequently, there is a need to utilize eco-friendly substances in the creation of superhydrophobic surfaces. As an example, Wang et al. [54] employed spray coating to apply PDMS-modified SiO_2-TiO_2 ternary composites to glass surfaces, creating a SiO_2-TiO_2@PDMS coating that demonstrated outstanding self-cleaning capabilities and photocatalytic efficiency. In comparison to fluorosilane, PDMS is a more economical option and provides several benefits in terms of environmental conservation and safety. The SiO_2-TiO_2@PDMS coating was subjected to heat treatment for 2 h at various temperatures, enabling modifications to its water-repellent properties and droplet adhesion characteristics based on the applied temperature. Water contact angle measurements for the coating, which demonstrated self-cleaning properties similar to the "lotus effect," were recorded as 147.8° following heat treatment at 25 °C and 152.2° after exposure to 400 °C (Figure 6.5c). These results suggest that high-temperature treatment enhances the coating's ability to repel. Interestingly, the SiO_2-TiO_2@PDMS coating maintained its exceptional water-repellent properties while demonstrating particular effectiveness in breaking down methyl orange, as evidenced when compared to a control sample (Figure 6.5c). Additionally, the coating exhibited resistance to liquids in acidic (pH = 2), alkaline (pH = 11), and organic dye solutions, demonstrating its remarkable durability even after being subjected to mechanical abrasion and UV light exposure.

6.3.7 Dip coating

Dip coating is a popular, cost-effective, and straightforward technique for applying TiO_2 nanoparticles (NPs) to glass surfaces [55]. This method involves submerging a glass substrate into a solution containing the photocatalytic material, resulting in the formation of a thin film. The TiO_2 precursor solution for dip coating is typically prepared using the sol-gel method [56]. Prior to coating, it is crucial to clean the glass substrate using plasma treatment or by washing it with specific solutions, such as acids, organic solvents, or neutral detergents [57]. Once the precursor solution and glass substrate are prepared, the dip coating process commences. Using specialized equipment, the substrate is immersed in the solution for a predetermined time and then withdrawn at a controlled rate. The coating thickness is influenced by the immersion duration and the number of dipping cycles. Following deposition, the films are dried and annealed to encourage TiO_2 crystallization. It is important to note that multiple precursor solutions can be utilized to create TiO_2-based heterogeneous photocatalytic coatings (Figure 6.5d).

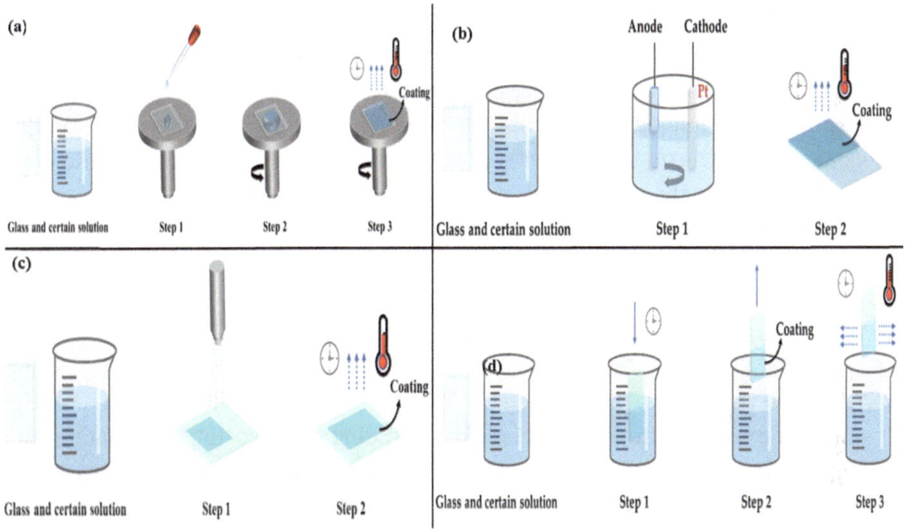

Figure 6.5: Schematic illustrations of four different coating processes: (a) spin coating, (b) electrodeposition (with the working electrode serving as the anode), (c) traditional spray coating, and (d) dip coating. Each diagram outlines the sequential steps involved in the respective coating method. This figure has been reproduced with permission from MDPI Publications [37].

The research indicates that the quantity and crystal structure of TiO_2 are critical factors in determining the photocatalytic efficiency of the resulting coatings [58]. While increasing the number of dip coating cycles only increases the coating thickness, it does not alter the crystal phase, which is instead influenced by the annealing temperatures during post-thermal processing. Furthermore, glass substrates often contain sodium ions, and if the photocatalytic layer is insufficiently thick, sodium migration from the glass can adversely affect the photocatalytic activity. Consequently, when employing dip coating techniques to produce photocatalytic coatings, it is vital to meticulously regulate both the frequency of dip coating cycles and the annealing processes. By fine-tuning these variables, it becomes possible to create the intended TiO_2 coating with suitable thickness and crystal structure.

6.4 Characterization of TiO_2 coating

Analyzing titanium dioxide (TiO_2) coatings is essential for comprehending their characteristics and efficacy in various applications, such as photocatalysis, self-cleaning surfaces, and optical devices. Multiple techniques are employed to evaluate the structural, morphological, and chemical attributes of TiO_2 coatings.

6.4.1 X-ray diffraction (XRD)

A fundamental method for analyzing titanium dioxide (TiO_2) coatings is X-ray diffraction (XRD). This technique provides insights into the crystalline structure, phase composition, and other structural characteristics of the material. The application of XRD is essential in evaluating the quality of TiO_2 films produced through various manufacturing methods, such as spray coating, dip coating, and chemical vapor deposition.

6.4.1.1 Key aspects of XRD in TiO_2 coating characterization

6.4.1.1.1 XRD testing for TiO_2 coatings

i. **Sample preparation:** Various methods, including spray coating, dip coating, and chemical vapor deposition (CVD), are employed to apply TiO_2 coatings onto appropriate surfaces such as glass, silicon, or metal. Following the application process, the coated samples may be subjected to annealing or other post-application treatments to improve their characteristics.
ii. **XRD measurement:** The coated sample is subsequently inserted into an XRD device. A beam of X-rays is aimed at the prepared sample, and it scatters at particular angles upon encountering the crystalline structure of TiO_2. The diffraction pattern that results is documented as a graph of intensity against angle (2θ).
iii. **Phase identification:** The XRD analysis shows clear peaks that indicate various crystalline forms of TiO_2, mainly anatase and rutile. Scientists can determine the phases present in the coating by comparing these detected peaks with standardized reference patterns found in resources like the International Centre for Diffraction Data (ICDD) [46].
iv. **Crystallinity assessment:** The crystallinity of the TiO_2 coating can be assessed by examining the clarity and strength of the diffraction peaks. More pronounced

peaks typically indicate a higher degree of crystallinity, which is often associated with improved photocatalytic performance [59].

v. **Grain size calculation**: The Scherrer equation can be employed to evaluate the expansion of diffraction peaks, providing an estimate of the average size of crystallites within the coating. Coatings with smaller crystallite dimensions typically exhibit greater surface areas, which can enhance their photocatalytic efficiency [60].

vi. **Texture analysis:** XRD analysis can also reveal information about the preferential alignment of crystallites within the coating, which may affect its performance characteristics.

6.4.2 Scanning electron microscopy

Scanning electron microscopy (SEM) is employed to examine the surface topography of titanium dioxide coatings at high magnification levels. This technique produces detailed images of the coating's surface features, including texture and distribution of particles, which are essential for assessing its photocatalytic effectiveness and self-cleaning capabilities.

i) Sample preparation

TiO_2 coatings are commonly applied to substrates utilizing spray, dip, or chemical vapor deposition processes before SEM examination. To ensure optimal imaging circumstances, the samples may need to be prepared in specific ways, such as cutting or mounting on a conductive stub. If the coatings are not conductive, a small layer of conductive material (such as gold or carbon) can be deposited on the surface to prevent charging during imaging.

ii) Scanning electron microscopic (SEM) imaging process

SEM employs a focused beam of electrons that moves across the surface of a specimen. As these electrons interact with the sample's atomic structure, they produce various signals, including secondary electrons, backscattered electrons, and X-rays. These diverse signals are then integrated to create detailed images that showcase the topography and compositional features of the sample's surface.

iii) Analysis of coating morphology

SEM allows for the observation of several key characteristics of TiO_2 coatings:

a) **Surface morphology***:* SEM images can reveal how consistent and homogeneous the coating surface is. For example, it can tell if the coating's texture is smooth or rough.

b) **Particle size and distribution:** The coating's TiO_2 particles can be examined for size and distribution. This knowledge is crucial for determining how these parameters affect photocatalytic activity.

c) **Coating thickness:** Cross-sectional SEM pictures can reliably quantify the thickness of TiO_2 coatings.
d) **Agglomeration:** SEM can detect agglomerates or clusters of TiO_2 particles that could impact the coating performance [61].

iv) Example findings

In studies involving TiO_2 coatings, SEM has been used to demonstrate various findings:

i) **Homogeneity:** TiO_2 coatings with specified deposition processes provide homogeneous thickness and consistent particle distribution over the substrate, according to research.
ii) **Photocatalytic efficiency:** The morphology seen in SEM pictures can be linked to photocatalytic performance; for example, rougher surfaces with more surface area may increase photocatalytic activity under UV light.

v) Images from studies

Here are some example images typically obtained from SEM analysis of TiO_2 coatings:

i) **Surface morphology:** SEM image reveals roughness and particle distribution on the surface of a TiO_2 coating.

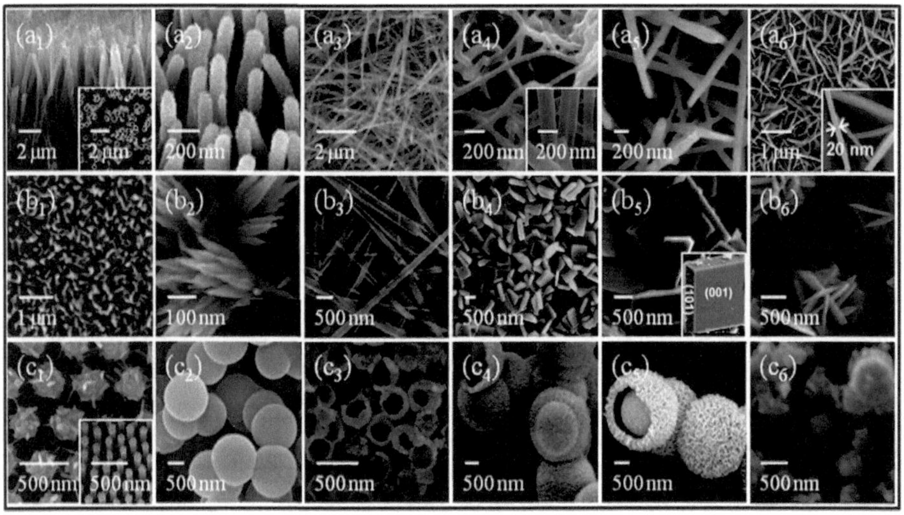

Figure 6.6: Scanning electron microscope (SEM) images showcase various TiO2 anatase nanostructures: (a1 and a2) depict nanotubes, (a3 and a4) illustrate nanofibers, (a5) displays nanorods, (a6) presents nanorods with nanoflakes, (b1) shows nanorod arrays, (b2) exhibits nanoribbons, (b3) portrays nanowires, (b4–b6) demonstrate nanosheets, (c1) reveals a 3D network, (c2) illustrates nanospheres, (c3) depicts nanocups, (c4 and c5) show nanospheres, and (c6) presents nanoparticles. Higher magnification images are provided in insets (a1, a4, a6, b5, and c1). These images are reproduced with permission from the Royal Society of Chemistry [63].

ii) **Cross-sectional view:** Cross-sectional SEM image of a TiO_2 coating on a substrate, showing thickness and stacking.
iii) **Agglomeration analysis:** This image shows agglomerated TiO_2 particles within a coating, which may affect its photocatalytic characteristics [62].

6.4.3 Transmission electron microscopy

Transmission electron microscopy (TEM) is utilized to examine the nanoscale features of TiO_2 coatings, including their microstructure, morphology, and crystallographic properties. The technique produces high-resolution images that reveal intricate details about the TiO_2 particles and their structural organization within the coating. This paper presents a comprehensive overview of TEM's application in the characterization of TiO_2 coatings:

i) Sample preparation
To conduct a TEM investigation, TiO_2 coatings must be thin enough to allow electrons to pass. This often entails preparing ultra-thin slices of coated samples, which are often less than 100 nanometers thick. To attain the desired thickness, techniques like focused ion beam (FIB) milling and ultramicrotomy are routinely used. The samples are then laid out on an appropriate grid for imaging.

ii) TEM imaging process
A high-energy electron beam passes through the material in TEM. As electrons interact with the sample, they scatter, producing a variety of signals that can be detected to create images. The photos reveal the interior structure and morphology of the TiO_2 coatings.

iii) Analysis of coating morphology
TEM provides detailed insights into several key aspects of TiO_2 coatings:
Particle size and distribution: TEM enables precise measurement of individual TiO_2 nanoparticle sizes and their distribution within the coating. This information is crucial for understanding how these factors affect photocatalytic activity.
Crystallographic structure: High-resolution TEM (HRTEM) can display lattice fringes, allowing researchers to identify the crystallographic orientation and phase of TiO_2, such as anatase or rutile.
Defects and interfaces: TEM can detect defects within the crystalline structure and examine the interfaces between different materials in composite coatings.
Morphological features: The morphology seen in TEM images offers insights into how processing conditions influence the physical structure of TiO_2 coatings.

iv) Example findings
TEM has been utilized in various studies of TiO_2 coatings to reveal significant results:

Nanostructure analysis: Research has demonstrated that TiO_2 nanoparticles can either form agglomerates or exhibit a uniform distribution, depending on the deposition method employed.
Phase identification: HRTEM images can verify the presence of specific crystalline phases by displaying distinct lattice spacings that correspond to anatase or rutile structures.

6.4.4 Contact angle measurement of TiO_2 coating

Contact angle measurement is an essential technique for evaluating the wettability of titanium dioxide (TiO_2) coatings. This property is especially significant in applications like self-cleaning surfaces, where either super hydrophilicity or super hydrophobicity is required. The interaction between a liquid droplet and a solid surface can be measured using the contact angle, which offers valuable information about the coating's surface energy and its water-attracting (hydrophilic) or water-repelling (hydrophobic) properties:

i) Understanding contact angle measurement
Definition: The contact angle serves as a method to evaluate the interaction between a solid surface and a liquid droplet. This measurement provides insights into the surface energy of the coating and reveals whether it exhibits hydrophilic (water-attracting) or hydrophobic (water-repelling) characteristics.
Hydrophilic surface: A contact angle of less than 90° indicates the surface is hydrophilic, meaning water spreads easily on it.
Hydrophobic surface: When the contact angle exceeds 90°, it indicates a hydrophobic surface, causing water to form droplets instead of spreading out.
Superhydrophilic surface: When the contact angle approaches 0°, it indicates extreme hydrophilicity, causing water droplets to fully spread across the surface.

ii) Importance of contact angle in TiO_2 coatings
Self-cleaning applications: Surfaces with self-cleaning capabilities often utilize TiO_2 coatings, which exhibit photocatalytic characteristics. The super hydrophilic nature of these surfaces allows water to spread out, effectively removing dirt and pollutants from the surface.
Photocatalytic activity: The wettability of TiO_2 coatings can influence their photocatalytic efficiency. A lower contact angle, indicating higher wettability, typically enhances the interaction between water and the coating, thereby improving the degradation of pollutants under UV light.

iii) Measurement techniques
Sessile drop method: A common technique involves placing a droplet of water on a TiO_2-coated surface, then using software to analyze a captured image and calculate the contact angle.

Wilhelmy plate method: This technique involves submerging a thin plate into a liquid to measure force changes, allowing dynamic contact angle measurements.
Goniometer: An instrument called a goniometer is utilized to determine the static contact angle by examining the profile of a droplet on a surface.

iv) Effects of UV irradiation on wettability
Research has shown that the wettability of TiO_2 coatings can be altered by UV light exposure:
Hydrophobic to hydrophilic transition: For example, a study revealed that a TiO_2 coating with an initial contact angle of nearly 150° could decrease to almost 0° under UV irradiation due to photocatalytic reactions that generate hydroxyl radicals, increasing surface energy and promoting hydrophilicity.
Durability and stability: The durability of these hydrophilic properties is vital for practical use. Studies have shown that TiO_2 coatings can retain their super hydrophilic characteristics even after multiple UV exposure cycles.

v) Example findings from research
One study reported that, after applying a TiO_{22} coating, the water contact angle significantly decreased from an initial value (e.g., 46.49°) to a much lower value (e.g., 14.89°) following a period of UV exposure. Another research highlighted the excellent self-cleaning properties of TiO_2 coatings due to their super hydrophilic nature, allowing water droplets to spread and effectively remove contaminants.

6.5 Application of TiO_2 coating

6.5.1 Photocatalytic applications

In recent decades, titanium dioxide (TiO_2) has emerged as a prominent photocatalyst, garnering substantial interest due to its robust chemical stability, widespread commercial availability, and exceptional catalytic characteristics (Figure 6.7b). Its versatility has been demonstrated across a broad spectrum of applications, encompassing energy conversion and storage, hydrogen generation, sensing technologies, and environmental solutions [64, 65]. The fundamental principle underlying the photocatalytic process in semiconductor materials like TiO_2 involves the creation of electron-hole pairs when exposed to light energy sufficient to overcome the band gap. This photo-excitation can subsequently trigger the formation of hydroxyl radicals in aqueous environments.

The catalytic efficiency of this system is primarily influenced by two competing processes: the recombination and trapping of charge carriers, as well as the competition between trapped carrier recombination and interfacial charge transfer. These factors collectively determine the overall effectiveness of the catalytic process. Consequently, enhancing the separation of charges and reducing the recombination of

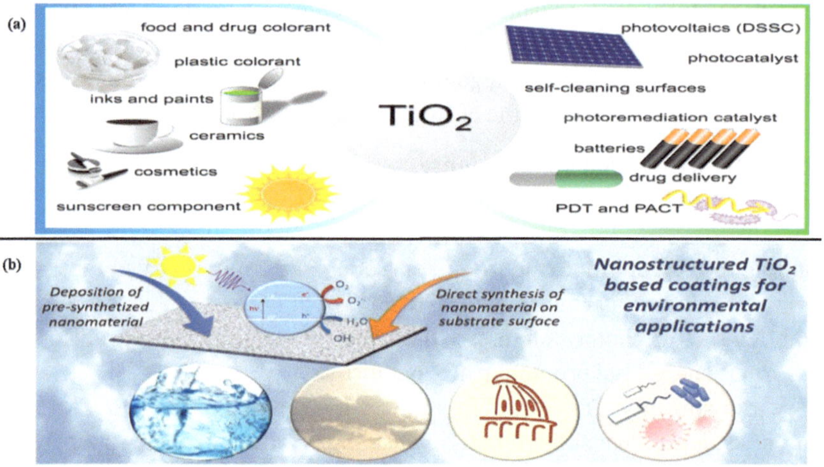

Figure 6.7: (a) Applications of TiO_2 coating (reproduced with permission from MDPI Publications [64]). (b) Photocatalytic applications (reproduced with permission from ELSEVIER Publications [32]).

charge carriers is essential for boosting the efficiency of charge transfer at interfaces. Nanostructured TiO_2 has shown superior performance compared to its bulk form, primarily due to its increased surface area-to-volume ratio. This characteristic significantly enhances the number of active sites available for adsorption and catalytic processes. Moreover, the band gap of TiO_2 nanoparticles varies with size, enabling the adjustment of electron-hole redox potentials and allowing for specific photochemical reactions. The smaller dimensions of the nanostructured semiconductor enhance the swift movement of light-generated charges to the catalyst's surface, thereby reducing the chances of harmful recombination within the nanostructure's interior. The limited effectiveness of TiO_2 materials under visible light has spurred research into photocatalysts that are active in this spectrum. Scientists have tackled this issue through various strategies, including: (i) developing hybrid nanostructured materials by combining TiO_2 with other semiconductors or plasmonic metal nanoparticles (NPs); (ii) introducing dopants such as metals or nitrogen; and (iii) integrating TiO_2 nanostructures with different carbon allotropes, like carbon nanotubes, graphene, or graphitic carbon nitride. These carbon-based materials serve a dual purpose: they boost TiO_2's photocatalytic performance by reducing the e^-/h^+ recombination rate and capturing photoexcited electrons from TiO_2's conduction band, while also enabling visible-light photocatalytic activity. This is achieved through photosensitization or doping effects, such as the formation of Ti-O-C bonds [62, 63]. Nanostructured TiO_2 materials are extensively researched photocatalysts in this field due to their size-dependent physical and chemical properties. These characteristics allow for the adjustment of their features to suit catalytic functions in environmental applications [67]. Such applications encompass the photocatalytic breakdown of contaminants in both air and water, surfaces that clean themselves, deactivation of microbes, and the degradation of pollutants. However, utilizing nanostructured TiO_2 in its dispersed nanopowder form comes with considerable disadvantages. These include difficulties in recovering and reusing the material, as well as environmental risks associated with the discharge of nanosized particles. For a considerable time, these problems have been significant obstacles in TiO_2 photocatalysis research [68–70]. Various strategies have been investigated to overcome these hurdles, including the attachment of nanostructured TiO_2 photocatalysts to stable and unreactive supports. In recent years, these technical constraints have been tackled by applying catalysts onto various substrates, chosen according to the intended use, such as glass, stainless steel, glass beads, fibers, silica, or textiles, to create an appropriate coating [71, 72]. Nevertheless, these coatings may encounter challenges such as the aggregation of nanostructured TiO_2, which diminishes photocatalytic effectiveness by substantially decreasing the number of active surface sites. Additional significant hurdles include managing coating thickness, geometry, and mechanical durability, which affect the material's practical use and applicability. While complex, the development and fabrication of TiO_2-based nanostructured coatings are crucial for environmental applications, including air and water purification, self-cleaning surfaces, and antimicrobial properties. This chapter summarizes current studies on nanostructured TiO_2-based coatings engineered for photocatalytic environmental uses [66]. It examines various methods for creating these coatings, ranging from direct synthesis of nanostructured TiO_2 on substrates to the application of pre-made TiO_2-based nanostructures. The review covers both traditional and innovative deposition techniques, emphasizing their features, benefits, and substrate suitability. Furthermore, it discusses commercially available nanostructured TiO_2-based products and their practical applications, such as water and air purification, self-cleaning surfaces, preservation of cultural heritage, and bacterial elimination. A particularly noteworthy example is the application of TiO_2 nanostructured coatings to deactivate SARS-CoV-2 on surfaces [73, 74].

6.5.2 Biomedical applications

For decades, titanium dioxide (TiO_2) has been a key component in various products, including pigments, paints, pharmaceuticals, cosmetics, and toothpaste [75–77]. Its significance grew substantially following the 1972 discovery of photocatalytic water splitting on a TiO_2 electrode when exposed to UV light [65]. Subsequently, TiO_2 materials have found applications across diverse sectors, spanning from energy to environmental solutions. These applications encompass photocatalytic splitting, photoelectrodes, supercapacitors, and dye-sensitized solar cells [78–80].

Advancements in nanotechnology have facilitated the production of diverse TiO_2 nanostructures, such as nanotubes, nanowires, nanorods, nanobelts, and nanoribbons [81–85]. These structures are characterized by their extensive surface areas and high levels of biocompatibility. The unique properties of these TiO_2 nanostructures have gen-

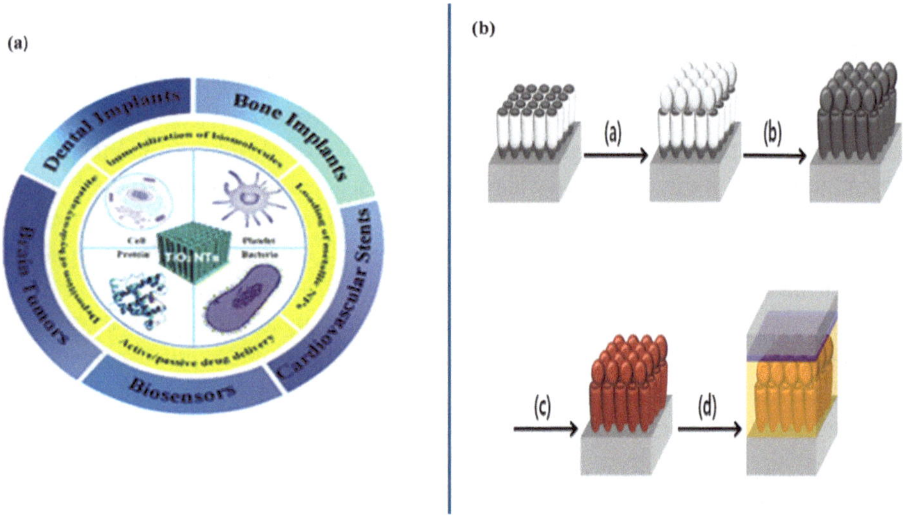

Figure 6.8: (a) Illustrations depicting TiO_2 applications in biomedicine [79]. (b) A comprehensive representation of the process for creating dye-sensitized solar cells (DSSCs) using freestanding TiO_2 nanotube arrays with either closed or open ends, enhanced with large TiO_2 nanoparticles (NPs) and carbon-based materials. The steps include: (a) applying a layer of large TiO_2 NPs; (b) generating carbon materials through chemical vapor deposition (CVD); (c) absorbing dye; and (d) assembling the DSSCs (republished with permission from MDPI Publications [80]).

erated considerable attention in the biomedical sector, primarily due to their capacity to incorporate functional components on their surfaces, which can influence biological reactions. When applied to the human body, these engineered nanomaterials can engage with the hierarchical nanostructures of biological systems. This interaction can replenish, stimulate, and maintain physiological responses while minimizing unwanted side effects [16]. For example, these materials can enhance the adsorption of proteins from the extracellular environment, such as fibronectin and vitronectin, creating conditions that favor early cell attachment to the material surface [86, 87]. The nanoscale topography and roughness of TiO_2 nanotubes (TNTs) closely resemble the dimensions of extracellular matrix signaling particles. This capability allows TNT surfaces to offer topographical signals for cellular recognition, thus influencing the expression of cell signaling molecules and further improving the integration of the biomaterial with surrounding healthy tissues [73]. Multiple studies have shown that nanostructures made of TiO_2 are exceptionally promising for use in implantable devices [88]. Implants, which are utilized to assist, substitute, or enhance damaged or diseased biological tissues, are frequently used in applications such as artificial hearts, synthetic blood vessels, vascular stents, and bone and dental structures. Implants primarily utilize three types of materials: metals, ceramics, and polymers. Metal-based implants excel in load-bearing applications due to their exceptional strength, rigidity, and durability [89]. In contrast, ceramic materials, including bio glass, possess a lower fracture resistance and higher elastic modulus compared to bone, rendering them unsuitable for meeting the in vivo mechanical requirements [90]. Although polymer implants readily bond with surrounding tissues, they carry the risk of causing tissue inflammation if monomers escape from the implant [91]. Over the past few decades, TNTs have garnered significant interest due to their exceptional biocompatibility, ability to promote bone growth, and capacity to integrate with bone tissue. Nevertheless, the widespread use of TNTs still faces considerable obstacles, particularly in terms of their antimicrobial properties and biological activity [92–94]. Researchers are currently investigating the potential of TNTs for various biomedical applications, including bone therapy, dental treatments, cardiovascular stent technology, and biosensor development (Figure 6.8).

6.5.3 TiO_2 in the food industry and cosmetics

The food and personal care industries have widely employed titanium dioxide powders as additives [75]. Among various white pigments, TiO_2 stands out due to its superior refractive index. As a result, it finds extensive use as a white colorant, color enhancer, and opacifying agent in numerous applications. TiO_2 in its microcrystalline state is utilized in various cosmetics and personal care items, ranging from toothpaste to hair dye products. It is regulated by the Cosmetics Regulation EC 1223/2009 Annex IV (which lists permitted colorants in cosmetic products, reference number: 143) [95] and has received approval from the US Food and Drug Administration (FDA, 21 CFR 73.2575) [3]. When in nanoform, TiO_2 is incorporated into sunscreens, day creams, lip balms, and foundations due to its capacity to shield against UVB, UVA, visible light, and infrared radiation [75]. TiO_2 is recognized as a physical UV filter and is included in the list of approved UV filters for cosmetic products [2, 4]. The US FDA has also authorized its use [5], with a maximum concentration of 25% in ready-to-use preparations. In the food industry, $TiO_2$is primarily utilized in its pure anatase and/or rutile forms, occasionally coated with minimal amounts of alumina and/or silica to enhance the product's technological properties. It is incorporated into various industrial food items, including confectionery, dairy products, cheese, fishery products, sauces, soups, and broths [10]. Among food products, chewing gum, sweets, and candies contain the highest concentration of TiO_2 pigments, with an average particle size of 110 nm and a wide size distribution [75]. Recent years have seen growing apprehension regarding the use of nanoparticles in food items, as evidenced by the increase in studies focused on quantifying TiO_2 in foodstuffs [95, 96]. Critical reviews have been conducted on research assessing the risks of titanium dioxide as a food additive [97, 98]. In 2016, the European Food Safety Authority (EFSA) reevaluated the safety of TiO_2 when used as a food additive [15]. Their findings indicated that both micro- and nano-sized forms of TiO_2 have negligible absorption in the gastrointestinal tract and do not pose a genotoxic concern [99].

6.7 Conclusion

In conclusion, titanium dioxide (TiO_2) coatings hold significant promise for the enhancement of biomedical devices, driven by their remarkable properties such as biocompatibility, chemical stability, and the ability to promote osseointegration. The application of various deposition techniques enables precise control over the surface characteristics, which can be tailored to optimize cellular responses and reduce infection risks. Furthermore, the exploration of doped TiO_2 variants presents opportunities to enhance antimicrobial activity and functionality. While the current advancements are encouraging, challenges related to large-scale manufacturing and long-term performance in physiological conditions must be addressed. Continuous advancements in research and development will be essential to surmount these challenges and fully exploit the capabilities of TiO_2 coatings in medical applications, ultimately enhancing patient care and extending the durability of biomedical devices.

References

[1] Wang, M., Zheng, Y., Yin, C., Dai, S., Fan, X., Jiang, Y., et al. Recent progress in antibacterial hydrogel coatings for targeting biofilm to prevent orthopedic implant-associated infections. Frontiers in Microbiology. 2023;14(December). *Front. Microbiol.* 14:1343202.

[2] Chen, X., & Selloni, A. (2014). Introduction: titanium dioxide (TiO_2) nanomaterials. *Chemical reviews*, 114(19), 9281–9282.

[3] Sun, S., Song, P., Cui, J., & Liang, S. Amorphous TiO2 nanostructures: Synthesis, fundamental properties and photocatalytic applications. *Catalysis Science and Technology*. 2019;*9*(16):4198–4215.

[4] Ahmad, W., Singh, A., Mishrwan, V. S., Joshi, S., & Rawat, A. Titanium dioxide nanomaterials: Synthesis and applications. *Asian Journal of Chemistry*. 2023;*35*(8):1770–1774.

[5] Çeşmeli, S., & Biray Avci, C. Application of titanium dioxide (TiO2) nanoparticles in cancer therapies. *Journal of Drug Targeting* [Internet]. 2019;*27*(7):762–766. Available from: http://dx.doi.org/10.1080/1061186X.2018.1527338

[6] Kamat, P. V. TiO 2 nanostructures: Recent physical chemistry advances. *Journal of Physical Chemistry C*. 2012;*116*(22):11849–11851.

[7] Rehman, F. U., Zhao, C., Jiang, H., & Wang, X. Biomedical applications of nano-titania in theranostics and photodynamic therapy. *Biomaterial Science*. 2016;*4*(1):40–54.

[8] Zeng, L., Ren, W., Xiang, L., Zheng, J., Chen, B., & Wu, A. Multifunctional Fe3O4-TiO2 nanocomposites for magnetic resonance imaging and potential photodynamic therapy. *Nanoscale*. 2013;*5*(5):2107–2113.

[9] Sébastien, S. Les soignants face à la mort. *Revue de l'Infirmiere*. 2012;*1*(180):39–41.

[10] Inoue, T., Fujishima, A., Konishi, S., & Honda, K. Photoelectrocatalytic reduction of carbon dioxide in aqueous suspensions of semiconductor powders [3]. Vol. *277*, *Nature*. 1979. p. 637–638.

[11] Ahmari, H., Heris, S. Z., & Khayyat, M. H. The effect of titanium dioxide nanoparticles and UV irradiation on photocatalytic degradation of Imidaclopride. *Environmental Technology (United Kingdom)* [Internet]. 2018;*39*(4):536–547. Available from: http://dx.doi.org/10.1080/09593330.2017.1306115

[12] Choi, H. C., Jung, Y. M., & Kim, S. B. Size effects in the Raman spectra of TiO2 nanoparticles. *Vibrational Spectroscopy*. 2005;*37*(1):33–38.

[13] Li, Y., Zhang, S., Yu, Q., & Yin, W. The effects of activated carbon supports on the structure and properties of TiO_2 nanoparticles prepared by a sol-gel method. *Applications of Surface Science*. 2007;*253*(23):9254–9258.
[14] Camps, I., Borlaf, M., Colomer, M. T., Moreno, R., Duta, L., Nita, C., et al. Structure-property relationships for Eu doped TiO_2 thin films grown by a laser assisted technique from colloidal sols. *RSC Advances*. 2017;*7*(60):37643–37653.
[15] Hoang, V. V., Zung, H., & Trong, N. H. B. Structural properties of amorphous TiO2 nanoparticles. *The European Physical Journal D-Atomic, Molecular, Optical and Plasma Physics*. 2007;*44*(3):515–524.
[16] Barshilia, H. C., & Gupta, N. Superhydrophobic polytetrafluoroethylene surfaces with leaf-like micro-protrusions through Ar + O_2 plasma etching process. *Vacuum*. 2014;*99*:42–48.
[17] Simons, P. Y., & Dachille, F. The structure of TiO_2 II, a high-pressure phase of TiO_2 . *Acta Crystallographica*. 1967;*23*(2):334–336.
[18] Kang, Xiaolan, Sihang Liu, Zideng Dai, Yunping He, Xuezhi Song, and Zhenquan Tan. 2019. "Titanium Dioxide: From Engineering to Applications" *Catalysts* 9, no. 2: 191.
[19] Selloni, A. Anatase shows its reactive side. *Nature Materials*. 2008;*7*(8):613–615.
[20] Gupta, S. M., & Tripathi, M. A review of TiO_2 nanoparticles. *Chinese Science Bulletin*. 2011;*56*(16):1639–1657.
[21] Tripathi, A. K., Singh, M. K., Mathpal, M. C., Mishra, S. K., & Agarwal, A. Study of structural transformation in TiO_2 nanoparticles and its optical properties. *Journal of Alloys and Compounds* [Internet]. 2013;*549*:114–120. Available from: http://dx.doi.org/10.1016/j.jallcom.2012.09.012
[22] Zhang, Q., Gao, L., & Guo, J. Effects of calcination on the photocatalytic properties of nanosized TiO2 powders prepared by TiCl4 hydrolysis. *Applied Catalysis B Environmental*. 2000;*26*(3):207–215.
[23] Govindasamy, G., Murugasen, P., & Sagadevan, S. Investigations on the synthesis, optical and electrical properties of TiO_2 thin films by Chemical Bath Deposition (CBD) method. *Materials Research*. 2016;*19*(2):413–419.
[24] Mo, S. D., & Ching, W. Y. Electronic and optical properties of three phases of titanium dioxide: Rutile, anatase, and brookite. *Physical Review B*. 1995;*51*(19):13023–13032.
[25] Shaikh, S. F., Mane, R. S., Min, B. K., Hwang, Y. J., & Joo, O. S. D-sorbitol-induced phase control of TiO_2 nanoparticles and its application for dye-sensitized solar cells. *Scientific Reports*. 2016;*6* (December 2015):1–10.
[26] Ali, I., Suhail, M., Alothman, Z. A., & Alwarthan, A. Recent advances in syntheses, properties and applications of TiO2 nanostructures. *RSC Advances*. 2018;*8*(53):30125–30147.
[27] Fujishima, A., Zhang, X., & Tryk, D. A. TiO_2 photocatalysis and related surface phenomena. *Surface Science Reports*. 2008;*63*(12):515–582.
[28] Nakata, K., & Fujishima, A. TiO 2 photocatalysis: Design and applications. *Journal of Photochemistry and Photobiology C Photochemistry Reviews*. 2012;*13*(3):169–189.
[29] Carp, O., Huisman, C. L., & Reller, A. Photoinduced reactivity of titanium dioxide. *Progress in Solid State Chemistry*. 2004;*32*(1–2):33–177.
[30] Auvinen, S., Alatalo, M., Haario, H., Jalava, J. P., & Lamminmäki, R. J. Size and shape dependence of the electronic and spectral properties inTiO_2 nanoparticles. *Journal of Physical Chemistry C*. 2011;*115*(17):8484–8493.
[31] Ikeda, K., Sakai, H., Baba, R., Hashimoto, K., & Fujishima, A. Photocatalytic reactions involving radical chain reactions using microelectrodes. *Journal of Physical Chemistry B*. 1997;*101*(14):2617–2620.
[32] Dell'Edera, M., Lo Porto, C., De Pasquale, I., Petronella, F., Curri, M. L., Agostiano, A., et al. Photocatalytic TiO_2-based coatings for environmental applications. *Catalysis Today* [Internet]. 2021;*380*(May):62–83. Available from: https://doi.org/10.1016/j.cattod.2021.04.023
[33] Equipment CE. Spin-coating overview. 2022; Available from: https://www.costeffectiveequipment.com/service-support/technical-resources/spin-coating-theory/

[34] Mustafa, H. A. M., & Jameel, D. A. Modeling and the main stages of spin coating process: A review. *Journal of Applied Science and Technology Trends*. 2021;*2*(02):119–123.

[35] Sahu, N., Parija, B., & Panigrahi, S. Fundamental understanding and modeling of spin coating process: A review. *Indian Journal of Physics*. 2009;*83*(4):493–502.

[36] Peeters, H., Keulemans, M., Nuyts, G., Vanmeert, F., Li, C., Minjauw, M., et al. Plasmonic gold-embedded TiO2 thin films as photocatalytic self-cleaning coatings. Applied Catalysis B Environmental 2020;267(October 2019). Applied Catalysis B: Environmental 267 (2020) 118654

[37] Tian, S., Feng, Y., Zheng, Z., & He, Z. TiO2-based photocatalytic coatings on glass substrates for environmental applications. *Coatings*. 2023;*13*(8):1–28.

[38] Mazur, M., Wojcieszak, D., Kaczmarek, D., Domaradzki, J., Song, S., Gibson, D., et al. Functional photocatalytically active and scratch resistant antireflective coating based on TiO_2 and SiO_2. *Applications of Surface Science*. 2016;*380*:165–171.

[39] Ding, R., Wang, K., Hong, K., Zhang, Y., & Cui, Y. Hierarchical core-shell tungsten oxide/TiO2 nanowires as an effective photocatalyst. *Chemical Physics Letters*. 2019;*714*(July 2018):156–159.

[40] Gondal, M. A., Ilyas, A. M., Fasasi, T. A., Dastageer, M. A., Seddigi, Z. S., Qahtan, T. F., et al. Synthesis of green TiO_2 /ZnO/CdS hybrid nano-catalyst for efficient light harvesting using an elegant pulsed laser ablation in liquids method. *Applications of Surface Science* [Internet]. 2015;*357*:2217–2222. Available from: http://dx.doi.org/10.1016/j.apsusc.2015.09.213

[41] Bricchi, B. R., Ghidelli, M., Mascaretti, L., Zapelli, A., Russo, V., Casari, C. S., et al. Integration of plasmonic Au nanoparticles in TiO_2 hierarchical structures in a single-step pulsed laser co-deposition. *Materials and Design* [Internet]. 2018;*156*:311–319. Available from: https://doi.org/10.1016/j.matdes.2018.06.051

[42] Kadri, Laid, Abdelkader Abderrahmane, Georgiana Bulai, Aurelian Carlescu, Corneliu Doroftei, Iuliana Motrescu, Silviu Gurlui, Liviu Leontie, and Mohamed Adnane. 2023. "Optical and Structural Analysis of TiO2–SiO2 Nanocomposite Thin Films Fabricated via Pulsed Laser Deposition Technique" Nanomaterials 13, no. 10: 1632.

[43] Ryan, S. W. "United States Patent Date of Patent." Syst. Method Program. A Weigh. Scale Usinga Key Signal Enter A Program. Mode 1 (2009): 14.

[44] John, R., Joseph, E., Cameron K, W. P. E., & Ralph E, V. A. E. United States Patent (19). 1975;(19).

[45] Lin, L., Deng, B., Sun, J., Peng, H., & Liu, Z. Bridging the gap between reality and ideal in chemical vapor deposition growth of graphene. *Chemical Reviews*. 2018;*118*(18):9281–9343.

[46] Kang, S., Mauchauffé, R., You, Y. S., & Moon, S. Y. Insights into the role of plasma in atmospheric pressure chemical vapor deposition of Titanium dioxide thin films. *Scientific Reports*. 2018;*8*(1):1–13.

[47] Liu, Xusong, Gang Wang, Hui Zhi, Jing Dong, Jian Hao, Xiang Zhang, Jing Wang, Danting Li, and Baosheng Liu. 2022. "Synthesis of the Porous ZnO Nanosheets and TiO2/ZnO/FTO Composite Films by a Low-Temperature Hydrothermal Method and Their Applications in Photocatalysis and Electrochromism" Coatings 12, no. 5: 695.

[48] Alpatov, S. S., Vasiliev, F. A., Aleshina, V. K., Vagramyan, T. A., & Semenikhin, O. A. Lithium electrodeposition in the presence of surfactants. *Russian Journal of Electrochemistry*. 2024;*60*(5):355–364.

[49] Lokhande, C. D. and Pawar, S. H.. "Electrodeposition of Thin Film Semiconductors". *Volume 111, Number 1 16 January*, edited by Görlich, Berlin, Boston: De Gruyter, 1989, pp. 17–40

[50] Wen, K., Liu, M., Liu, X., Deng, C., & Zhou, K. Deposition of photocatalytic TiO_2 coating by modifying the solidification pathway in plasma spraying. *Coatings*. 2017;*7*(10):1–9.

[51] Šuligoj, Andraž, Olena Pliekhova, Nives Vodišek, Mohor Mihelčič, Angelja K. Surca, Roman Kunič, Barbara Šubic, Jernej Starman, Aleš Ugovšek, and Urška Lavrenčič Štangar. 2019. "Field Test of Self-Cleaning Zr-Modified-TiO_2-SiO_2 Films on Glass with a Demonstration of Their Anti-Fogging Effect" Materials 12, no. 13: 2196

[52] Dundar, Ibrahim, Marina Krichevskaya, Atanas Katerski, Malle Krunks, and Ilona Oja Acik. 2019. "Photocatalytic Degradation of Different VOCs in the Gas-Phase over TiO_2 Thin Films Prepared by Ultrasonic Spray Pyrolysis" *Catalysts* 9, no. 11: 915.

[53] Zong, L., Wu, Y., Li, X., & Jiang, B. The preparation of superhydrophobic photocatalytic fluorosilicone/ SiO_2–TiO_2 coating and its self-cleaning performance. *Journal of Coatings Technology and Research*. 2021;*18*(5):1245–1259.

[54] Wang, X., Ding, H., Sun, S., Zhang, H., Zhou, R., Li, Y., et al. Preparation of a temperature-sensitive superhydrophobic self-cleaning SiO_2-TiO_2@PDMS coating with photocatalytic activity. *Surf Coatings Technol* [Internet]. 2021;*408*(December 2020):126853. Available from: https://doi.org/10.1016/j.surfcoat.2021.126853

[55] Lu, Bing-Jyh, Keng-Ta Lin, Yi-Ming Kuo, and Cheng-Hsien Tsai. 2021. "Preparation of High-Transparency, Super hydrophilic Visible Photo-Induced Photocatalytic Film via a Rapid Plasma-Modification Process" Coatings 11, no. 7: 784

[56] Mittal, T. Self-cleaning smart photocatalytic coatings for water treatment. *Materials Today: Proceedings* [Internet]. 2023;*78*:891–894. Available from: https://doi.org/10.1016/j.matpr.2022.12.094

[57] Najafidoust, A., Allahyari, S., Rahemi, N., & Tasbihi, M. Uniform coating of TiO_2 nanoparticles using biotemplates for photocatalytic wastewater treatment. *Ceramics International*. 2020;*46*(4):4707–4719.

[58] Ohya, Yutaka, Jyunji Mishina, Tatsuya Matsuda, Takayuki Ban, and Yasutaka Takahashi. "Crystallization and Microstructure Development of Sol–Gel-Derived Titanium Dioxide Thin Films with Single and Multiple Layers." *Journal of the American Ceramic Society* 82, no. 10 (1999): 2601–2606.

[59] Rosales, A., Maury-Ramírez, A., Gutiérrez, R. M. D., Guzmán, C., & Esquivel, K. SiO_2@TiO_2 coating: Synthesis, physical characterization and photocatalytic evaluation. *Coatings*. 2018;*8*(4):1–13.

[60] Guin, D., Manorama, S. V., Latha, J. N. L., & Singh, S. Photoreduction of silver on bare and colloidal TiO_2 nanoparticles/nanotubes: Synthesis, characterization, and tested for antibacterial outcome. *Journal of Physical Chemistry C*. 2007;*111*(36):13393–13397.

[61] Iesalnieks, Mairis, Raivis Eglītis, Tālis Juhna, Krišjānis Šmits, and Andris Šutka. 2022. "Photocatalytic Activity of TiO2 Coatings Obtained at Room Temperature on a Polymethyl Methacrylate Substrate" *International Journal of Molecular Sciences* 23, no. 21: 12936.

[62] Yang, M., Liu, G., Zhao, W., Meng, F., Lao, Y., & Liang, Y. Preparation and properties of TiO_2 photocatalytic self-cleaning coating material for glass. *International Journal of Low-Carbon Technologies*. 2023;*18*(March):322–330.

[63] Verma, R., Gangwar, J., & Srivastava, A. K. Multiphase TiO2 nanostructures: A review of efficient synthesis, growth mechanism, probing capabilities, and applications in bio-safety and health. *RSC Advances*. 2017;*7*(70):44199–44224.

[64] Joseph, J. Titanium dioxide nanoparticles: Prospects and applications in nanotheranostics. *Inorg Nanosyst Theranostic Nanosyst* Vol *2*. 2023;253–278.

[65] Pelaez, M., Nolan, N. T., Pillai, S. C., Seery, M. K., Falaras, P., Kontos, A. G., et al. A review on the visible light active titanium dioxide photocatalysts for environmental applications. *Applied Catalysis B Environmental* [Internet]. 2012;*125*:331–349. Available from: http://dx.doi.org/10.1016/j.apcatb.2012.05.036

[66] Panniello, A., Curri, M. L., Diso, D., Licciulli, A., Locaputo, V., Agostiano, A., et al. Nanocrystalline TiO 2 based films onto fibers for photocatalytic degradation of organic dye in aqueous solution. *Applied Catalysis B Environmental* [Internet]. 2012;*121–122*:190–197. Available from: http://dx.doi.org/10.1016/j.apcatb.2012.03.019

[67] Petrella, A., Mascolo, G., Murgolo, S., Petruzzelli, V., Ranieri, E., Spasiano, D., et al. Photocatalytic oxidation of organic micro-pollutants: Pilot plant investigation and mechanistic aspects of the degradation reaction. *Chemical Engineering Communications*. 2016;*203*(10):1298–1307.

[68] Cristina Yeber, M., Rodríguez, J., Freer, J., Durán, N., & Mansilla H., D. Photocatalytic degradation of cellulose bleaching effluent by supported TiO_2 and ZnO. *Chemosphere*. 2000;*41*(8):1193–1197.

[69] Horikoshi, S., & Serpone, N. Can the photocatalyst TiO_2 be incorporated into a wastewater treatment method? Background and prospects. *Catalysis Today* [Internet]. 2020;*340* (November 2018):334–346. Available from: https://doi.org/10.1016/j.cattod.2018.10.020
[70] Anwer, H., Mahmood, A., Lee, J., Kim, K. H., Park, J. W., & Yip, A. C. K. Photocatalysts for degradation of dyes in industrial effluents: Opportunities and challenges. *Nano Research*. 2019;*12*(5):955–972.
[71] Srikanth, B., Goutham, R., Badri Narayan, R., Ramprasath, A., Gopinath, K. P., & Sankaranarayanan, A. R. Recent advancements in supporting materials for immobilised photocatalytic applications in waste water treatment. *Journal of Environmental Management* [Internet]. 2017;*200*:60–78. Available from: http://dx.doi.org/10.1016/j.jenvman.2017.05.063
[72] Xing, Z., Zhang, J., Cui, J., Yin, J., Zhao, T., Kuang, J., et al. Recent advances in floating TiO_2-based photocatalysts for environmental application. *Applied Catalysis B Environmental* [Internet]. 2018;*225* (October 2017):452–467. Available from: https://doi.org/10.1016/j.apcatb.2017.12.005
[73] Hashimoto, K., Irie, H., & Fujishima, A. TiO 2 photocatalysis: A historical overview and future prospects. *Japanese Journal of Applied Physics, Part 1: Regular Papers, Short Notes & Review Papers*. 2005;*44*(12):8269–8285.
[74] Bachler, G., Von Goetz, N., & Hungerbuhler, K. Using physiologically based pharmacokinetic (PBPK) modeling for dietary risk assessment of titanium dioxide (TiO2) nanoparticles. *Nanotoxicology*. 2015;*9*(3):373–380.
[75] Weir, A., Westerhoff, P., Fabricius, L., Hristovski, K., & Von Goetz, N. Titanium dioxide nanoparticles in food and personal care products. *Environmental Science Technology*. 2012;*46*(4):2242–2250.
[76] Zhang, X. Y., Li, H. P., Cui, X. L., & Lin, Y. Graphene/TiO2 nanocomposites: Synthesis, characterization and application in hydrogen evolution from water photocatalytic splitting. *Journal of Materials Chemistry*. 2010;*20*(14):2801–2806.
[77] Sun, Wen-Tao, Yuan Yu, Hua-Yong Pan, Xian-Feng Gao, Qing Chen, and Lian-Mao Peng. "CdS quantum dots sensitized TiO_2 nanotube-array photoelectrodes." *Journal of the American Chemical Society* 130, no. 4 (2008): 1124–1125.
[78] Jennings, J. R., Ghicov, A., Peter, L. M., Schmuki, P., & Walker, A. B. Dye-sensitized solar cells based on oriented TiO_2 nanotube arrays: Transport, trapping, and transfer of electrons. *Journal of the American Chemical Society*. 2008;*130*(40):13364–13372.
[79] Wang, G., Song, D., Qiao, Y., Cheng, J., Liu, H., Jiang, J., et al. Developing super-hydrophobic and corrosion-resistant coating on magnesium-lithium alloy via one-step hydrothermal processing. *Journal of Magnesium and Alloys*. 2023;*11*(4):1422–1439.
[80] Rho, Won-Yeop, Da Hyun Song, Sang Hun Lee, and Bong-Hyun Jun. 2017. "Enhanced Efficiency in Dye-Sensitized Solar Cells by Electron Transport and Light Scattering on Freestanding TiO_2 Nanotube Arrays" *Nanomaterials* 7, no. 10: 345.
[81] Zhao, Z., Tian, J., Sang, Y., Cabot, A., & Liu, H. Structure, Synthesis, and Applications of TiO_2 Nanobelts. *Advances in Materials*. 2015;*27*(16):2557–2582.
[82] Santara, B., Giri, P. K., Imakita, K., & Fujii, M. Evidence of oxygen vacancy induced room temperature ferromagnetism in solvothermally synthesized undoped TiO_2 nanoribbons. *Nanoscale*. 2013;*5*(12):5476–5488.
[83] Wang, G., Wang, H., Ling, Y., Tang, Y., Yang, X., Fitzmorris, R. C., et al. Hydrogen-treated TiO2 nanowire arrays for photoelectrochemical water splitting. *Nano Letters*. 2011;*11*(7):3026–3033.
[84] Mor, G. K., Varghese, O. K., Paulose, M., Shankar, K., & Grimes, C. A. A review on highly ordered, vertically oriented TiO_2 nanotube arrays: Fabrication, material properties, and solar energy applications. *Solar Energy Materials and Solar Cells*. 2006;*90*(14):2011–2075.
[85] Feng, X., Shankar, K., Varghese, O. K., Paulose, M., Latempa, T. J., & Grimes, C. A. Vertically aligned single crystal TiO_2 nanowire arrays grown directly on transparent conducting oxide coated glass: Synthesis details and applications. *Nano Letters*. 2008;*8*(11):3781–3786.

[86] Lavenus, S., Ricquier, J. C., Louarn, G., & Layrolle, P. Cell interaction with nanopatterned surface of implants. *Nanomedicine*. 2010;*5*(6):937–947.
[87] Anselme, K. Osteoblast adhesion on biomaterials. *Biomaterials*. 2000;*21*(7):667–681.
[88] Minagar, S., Berndt, C. C., Wang, J., Ivanova, E., & Wen, C. A review of the application of anodization for the fabrication of nanotubes on metal implant surfaces. *Acta Biomaterial* [Internet]. 2012;*8*(8):2875–2888. Available from: http://dx.doi.org/10.1016/j.actbio.2012.04.005
[89] Geetha, M., Singh, A. K., Asokamani, R., & Gogia, A. K. Ti based biomaterials, the ultimate choice for orthopaedic implants – A review. *Progress in Materials Science* [Internet]. 2009;*54*(3):397–425. Available from: http://dx.doi.org/10.1016/j.pmatsci.2008.06.004
[90] Minagar, S., Wang, J., Berndt, C. C., Ivanova, E. P., & Wen, C. Cell response of anodized nanotubes on titanium and titanium alloys. *Journal of Biomedical Materials Research Part A*. 2013;*101*(9):2726–2739.
[91] Kokubo, T., Kim, H. M., & Kawashita, M. Novel bioactive materials with different mechanical properties. *Biomaterials*. 2003;*24*(13):2161–2175.
[92] Park, J., Bauer, S., Schlegel, K. A., Neukam, F. W., Von Mark, K. D., & Schmuki, P. TiO_2 nanotube surfaces: 15 nm – An optimal length scale of surface topography for cell adhesion and differentiation. *Small*. 2009;*5*(6):666–671.
[93] Li, M., Mondrinos, M. J., Chen, X., Gandhi, M. R., Ko, F. K., & Lelkes, P. I. Elastin blends for tissue engineering scaffolds. *Journal of Biomedical Materials Research, Part A* [Internet]. 2006;*79*(4):963–973. Available from: http://www.ncbi.nlm.nih.gov/pubmed/16948146
[94] Lai, Y., Lin, L., Pan, F., Huang, J., Song, R., Huang, Y., et al. Bioinspired patterning with extreme wettability contrast on TiO_2 nanotube array surface: A versatile platform for biomedical applications. *Small*. 2013;*9*(17):2945–2953.
[95] UNION, PEAN. "Regulation (EC) No 1223/2009 of the european parliament and of the council." *Official Journal of the European Union L* 342 (2009): 59.
[96] Peters, R. J. B., Van Bemmel, G., Herrera-Rivera, Z., Helsper, H. P. F. G., Marvin, H. J. P., Weigel, S., et al. Characterization of titanium dioxide nanoparticles in food products: Analytical methods to define nanoparticles. *Journal of Agricultural and Food Chemistry*. 2014;*62*(27):6285–6293.
[97] Yang, Y., Doudrick, K., Bi, X., Hristovski, K., Herckes, P., Westerhoff, P., et al. Characterization of food-grade titanium dioxide: The presence of nanosized particles. *Environmental Science Technology*. 2014;*48*(11):6391–6400.
[98] Winkler, H. C., Notter, T., Meyer, U., & Naegeli, H. Critical review of the safety assessment of titanium dioxide additives in food. *Journal of Nanobiotechnology* [Internet]. 2018;*16*(1):1–19. Available from: https://doi.org/10.1186/s12951-018-0376-8
[99] Hwang, J. S., Yu, J., Kim, H. M., Oh, J. M., & Choi, S. J. Food additive titanium dioxide and its fate in commercial foods. *Nanomaterials*. 2019;*9*(8):1–19.

Harish Singh Dhami, Srinivasan Chandrasekar

Chapter 7
Surface topographies in the manufacturing of biomedical implants

Abstract: Surface topography is a critical factor in the design and performance of biomedical implants, influencing cell adhesion, migration, and proliferation. The correlation between surface texture at various length scales and biocompatibility has become an active area of research, as it directly impacts the interaction between cells and implant materials. Traditional manufacturing techniques, such as cutting, turning, and multipoint material removal processes like abrasive finishing and sandblasting, create distinct surface characteristics that play a significant role in the overall functionality of implants. However, recent advancements in free-form fabrication, particularly laser-based additive manufacturing, have introduced new possibilities for creating patient-specific implants with customized properties. Despite these benefits, additive manufacturing methods often result in inherent surface textures that require further post-processing to optimize the surface for final applications. This chapter explores the emergence of surface roughness in various manufacturing processes and discusses methods to control surface topography for enhanced biocompatibility and performance of biomedical implants.

Keywords: Surface topography, biomedical implants, surface finishing, additive manufacturing

7.1 Introduction

The emergence of **surface topography** in the manufacturing of **biomedical implants** and its evolution during performance are pivotal questions addressed in this chapter. The former question pertains primarily to the type of manufacturing processes employed in implant production, necessitating a thorough investigation of various methodologies. The latter concerns the dynamic tribological factors that come into play during the implant's service life. Although these aspects are seemingly interconnected, they warrant distinct consideration due to their fundamentally different mechanistic underpinnings. This chapter focuses on the evolution of surface topography during the manufacturing of biomedical implants. Surface topography, encom-

Harish Singh Dhami, Srinivasan Chandrasekar, Department of Industrial Engineering, Purdue University, West Lafayette, e-mail: dhamih@purdue.edu

https://doi.org/10.1515/9783111571423-007

passing parameters such as size, shape, and texture, plays a crucial role in modulating cell behavior, including adhesion, reaction, migration, and proliferation [1]. Consequently, a range of treatments is applied to modify the surface morphology of biomedical implants to investigate the cell-material interface in vitro. Numerous studies have explored the relationship between surface characteristics and cellular integration in biomedical implants with a consensus that surface morphology significantly influences cellular responses, particularly during the initial stages of growth [2–5].

The impact of implant surface topography on biological functions, including cell and tissue adhesion and growth, remains a key area of research. Surfaces with random or regular textures exhibit different influences on performance and product life, emphasizing the importance of preparing surfaces appropriately [6]. Preparation of surfaces for such studies typically involves machining, abrasive polishing, acid etching, and electro-erosion. Materials such as titanium (Ti)- and magnesium (Mg)-based alloys and their coatings are extensively used in implant applications due to their nontoxic nature and biocompatibility [7–10]. These metals offer high specific strength and an elastic modulus comparable to bone, making them ideal for such applications. Surface modification through coatings, applied via chemical vapor deposition or physical vapor deposition, enhances wear resistance, cell adhesion, and corrosion behavior [11, 12]. Rough surfaces produced by sandblasting have been shown to promote greater cell attachment compared to smooth surfaces. Recent efforts have focused on developing bone cell culture models to facilitate in vitro formation of cell-material interfaces [13, 14]. Computational modeling of surface coating-based microtopographies in two-dimensional (2D) and three-dimensional (3D), aimed at prosthetic implant applications, represents another area of interest [15]. Integrating these features within the design environment can provide detailed quantitative descriptions at the microscopic scale, enabling high-precision engineering design and manufacturing of biomedical implants.

The demand for biomedical implants, consequently its manufacturing industry, has seen a surge over past decades, driven by advancements in manufacturing technologies and the development of novel biocompatible materials [16, 17]. This growth has fostered a new research domain within the materials and manufacturing community. Traditional manufacturing processes such as casting, forging, and machining have been primarily utilized for producing such components [18]. However, these processes impose limitations on achieving specific surface roughness and topographies. Additive manufacturing, with its capability for selective deposition, has emerged as a significant advancement, allowing the production of customized components tailored to individual needs [19]. Nevertheless, postprocessing is often required to achieve the desired surface topography. Therefore, it is essential to investigate the factors governing surface roughness and topography across different manufacturing processes used in biomedical implant studies, ultimately contributing to practical applications.

The manufacturing process employed and the underlying mechanism of material removal or plastic deformation govern the resulting surface topography. For instance,

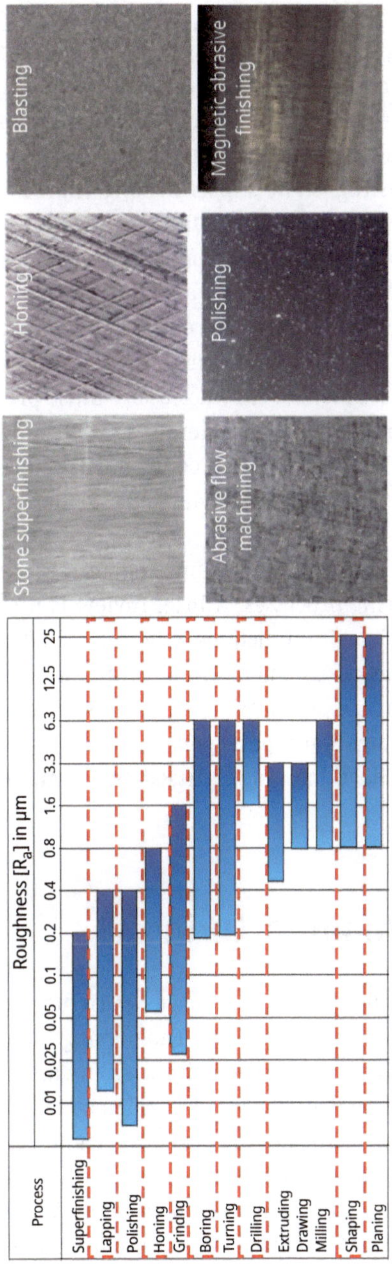

Figure 7.1: Typical roughness values obtained by different manufacturing and finishing processes (left panel). (Right panel) Surface textures generated by different abrasive fine-finishing methods (adapted from Refs. [6, 20]).

a range of surface finish can be achieved in casting depending upon the casting method, mold, and material used. The resulting finish is essentially being replica of the mold surface. Processes involving single- or multipoint cutting tools can have multiple attributes contributing toward the final surface finish such as depth of cut and plastic flow mode controlling material removal [21]. On the other hand, abrasive-based surface finishing processes (e.g., sandblasting, stone superfinishing, blasting, and grinding) pose a higher degree of complexity, wherein the resulting surface is ensemble of stochastic unit phenomena. Thus, a variety of surface topographies can be achieved, contingent upon the specific manufacturing methods and finishing processes. An illustration of surface roughness achieved through different manufacturing techniques is presented in Figure 7.1 (adapted from Ref. [6]), highlighting the limitations in achieving desired roughness. The processes highlighted (boxed) are those predominantly employed in the manufacturing of implants. Consequently, the selection of appropriate manufacturing processes is crucial. Furthermore, controlling surface roughness for specific applications is a critical consideration that warrants further attention.

We begin by discussing general surface characteristics and examining the surface topographies resulting from various manufacturing processes. A brief review is provided on the influence of surface roughness on biological functions, followed by a discussion of standard surface-cell culture studies. Subsequently, conventional and emerging manufacturing methods, as well as surface preparation techniques for biomedical implants are explored. Finally, the fundamental aspects governing the emergence of surface roughness for engineered applications are discussed and summarized.

7.2 Surface characteristics in manufacturing

In practice, all real surfaces comprise peaks and valleys at the microscale, which may be irregular or uniform, and are often imperceptible to the naked eye. Quantifying these features is essential for engineering applications to ensure repeatability, even in ensemble contexts. Surface characteristics can be complex and are typically classified by length scale into roughness, waviness, and form, as schematically illustrated in Figure 7.2 (adapted from Ref. [22]). Roughness refers to the smallest scale surface irregularities, inherent to the manufacturing process, and is generally superimposed on waviness. Waviness results from factors such as machine tool vibration, chatter, workpiece deflection, or material strain. In contrast, form represents the overall shape of the part, disregarding small-scale roughness and waviness.

Let us see this from a practical example. Figure 7.3 illustrates a 3D optical profile of surfaces exhibiting varying roughness, waviness, and form. In Figure 7.3(a), the roughness of a typical rolled sample (Al6061) is shown, with a root mean square

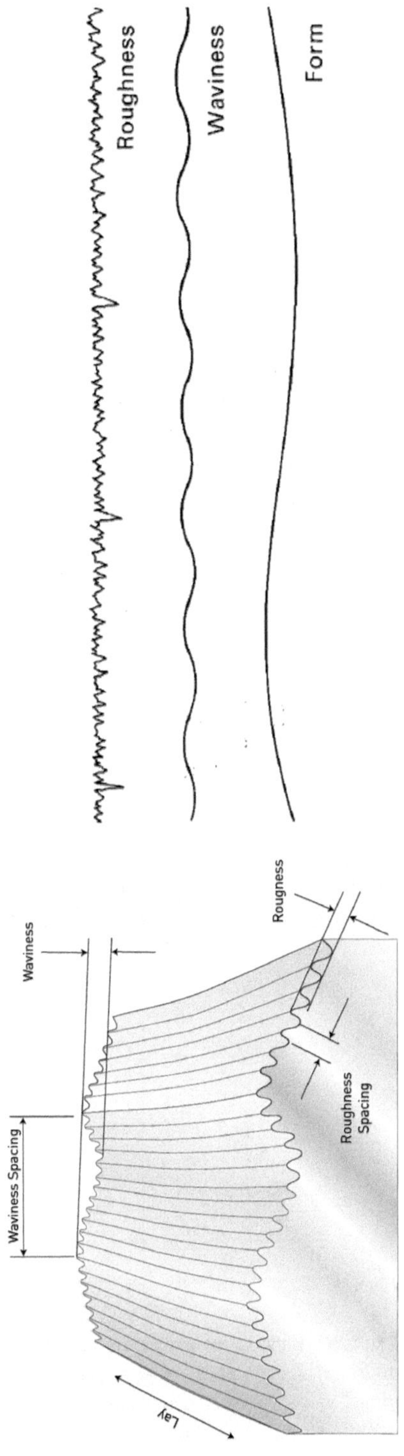

Figure 7.2: Roughness and waviness are the constituents of surface texture, and a surface profile represents the combined effects of roughness, waviness, and form (adapted from Ref. [20, 22]).

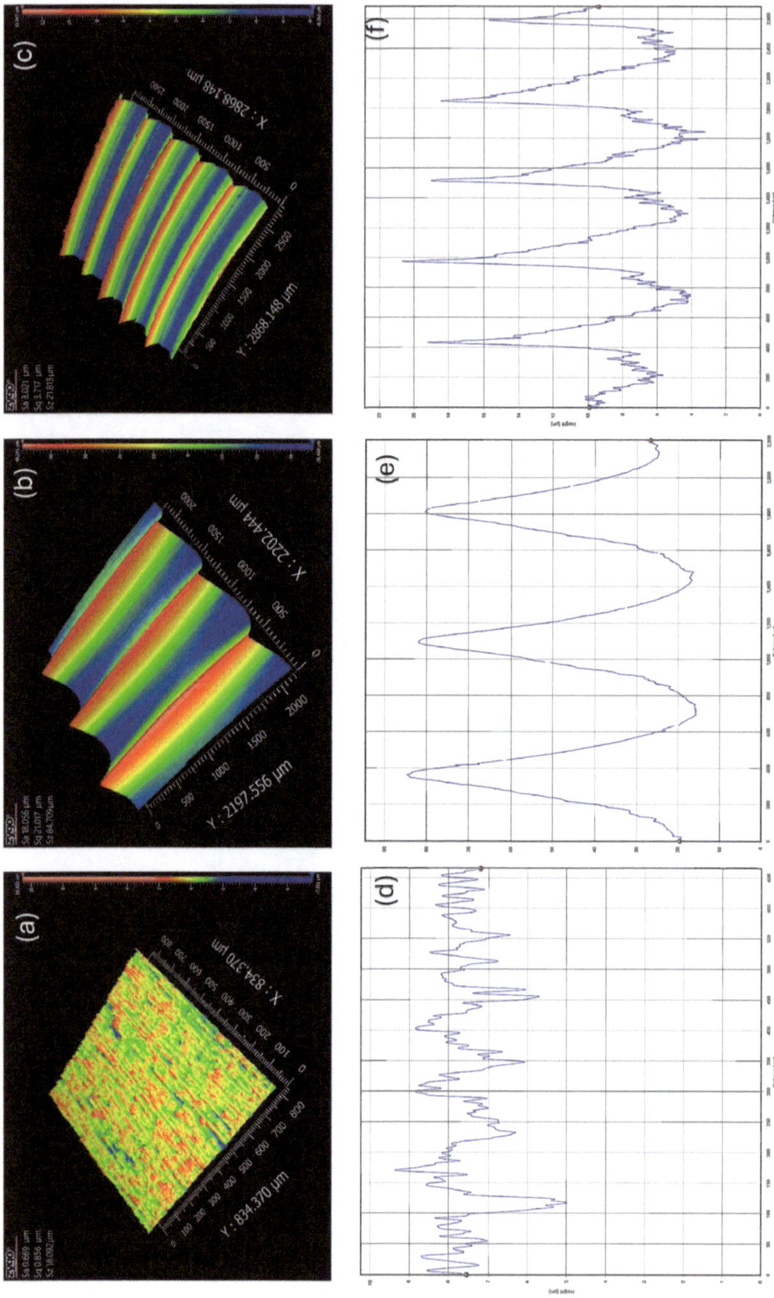

Figure 7.3: The optical profilometry results of three different surfaces, showing the roughness and waviness of (a) a rolled aluminum 6061 samples, (b) a turned surface, and (c) vertical milled surface. The corresponding linear profiles, displaying the roughness profile values, are shown in (d), (e), and (f), respectively.

(RMS) height of ~0.8 µm. Figure 7.3(b) presents a surface obtained through turning, highlighting the waviness, while Figure 7.3(c) depicts a milled surface. The latter two images represent larger length scales and show the imposed roughness between the peaks and valleys of the waviness. Notably, there is a significant difference in roughness between successive peaks in the turning and milling processes. The results are enlisted in Table 7.1, indicating the maximum RMS height in case of turning, followed by end milling and rolling. Beyond peaks and valleys, lay is another critical characteristic inherent to machined surfaces, characterized by its distinct direction. Lay can be described as parallel, perpendicular, crossed, multidirectional, or circular. However, some surfaces, such as those from casting or sintering, exhibit granular features that are challenging to characterize meaningfully.

Table 7.1: Optical profilometry data depicting surface roughness resulting from different manufacturing methods presented in Figure 7.3.

Parameter	(a) Rolled	(b) Turned	(c) End-milled
Peak height (µm)	3.24	48.24	14.11
Valley depth (µm)	7.71	36.47	8.92
Maximum peak to valley height (µm)	10.96	84.71	23.03
Arithmetic mean height (µm)	0.62	18.06	2.84
Root mean square height (µm)	0.83	21.02	3.56

While a 2D description may not adequately capture the complete characterization of a surface, this limitation will be further addressed in subsequent sections. These surface characteristics are equally important in the context of biomedical implant applications. Section 7.3 explores how surface topography influences cell and tissue response, hydrophobicity, and adhesion in biomedical implants.

7.3 Role of surface topography in biomedical implants

Typically, the size of human osteoblasts (~a few microns) is considerably smaller than the surface features produced by machining or abrasive polishing. Machining tends to create regular topographies, whereas abrasive polishing results in more random (chaotic) surface features. These differences in surface topography significantly influence cell adhesion and growth. Studies have shown that chaotic surfaces can lead to reduced cell adhesion and proliferation due to poor contact with focal adhesion plaques [23]. In a study carried out by Anselme and Bigerelle to check the efficiency in discriminating the long-term adhesion of human osteoblast on pure titanium substrate with different surface morphologies prepared using sandblasting, polishing,

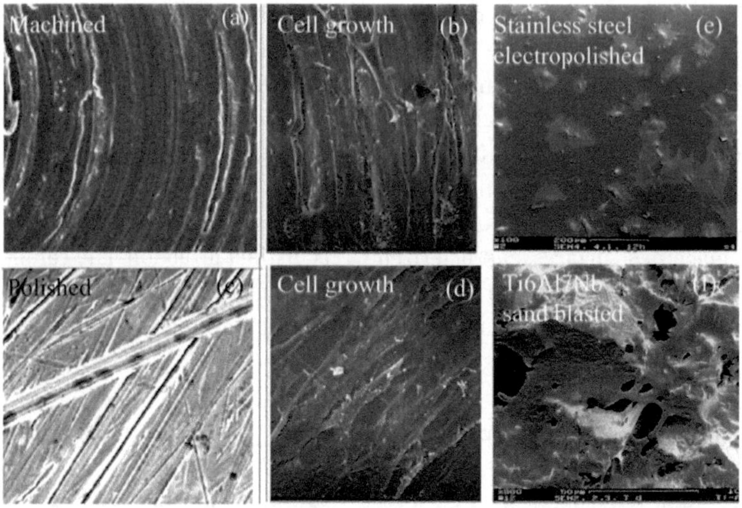

Figure 7.4: Scanning electron microscope (SEM) images of Ti6Al4V surfaces: (a) a machined surface and (b) a surface polished with silicon carbide paper. Corresponding cell growth on these surfaces after 14 days in culture is shown in (c) and (d), respectively (adapted from Ref. [24]). Additionally, cell adherence is depicted on: (e) a smooth electro-polished stainless steel surface with flat adherent cells after 12 h in vitro, and (f) a similar study conducted on a sandblasted Ti6Al7Nb surface (adapted from Ref. [25]).

electro-etching, acid etching, and machining. The results demonstrated that a higher degree of spread is the case of surfaces with low roughness amplitude compared to the rough surface [4].

A scale-independent surface order parameter, based on the statistical bootstrap method rather than average roughness (R_a), is proposed to characterize surfaces. It is suggested that better cell adhesion can be achieved on isotropic surfaces with low order, provided the order is quantified at a scale significant to the size of the cell. Figure 7.4(a)–(d) illustrates the results using a scanning electron microscope, showing cell growth after 14 days on machined and polished titanium substrates [24]. The cell growth on these surfaces appears uniform, with several filamentous extensions that do not conflate.

In another cell culture study by Schmidt et al. [25], different material substrates such as pure titanium, Ti6Al4V, Ti6Al7Nb, and stainless steel provide morphological evolution of cell adherence for orthopedic applications. These surfaces are prepared using electropolishing, vibratory grinding, and sand blasting. Scanning electron microscopic-based time-evolution analysis revealed that the cell growth on smoother surfaces occurred in a flattened, fibroblast-like manner. In contrast, titanium alloys with rough surface characteristics, prepared using sandblasting, induced a 3D growth characterized by the formation of multilayered cell structures with increased cellular depth and complexity, as shown in Figure 7.4(e) and (f). These studies indicate a strong correlation between implant surface roughness and cell culture, suggesting that rougher surfaces promote enhanced cell proliferation and 3D growth, while smoother surfaces encourage flattened, fibroblast-like cell growth. This highlights the importance of tailoring surface roughness to optimize cellular responses and improve implant integration. In addition to roughness, surface features like grooves and ridges (waviness), and curvature (form) play a significant role in protein folding, accumulation, and orientation, as well as in receptor-ligand accessibility [26]. Therefore, manufacturing and surface preparation of biomedical implants are crucial in providing a natural environment for osteoblast and as a starting point for successful osseointegration of an implant. Section 7.4 reviews various methods used in the manufacturing of biomedical implants, highlighting their advantages and limitations.

7.4 Manufacturing and surface preparation

Conventionally, a significant portion of implants are produced using casting, sintering, and machining processes, with subsequent surface modifications achieved through coatings or abrasive finishing techniques [27]. Consequently, understanding the evolution of surface roughness in as-produced components is crucial. As discussed in Section 7.1, and illustrated in Figure 7.1 (left panel), a quantitative representation of the roughness (R_a) resulting from various manufacturing and surface finishing operations is provided, with particular emphasis on processes relevant to the production and preparation of biomedical implants. It is important to note that the parameter R_a alone may be misleading and does not offer a comprehensive description of a surface. It solely reflects the average height of the peaks and valleys along a given direction. The full surface characteristics, shaped by different manufacturing and finishing processes, may exhibit area-dependent features, as shown clearly in Figure 7.1 (right panel), which presents optical images of surface textures derived from various finishing operations, including stone-based superfinishing, sandblasting, abrasive flow machining, honing, polishing, and magnetic abrasive finishing. The average surface roughness parameter (S_a), which represents the average peak-to-valley height distributed over a 3D surface, provides a more comprehensive understanding of the resulting surface. Surfaces produced through casting or forging often exhibit variable roughness, which can vary significantly due to the interface phenomena occurring at the two contacting surfaces.

Casting is one of the most widely used methods for producing complex structures from both ferrous and nonferrous metals, including those used in biomedical implants. The process involves the use of predesigned mold, which can be either temporary or permanent, and is typically fabricated from refractory-based materials (e.g., sand or clay) for high-melting-point alloys or metals (such as steel or nickel-based alloys), or from other materials suited for low-melting-point alloys. During casting, liquid metal is poured into the mold cavity, with the assumption that the surface topography of the mold will be replicated. However, variations in the resulting surface can

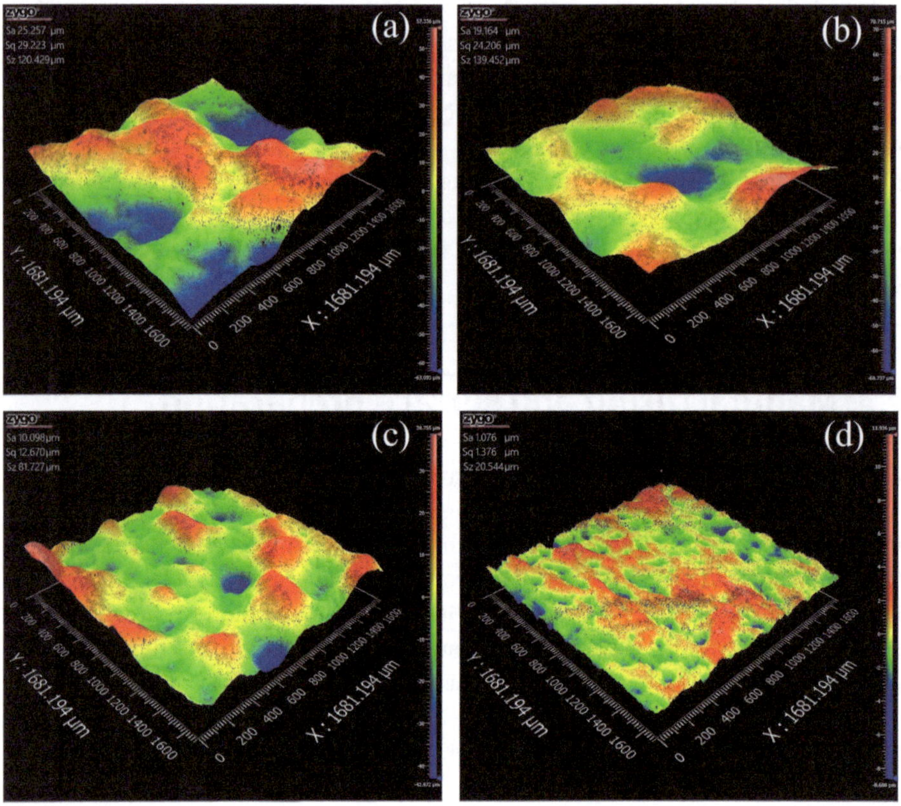

Figure 7.5: Optical profilometry results of standard surfaces produced via different casting methods: (a) temporary sand casting, (b) permanent sand casting, (c) investment casting, and (d) die casting.

occur due to factors such as shrinkage and the viscosity of the poured metal. The type of mold used influences the resulting surface roughness, which can be observed in the optical profilometry results of four standard surfaces created using different molds (see Figure 7.5). The surface produced by a temporary sand mold exhibits the highest roughness, with an average surface roughness (S_a) of ~25 µm, compared to ~20 µm for surfaces produced using a permanent mold (see Figure 7.5(a) and (b)). Investment casting results in surfaces with an average roughness of ~10 µm, as shown in Figure 7.5(c). The smoothest surfaces achievable via casting are obtained using metal molds, commonly referred to as die casting, which can yield a surface profile with an S_a of approximately 1 µm, as demonstrated in Figure 7.5(d). Therefore, selecting the appropriate casting method is crucial, depending on the required surface roughness and the material limitations. Additionally, free-form fabrication techniques, such as additive manufacturing, offer an alternative approach where the resulting surface is not influenced by material-die interactions.

Additive manufacturing is an emerging technology with transformative potential for the production of biomedical implants, largely due to its capability for patient-specific customization. The integration of tomography further enhances this potential by enabling the recreation of 3D designs tailored to individual patients. The selective deposition of material allows for spatially tailored functionality. Additive manufacturing supports a wide range of biocompatible materials, including metals, polymers, and ceramics. Figure 7.6(a) presents a schematic representation of metal-based laser-directed energy deposition, where a high-power infrared laser beam is used to melt and deposit incoming powder onto a substrate. Material is deposited in tracks, and successive overlapping tracks form a solid part.

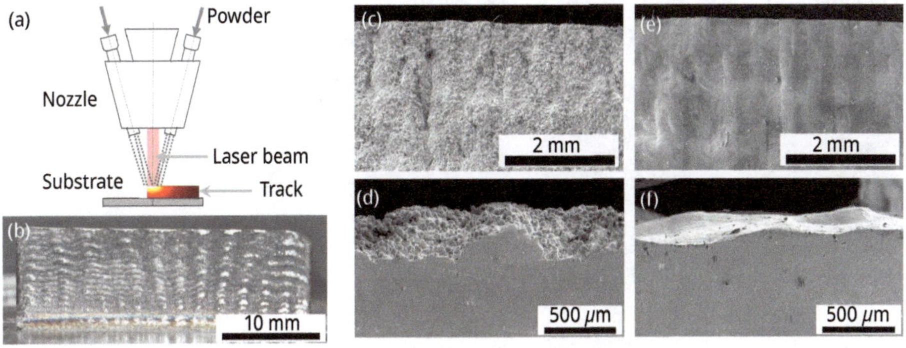

Figure 7.6: Schematic showing the material deposition in laser-directed energy deposition (DED) additive manufacturing technology (a), and (b) a multitrack solid structure printed using DED showing the waviness on the surface. (c) and (e) Top view of as-deposited surface with satellite powders, cleaned surface, and respective cross section on (d) and (f).

The layer-by-layer deposition process is inherent to additive manufacturing results in surface waviness, which can be minimized to some extent but cannot be eliminated. Figure 7.6(b) illustrates a printed part where the waviness from successive layers is visible. Similarly, processes that use powder as a feedstock tend to produce surfaces with satellite powder particles, contributing to surface roughness, as shown in Figure 7.6(c) and (d). However, this surface roughness can be modified through postprocessing techniques such as laser cleaning, sandblasting, and abrasive polishing. Figure 7.6(e) and (f) provides an example where the printed surface has been polished using a laser.

Postprocessing or finishing is an additional step often required for both conventionally manufactured parts (such as those produced by casting, forging, or machining) and laser-based additive manufacturing. This step typically involves the use of fine abrasive tools, such as those used in surface grinding or honing for regular shapes, or free-flowing abrasive slurries for more complex structures. The resulting surface topography is primarily influenced by factors such as the size and hardness of the abrasive grit. Furthermore, the surface texture is determined by the specific finishing op-

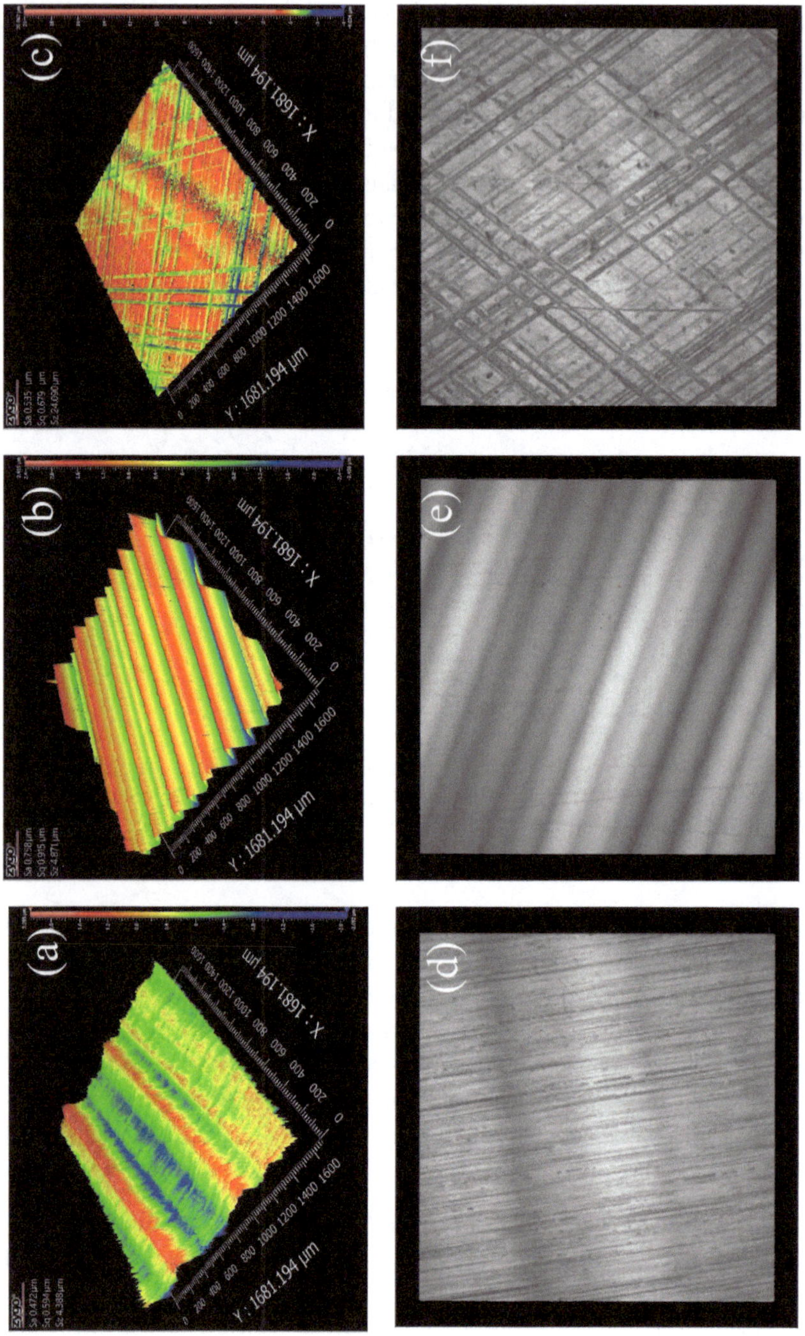

Figure 7.7: Standard surface topographies resulting from different finishing processes: (a) surface grinding, (b) honing, and (c) Blanchard grinding, with corresponding optical images in (d), (e), and (f), respectively.

erations employed. For example, processes such as surface grinding or honing tend to produce regular patterns, as observed in the optical profilometry results shown in Figure 7.7(a) and (b), with corresponding optical images in (d) and (e). It is important to note that the direction of the resulting surface patterns (or lays) depends on the tool path used during these operations. Blanchard grinding, which utilizes a rotating magnetic base to hold the workpiece and an abrasive wheel rotating on a fixed axis, produces cross-lay marks, as shown in Figure 7.7(f) and the corresponding profilometry image in Figure 7.7(c). In contrast, processes involving free-flowing abrasives typically yield a granulated finish. The emergence and evolution of surface roughness in these manufacturing and finishing processes is an important area of investigation that requires a deeper understanding of the underlying mechanisms. This knowledge is crucial for understanding surface topographies across different processes and for the selection and customization of processes to achieve the desired functional properties. A more comprehensive discussion of these aspects is mentioned in Section 7.5.

7.5 Emergence and control of surface roughness

The emergence and evolution of surface topography in different manufacturing processes is a complex question that requires careful consideration. It can be broadly categorized into two main groups: processes involving material removal (such as machining and abrasive processing) and those that do not involve material removal, such as casting, forging, and sheet metal forming. Strictly speaking, large-scale plastic deformation and, to some extent, ductile fracture play a critical role in the evolution of surface roughness. To fully understand the morphology and contours of surfaces, it is essential to investigate several factors, including their deformation behavior under pressure, the adhesion characteristics of solids, the mechanical properties of the interface, the influence of surface films, and the energy dissipation associated with surface deformation during sliding.

The resulting surface roughness is closely linked to the specific manufacturing processes employed, which can typically be characterized based on the fundamental mechanisms of material removal or plastic deformation. For instance, in bulk deformation processes, such as forging, where a hard tool is pressed against a softer material, the asperities of the harder material are expected to be replicated onto the softer counterpart. However, it is noteworthy that the asperities on the softer material often remain unchanged, even under pressures exceeding the yield stress of the material [28]. It continues until the bulk plastic deformation region expands, eventually flattening the asperities. In sheet metal forming, surface texture develops because of grain reorientation, rotation, or sliding. This leads to a bubbly surface appearance, commonly known as the orange peel effect. Thus, resulting roughness is a function of initial grain size or microstructure in general. A comprehensive discussion on the emergence of surface roughness in engineered surfaces is provided in Ref. [29].

In contrast, abrasive-based finishing operations involve numerous unit actions occurring at each abrasive-workpiece contact. The complexity of understanding surface topography evolution in these processes necessitates an in-depth analysis of the unit interactions responsible for material removal (chip formation) or plastic deformation. The surface finish achieved is directly influenced by the depth of cut (or uncut chip thickness), which is generally smaller in abrasive finishing processes. Consequently, the specific energy expended per unit of material removed to achieve a smoother surface is significantly high [6]. It is noteworthy that the average peak-to-valley height during abrasive finishing processes is a monotonic function of time. The final steady-state surface roughness (R_a) may be higher or lower than the initial workpiece surface, depending on the abrasive grit size. Furthermore, the resulting surface topography is typically independent of the microstructural characteristics. This independence allows for the modeling of surface roughness evolution, thereby enabling the design of processes aimed at achieving targeted surface topographies, particularly for biomedical applications.

7.6 Summary

Surface topography plays a critical role in biomedical implants, as it directly influences cell adhesion and proliferation. Investigating the correlation between surface texture at various length scales and biocompatibility remains an active area of research. Traditionally, these surfaces are prepared using single-tool operations such as cutting and turning, or through multipoint material removal methods such as abrasive finishing or sandblasting. Understanding the emergence and characteristics of the surface in these processes is crucial for effective surface preparation. Free-form fabrication methods, such as laser-based additive manufacturing, offer a promising avenue for customizing and producing parts based on patient-specific requirements. However, this process inherently introduces surface textures that necessitate additional post-processing steps before the final application. This approach opens opportunities to fabricate parts with tailored properties using a wide range of materials.

References

[1] Köunönen, M., Hormia, M., Kivilahti, J., Hautaniemi, J., & Thesleff, I. (1992). Effect of surface processing on the attachment, orientation, and proliferation of human gingival fibroblasts on titanium. *Journal of Biomedical Materials Research*, *26*(10), 1325–1341.

[2] Li, D., Liu, B., Wu, J., & Chen, J. (2001). Bone interface of dental implants cytologically influenced by a modified sandblasted surface: A preliminary: In vitro: Study. *Implant Dentistry*, *10*(2), 132–138.

[3] Anselme, K., Ploux, L., & Ponche, A. (2010). Cell/material interfaces: Influence of surface chemistry and surface topography on cell adhesion. *Journal of Adhesion Science and Technology*, *24*(5), 831–852.

[4] Anselme, K., & Bigerelle, M. (2005). Topography effects of pure titanium substrates on human osteoblast long- term adhesion. *Acta Biomaterialia*, *1*(2), 211–222.

[5] Sul, Y. T., Kang, B. S., Johansson, C., Um, H. S., Park, C. J., & Albrektsson, T. (2009). The roles of surface chemistry and topography in the strength and rate of osseointegration of titanium implants in bone. *Journal of Biomedical Materials Research Part A: An Official Journal of the Society for Biomaterials, the Japanese Society for Biomaterials, and the Australian Society for Biomaterials and the Korean Society for Biomaterials*, *89*(4), 942–950.

[6] Hashimoto, F., Yamaguchi, H., Krajnik, P., Wegener, K., Chaudhari, R., Hoffmeister, H. W., et al. (2016). Abrasive fine-finishing technology. *CIRP Annals*, *65*(2). 597–620.

[7] Zhang, T., Wang, W., Liu, J., Wang, L., Tang, Y., & Wang, K. (2022). A review on magnesium alloys for biomedical applications. *Frontiers in Bioengineering and Biotechnology*, *10*, 953344.

[8] Riaz, U., Shabib, I., & Haider, W. (2019). The current trends of Mg alloys in biomedical applications – A review. *Journal of Biomedical Materials Research Part B: Applied Biomaterials*, *107*(6), 1970–1996.

[9] Niinomi, M. (2008). Mechanical biocompatibilities of titanium alloys for biomedical applications. *Journal of the Mechanical Behavior of Biomedical Materials*, *1*(1), 30–42.

[10] Rack, H., & Qazi, J. (2006). Titanium alloys for biomedical applications. *Materials Science and Engineering: C*, *26*(8), 1269–1277.

[11] Palani, S., Michael, E. G., Desta, M., Atnaw, S. M., Banoth, R., & Kolanji, S. (2022). Physical vapor deposition coating process in biomedical applications: An overview. *Sustainable Advanced Manufacturing and Materials Processing*, 1(1) 67–93.

[12] Verma, P. K., Singh, S., Kapoor, M., & Singh, S. (2024). A review on the surface topography and corrosion behavior of Mg-alloy coatings for biomedical implants. *Results in Surfaces and Interfaces*, 15, 100227.

[13] Ellermann, E., Meyer, N., Cameron, R. E., & Best, S. M. (2023). In vitro angiogenesis in response to biomaterial properties for bone tissue engineering: A review of the state of the art. *Regenerative Biomaterials*, *10*, rbad027.

[14] Bicho, D., Pina, S., Oliveira, J. M., & Reis, R. L. (2018). In vitro mimetic models for the bone-cartilage interface regeneration. *Osteochondral Tissue Engineering: Challenges, Current Strategies, and Technological Advances*, 1059, 373–394.

[15] Senin, N., & Groppetti, R. (2005). Surface microtopography design and manufacturing through topography descriptors: An application to prosthetic implant surfaces. *Computer-Aided Design*, *37*(11), 1163–1175.

[16] Chen, R., Chang, R. C., Tai, B., Huang, Y., Ozdoganlar, B., Li, W., et al. (2020). Biomedical manufacturing: A review of the emerging research and applications. *Journal of Manufacturing Science and Engineering*, *142*(11). 110807.

[17] Emelogu, A., Marufuzzaman, M., Thompson, S. M., Shamsaei, N., & Bian, L. (2016). Additive manufacturing of biomedical implants: A feasibility assessment via supply-chain cost analysis. *Additive Manufacturing*, *11*, 97–113.

[18] Niinomi, M. (2019). *Metals for Biomedical Devices*, Woodhead Publishing, Sawston, Cambridge, United Kingdom.

[19] Murr, L. E., Gaytan, S., Medina, F., Lopez, H., Martinez, E., Machado, B., et al. (2010). Next-generation biomedical implants using additive manufacturing of complex, cellular and functional mesh arrays. *Philosophical Transactions of the Royal Society A: Mathematical, Physical and Engineering Sciences*, *368* (1917). 1999–2032.

[20] Dagnall, H. (1986). *Exploring Surface Texture*, Rank Taylor Hobson, Leicester, UK, 20 (edition 2).

[21] Udupa, A., Sugihara, T., Viswanathan, K., & Chandrasekar, S. (2019). Altering the stability of surface plastic flow via mechanochemical effects. *Physical Review Applied*, *11*(1), 014021.

[22] Hobson, T. (2003). Exploring surface texture: A fundamental guide to the measurement of surface finish. *Leicester–Inglaterra, 4th Ed, Taylor Hobson Ltda*, *100*, 9.
[23] Anselme, K., & Bigerelle, M. (2011). Role of materials surface topography on mammalian cell response. *Inter- National Materials Reviews*, *56*(4), 243–266.
[24] Bigerelle, M., Anselme, K., Dufresne, E., Hardouin, P., & Iost, A. (2002). An unscaled parameter to measure the order of surfaces: A new surface elaboration to increase cells adhesion. *Biomolecular Engineering*, *19*(2–6), 79–83.
[25] Schmidt, C., Kaspar, D., Sarkar, M. R., Claes, L. E., & Ignatius, A. A. (2002). A scanning electron microscopy study of human osteoblast morphology on five orthopedic metals. *Journal of Biomedical Materials Research*, *63*(3), 252–261.
[26] Harvey, A. G., Hill, E. W., & Bayat, A. (2013). Designing implant surface topography for improved biocompatibility. *Expert Review of Medical Devices*, *10*(2), 257–267.
[27] Abellán-Nebot, J. V., Siller, H. R., Vila, C., & Rodríguez, C. A. (2012). An experimental study of process variables in turning operations of Ti–6Al–4V and Cr–Co spherical prostheses. *The International Journal of Advanced Manufacturing Technology*, *63*, 887–902.
[28] Bowden, F. P., & Tabor, D. (2001). *The Friction and Lubrication of Solids* vol. 1, Oxford university press, Oxford, UK.
[29] Aghababaei, R., Brodsky, E. E., Molinari, J. F., & Chandrasekar, S. (2022). How roughness emerges on natural and engineered surfaces. *MRS Bulletin*, *47*(12), 1229–1236.

Abhinay Thakur*, Valentine C. Anadebe, Abdelkader Zarrouk, Omar Dagdag, Ashish Kumar, and Virat Khanna

Chapter 8
Nanocoating for medical devices

Abstract: Nowadays, nanocoating is a revolutionary technology that has been introduced to the medical devices field to provide advanced solutions for biocompatibility, antimicrobial, and overall performance. This chapter focuses on the formation of nanocoating and its applications in common medical tools and devices, such as implants, instruments used in surgeries, and diagnostic equipment. The properties of nanomaterials include enormous surface area, enhanced mechanical force, and the ability to modify substrate surfaces and make them ideal nanomaterials for coating. It presents the nanoparticles in medical coatings as nanoparticles, nanostructured surfaces and nanocomposites and focuses on the interactions of these nanomaterials for controlling biologic response and preventing device-related infections and promoting tissue integration. The chapter presents the challenges of the regulatory and safety concerns of nanocoating in medical applications being on the biocompatibility examination and toxicity implications. Lastly, the chapter explores the future trends in nanocoating technology, including the integration of smart materials that can adapt to surrounding conditions and the opportunities of delivering personalized drugs through targeted coatings.

Keywords: Nanocoatings, medical, antimicrobial, nanoparticles, toxicity

***Corresponding author: Abhinay Thakur**, Division of Research and Development, Lovely Professional University, Phagwara 144411, Punjab, India, e-mail: thakurabhinay96@gmail.com

Valentine C. Anadebe, Department of Chemical Engineering, Alex Ekwueme Federal University Ndufu Alike, P. M. B. 1010, Abakaliki, Ebonyi State, Nigeria

Abdelkader Zarrouk, Laboratory of Materials, Nanotechnology and Environment, Faculty of Sciences, Mohammed V University in Rabat, Av. Ibn Battouta, P.O. Box 1014, Agdal-Rabat, Morocco; Research Centre, Manchester Salt & Catalysis, Unit C, 88–90 Chorlton Road, M15 4AN Manchester, United Kingdom

Omar Dagdag, Department of Mechanical Engineering, Gachon University, Seongnam 13120, Republic of Korea

Ashish Kumar, Department of Science, Technology and Technical Education, Nalanda College of Engineering, Bihar Engineering University, Government of Bihar, 803108 Bihar, India

Virat Khanna, University Centre for Research and Development, Chandigarh University, Mohali 140413, Punjab, India

https://doi.org/10.1515/9783111571423-008

8.1 Introduction

Nanocoatings have brought new changes and opportunities for acceptance in the medical device market by enabling the improvements of biocompatibility of the coatings, antimicrobial properties, and device performance. Recent progress in nanotechnology has led to nanocoatings that offer improved functionality and safety due to the fabrication of medical devices at the nanofabrication level [1–3]. As the use of high-performance medical devices continues to rise, many devices are specifically designed to incorporate nanocoatings. Nanocoatings are thin films of nanomaterials that are deposited on the medical tools to impart desired characteristics, such as increased efficiency, mechanical strength, and resistance to microbial colonization.. By controlling nanomaterials, scientists can exploit the unique properties inherent to materials at the nanoscale, allowing for the design of coatings tailored to specific healthcare applications [4, 5]. The unique properties of nanomaterials, such as high surface-to-volume ratio, the ability to introduce versatile functional groups, and the chemical reactivity, make them ideal for use in medical coatings. Nanocoatings play a critical role in medical devices and cannot be overstated.. Medical technology is an industry that has encountered various challenges concerning device-associated infection, material breakdown, and biocompatibility. These and many other problems are not well solved by traditional coatings, which subsequently results in complications that can endanger patient lives and the effectiveness of the underlying medical devices [6]. Nanocoatings therefore present a possible solution to the problem by creating a layer that can be designed to work against bacterial adhesion and promote tissue growth and the general performance of the implant device. However, the use of nanocoatings it is also adds value by negating in part the need for the use of antibiotics and or any other pharmaceutical agents which are very essential in health facilities. It can be also noted that one of the most important characteristics of nanocoatings is their characteristics attributable to the size of the particles used in the creation of nanocoatings. At this scale, materials appear to have different physical and chemical characteristics than the large-scale materials. For example, nanoparticles have much larger surface area to volume ratio, increased reactivity, and provide better interaction with the biological systems. This property is especially important in biomedical devices, where the adventures between the device surface and tissue can determine the proper implantation or deployment of devices [7, 8]. Moreover, this nanoscale structure can also enhance a controlled delivery of therapeutic agents; thus, the materials can be multifunctional and can both provide physical support and release the therapeutic agents at the targeted site. Yet another important feature and property of nanomaterials is the tunability of their properties for certain applications. It is possible to alter the surface properties of nanoparticles to cause coatings to change properties in response to stimuli including pH or temperature as well as enhance biocompatibility with tissues. Figure 8.1 shows the stages and formation of biofilm integrating with nanoparticles on biomedical devices.

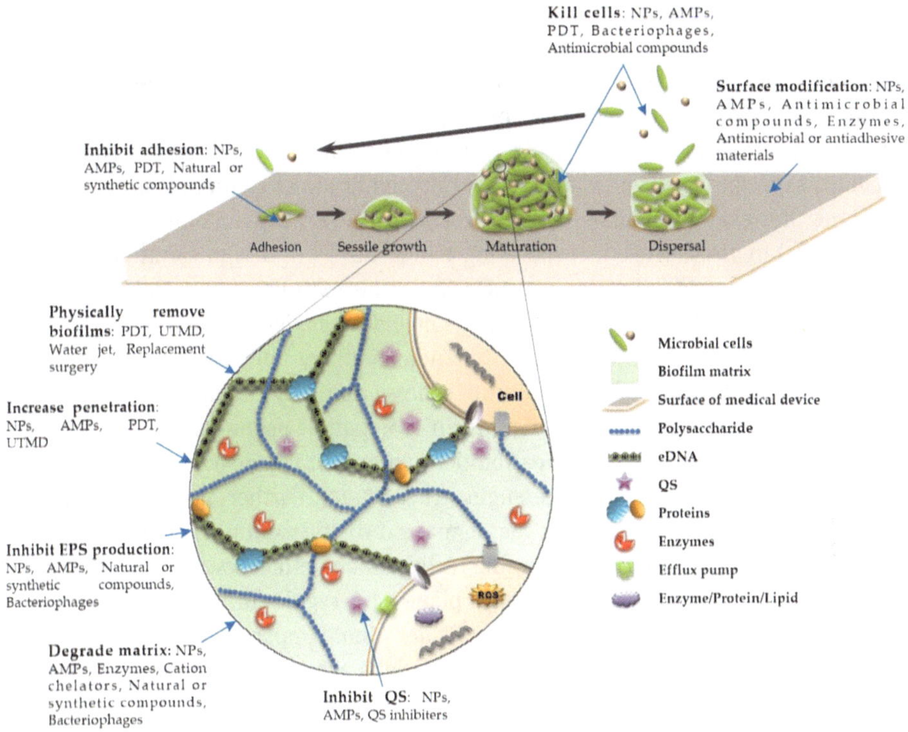

Figure 8.1: Phases of biofilm formation on medical devices and strategies for biofilm prevention and control. Adapted from Ref. [9] under CCBY 4.0.

Computer Numerical Control capabilities cannot be effectively accomplished using conventional materials, which is why nanocoatings are revolutionary when it comes to creating new smart medical devices that can be easily adjusted to meet the needs of patients. In addition, the integration of nanoparticles with antibacterial properties into the technology minimizes the chance of infections related to implanted medical devices and hence enhances the quality of services offered and the costs incurred. It is not just that nanocoating has now become a tool to improve the performance of medical devices by making them biocompatible and antimicrobial [10, 11]. They also enhance the mechanical properties of the devices, whose aspect is vital for the enhancement of performance in specific climates. For instance, nanocoatings can prevent wear and tear and thus ensure the survival of the joints put in place by means of implantation in healthcare practice. This is mainly evident in load-bearing applications where mechanical failure results in a number of adverse effects on patients. Making use of poor and susceptible material for a medical device can lead to its failure and require surgery to replace the faulty device and this can cause the patient discomfort and pain, but through nanocoating, the lifespan of such devices is greatly enhanced by adding a thin and effective shield.

Notably, the aesthetic properties of nanocoatings also have the potential to add value to the parent medical devices. The flexibility in achieving certain desired colors and smoothness of the coating is an added advantage in issues to do with aesthetic value, which are small but could eventually be a factor that would make patients or even the healthcare givers go for a certain make or model of the device. This aspect becomes especially critical in applications involving direct communication with the consumer, for example, dental implants, cosmetic surgery instruments and other similar medical devices. Thus, incorporating nanocoatings in design solutions not only enables the production of effective devices but is also attractive in visual aspects. We expect the future of nanocoating in the medical devices field to grow rapidly due to the advancement and discovery in the nanotechnology field. Sappers are actively working on the new material and coating process that they can apply to medical device enhancement. For instance, combining biocompatible polymers with nanomaterials results in coatings with biocompatibilities that encourage the adhesion and growth of cells enhancing healing in tissue engineering [12–14]. Further, nanotechnology, especially self-cleaning nanocoatings that can avoid the formation of biofilms and contaminants is excellent news in increasing the safety and efficiency of numerous medical applications. Remarkably, the part played by regulation on the concepts of nanocoatings deployment is worthy of consideration. Similar to any new method of applying technology in the medical field, questions arise about the use of nanocoatings in the development of medical devices. Current and future regulatory agencies are more concerned with the provision of standards required in the use of nanomaterials in healthcare products to gain assurance that the products used have undergone stringent tests, particularly regarding biocompatibility and toxicity. This process is important in receiving market approval to allow patients confidence in the safety of the devices in use. With more introduction of the material, there will be a need to draw top-notch collaboration between the researchers, manufacturers, and regulatory bodies in setting standards for nanocoatings in the devices. Table 8.1 provides a quick overview of various nanocoatings, highlighting their specific applications in medical devices along with their respective advantages and disadvantages [15–24].

Table 8.1: Overview of various nanocoatings used on medical devices, highlighting their specific applications, advantages, and disadvantages.

Nanocoating material	Medical device	Advantages	Disadvantages
Silver nanoparticles	Orthopedic implants	– Excellent antibacterial properties – Reduces risk of biofilm formation	– Potential cytotoxicity – May develop resistance in bacteria

Table 8.1 (continued)

Nanocoating material	Medical device	Advantages	Disadvantages
Chitosan coating	Catheters	– Biocompatible – Antimicrobial effects – Promotes cell adhesion	– Limited mechanical strength – Susceptible to enzymatic degradation
Hydroxyapatite	Bone grafts	– Promotes osteoconductivity – Enhances osseointegration	– Limited antibacterial properties – May be brittle under stress
Titanium dioxide (TiO_2)	Dental implants	– Biocompatible – Antimicrobial properties – Promotes cell attachment	– Limited effectiveness in wet environments – May require UV activation
Silk fibroin	Surgical sutures	– Biodegradable – Promotes tissue regeneration – Low inflammatory response	– Mechanical properties may not be suitable for all applications
Zinc oxide (ZnO)	Wound dressings	– Antimicrobial properties – Promotes wound healing	– Potential skin irritation – Stability issues in humid environments
Niosomes	Orthopedic implants	– Sustained drug release – Biocompatible – Reduced risk of infection	– Complex formulation process – Stability concerns under physiological conditions

8.2 Types of nanomaterials used in medical coatings

8.2.1 Nanoparticles

Nanoparticles are particulate materials with sizes ranging from 1 to 100 nm. As a result of their nanoscale dimensions, they have a high surface area-to-volume ratio, a higher catalytic activity, and novel optical and electrical response. These properties make nanoparticles with many applications in medical coatings, particularly in enhancing biocompatibility and antimicrobial action [25–27]. Antimicrobial properties are one of the key applications of nanoparticles in medical coatings. Silver, gold, and copper metallic nanoparticles are some of the most investigated particles because of their antimicrobial properties. Of these, silver nanoparticles are especially widespread; these nanoparticles possess antibacterial properties and are effective against a broad range of pathogens. This makes silver nanoparticles very important in providing coatings for medical applications, including catheters, implants, and surgical instruments, where there is an increased likelihood of acquiring an infection. The use of silver nanoparticles in the surface coatings of these devices would help the manufacturers minimize the incidences of those device-borne infections and therefore enhance the health of the patients. The second critical area that greatly benefits from the use of nanoparticles in medical coatings is the delivery of drugs [28, 29]. It was demonstrated that nanoparticles can be designed to incorporate the therapeutic agents, thus providing a time and targeted controlled-release delivery system. For example, polymeric nanoparticles may be programmed to deliver medicine only upon exposure to certain conditions like chemical or physical conditions like pH or temperature respectively. This property is especially useful when localized therapy is desired, for example in the treatment of cancer or constructing tissue scaffolds. Cohosting of nanoparticles has the potential of increasing the positive effect of the drugs while reducing their impact on the side effects, thus making treatment much more effective and safe. In addition, medical coatings can also be improved by means of nanoparticles in terms of mechanical characteristics. For instance, incorporation of silica or titanium dioxide nanoparticles in the polymeric coatings enhances their toughness, service life, and wear resistance. This improvement is significant in uses including orthopedic implants and dental structures since the firmness of the item is vital. Using nanoparticles offers enhanced mechanical properties such as improved strength and wear resistance which would translate to longer implant service life and decreased rate of implant revisions hence the quality life of the patients will be enhanced.

Moreover, nanoparticles possess mechanical and antimicrobial properties, which are supplemented by unique optical features that are suitable for use in medical diagnosis. For instance, gold nanoparticles are used in imaging procedures because of their capability to improve contrast in imaging techniques like CT and MRI [30, 31]. The drawback however, is that it can be challenging to obtain information about the location of a particular disease or tumor since it is not always easy to target a particular tissue or a cell through an imaging agent; however, this can be achieved by simply attaching targeting ligands to the gold nanoparticles. It is evident that the application of nanoparticles in medical coatings presents significant benefits, notwithstanding, there are difficulties as well. The first is the toxicity of nanoparticles; the size and higher reactivity in comparison with large-scale particles may cause negative effects on organisms. As with any material that is used in the treatment of patients, it is very important to ascertain that nanoparticles used in medical coatings do not have any adverse effects [32, 33]. Every day there are growing concerns about setting environmental standards for the proper application of nanoparticles especially in the medical field since they affect human lives which were not practically considered in the process of the development of nanoparticles.

8.2.2 Nanostructured surfaces

Another type of nanomaterial that received a lot of attention in the area of medical coatings is nanostructured surfaces. This is because these surfaces are defined by their topography and these parameters can; be designed at the nanoscale to control biological responses. Surface features characterizing solidity or porosity, roughness or smoothness, are pivotal in the manner organisms intersect with medical tools and how they modify cell attachment and growth. The first advantage of nanostructured surfaces for application in cells is that it improves cell adhesion rate as well as their growth [34–36]. The structural arrangement of the surface can be designed to produce functional micro- and nanoscale re-entrants that replicate the extracellular matrix (ECM), the external cellular environment. The success of any cell surface interaction depends on the topographical features of the surface; therefore, researchers can opt to incorporate surface characteristics that favor cell attachment and proliferation in areas such as tissue engineering and regenerative medicine. For instance, through the conversion of the nanostructures surfacing polymers and/or ceramics, one can design a scaffold for the growth of new tissues to help in the healing and regeneration of damaged tissues. In the context of cell adhesion, nanostructured surfaces can also affect the actions of immune cells as well [37]. This chapter examines the relationship that takes place between the immune system and the medical devices in order to support the implantation and other related medical equipment. Anti-inflammatory properties are shown to be achievable when nanostructured surfaces control immune responses improving the environment to promote healing. For example, specific surface treatments that prevent protein and cell attachment significantly allow for low inflammation and enhanced biocompatibility [38, 39]. This property becomes particularly important in cases of long-term implantation, in which the activity of the device and the corresponding host tissue may determine the effectiveness of the surgery.

However, nanostructured surfaces can also be used to improve the antimicrobial characteristics of medical coatings. The enhanced area in contact with the adjacent surface and the altered surface morphology make it harder for bacteria to stick and form a biofilm that causes subsequent infections. Perhaps, the most investigated approaches to produce nanostructured surfaces are laser ablation, electrospinning, and chemical etching. These techniques help to exert a good level of feature control and their ability to provide optimal antibacterial function yet biocompatibility can be achieved. The extendibility of nanostructure surfaces for use in medical coatings further offers the possibility for smart surfaces that can react to a change in environment. For instance, scientists are concerned with surfaces that are adaptive to changes in temperature, pH, or any specific biomolecules [40, 41]. Such activity can improve the capability of various types of medical devices, and provide the device with the functions of a specific drug delivery, or the function of cleaning itself. This kind of possibility of designing medical coatings with the ability to change with the surrounding environment was a major boost to the medical coating industry. Still, some problems are associated with the application of nanostructured surfaces, being one of the numerous beneficial methods. The deposition of nanostructures on substrate surfaces is relatively a complex process that may need expensive technologies and tools. However, one drawback with these surfaces is their stability in physiological conditions over the long term. Achieving these structures must be done in a way that retains functionality over time, especially in the context of their interaction with living organisms.

8.2.3 Nanocomposites

Nanocomposites can be defined as materials that consist of nanoparticles reinforced by a matrix material of polymer, metal, or ceramic. One of the great characteristics of nanocomposites is increased stiffness and improved response of the material when nanoparticles are incorporated into a matrix, making medical coatings an important application of these materials. Due to the size and structure of nanoparticles, the researchers can create coatings that possess enhanced mechanical, thermal, and biological performance [42–44]. Promising properties of nanocomposites such as improved mechanical properties and durability make them suitable for use in medical coatings. The reinforcement of the matrix material by incorporating nanoparticles like Silica, clay, or carbon nanotubes will enhance or improve on wear, impact, or deformation resistance of coatings. This enhancement is noteworthy in application that the mechanical properties of the device are very essential in the performance of the product such as orthopedic implants, dental materials, and surgical instruments [45–47]. Nanocomposites can help increase the mechanical properties of coatings prepared therefore increasing the life span of the devices without failure during usage. Beyond mechanical strength, there is the possibility that the nanocomposites confer antimicrobial characteristics to medical coatings. For example, by dispersing into polymer matrix silver nanoparticles, one can obtain a composite material that will, on one hand, provide structural support, and on the other – exhibit the antimicrobial effect. This bifunctional property is of particular value in cases where bacterial colonization is a concern, and methods for avoiding this event are also required to preserve the mechanical stability of the artificial limb. The authors believe that the matrix materials and nanoparticles are flexible entities that scientists are working on to find new optimal materials for particular medical uses of nanocomposites [48, 49].

Thermal properties are another important factor that forms the basis of the design of medical coatings. A nanocomposite can be specially tailored to enhance its thermal stability and thermal conductivity depending on the medical applications. For instance, thermally conductive nanocomposites can be applied in electronic medical tools including sensors and diagnostic applications where heat dissipation is critical for functionality. The integration of thermally conductive nanoparticles can improve the general performance of these devices so that they would properly work in the medical environment. Another important consideration of nanocomposites is related to their biocompatibility with the given application in the medical coating field. Currently, there is a great interest in the development of different matrix materials and nanoparticles for manufacturing bio-inspired composites with both high mechanical and biological performance. Chitosan and collagen are majorly preferred materials for matrices in natural polymers because of their biocompatibility for cell growth [50, 51]. When these natural polymers are blended with nanomaterials, scientists can develop nanocomposites that will enhance tissue adhesion and excellent healing properties alongside enhanced mechanical properties. In addition, nanocomposites have lucrative prospects as carriers for site-specific drug deployment. Thus, the co-encapsulation of therapeutic agents within nanoparticles can precede the preparation of coatings that are capable of controlled drug delivery. This capability is most useful in the use of drug-eluting stents and orthopedic implants where the concentration of the drug enables good therapeutic response while minimizing the side effects to the rest of the body. Nanocomposite coatings for drug delivery applications call for controlled drug release rates and compatibility between the drug, particles, and matrix.

8.3 Fabrication techniques for nanocoatings

8.3.1 Sol-gel processes

Sol-gel process is one of the well-known methods of nanocoating synthesis that is quite flexible. The technique involves the transformation of a sol (a colloidal suspension of fine solid particles in a liquid) into a gel phase and then into a solid mass. The sol-gel process is especially beneficial in the production of coatings with desired porosity, surface profile, and chemical content to be appropriate for medical use. The sol-gel process involves the preparation of a precursor solution identified as metal alkoxides or metal salts in a solvent. These precursors immobilize by hydrolysis and polycondensation into forming the three-dimensional web of inorganic polymers. The control over the hydrolysis and condensation reactions makes it possible to obtain nanostructured coatings possessing certain characteristics [52, 53]. For example, it is possible to control the rate of reaction such as pH, temperature and concentration of precursors thus it would be easy to control the characteristics of the final coating. In the field of coatings, one of the important benefits of the sol-gel process is to synthesize uniform films on metals, ceramics and polymers. The sol-gel technique can be employed depending on the deposition method; dip coating, spin coating, and spray coating. In the case of dip-coating, the substrate is totally dipped in the sol solution and then withdrawn at a certain speed and thus providing a uniform layer thickness. Spin-coating is a process of applying a small volume of sol on a substrate followed by spinning at high RPM in order to spread the solution uniformly on the interface. It is essential to note that both methods exhibit thin film thickness and uniformity control to high standards, and are particularly suitable for the medical device sector. This process also allows for the addition of bioactive elements into the nanocoatings successfully [54–56]. By doping sol-gel precursor solution with nanoparticles like silver or gold, it is possible to make coatings that have antibacterial properties. These bioactive coatings can be incorporated into different medical applications that include implants, surgery tools as well as catheters to enhance bioavailability and reduce surface infection. Hence, due to the special features such as tunability of the sol-gel process in terms of the type of materials used and incorporated property, sol-gel processing of nanocoatings has found great applicability in the medical field.

Also, the sol-gel process makes it possible to achieve good adhesion of resulting coatings to the substrate, which again is important for the stability and functionality of medical devices. The adhesion between the sol-gel coating and the substrate can be improved through modifications that must guarantee their durability during application. This feature is highly relevant in the areas like orthopedics implantations where mechanical properties play a key role in order to guarantee the patient's safety. The sol-gel process has specific benefits on the other hand; there are some issues with its application on coatings. Also, it has wide applicability in the mass production of snacks. The mentioned synthesis method of the sol-gel process can be still time-consuming; furthermore, getting homogenized coatings on large areas could still need further optimization of parameters. Apart from the sol-gel process, the drying and curing parameters also play a significant role in determining the properties of the prepared coatings and therefore call for proper regulation.

8.3.2 Chemical vapor deposition

Another well-known method that has garnered significant use in the process of creating nanocoatings including in the field of medicine is chemical vapor deposition (CVD). CVD techniques are based on chemical reactions carried out at relatively low pressure where thin films are produced from gaseous precursors by depositing them on a substrate. This process also provides an opportunity to form appreciable, dense and uniformly dense coatings for optimum adhesion that can be useful in many medical applications [57, 58]. The CVD process initiates with the injection of gaseous precursors into a reaction chamber bearing the substrate. Such precursors include organometallic compounds, metal halides and other volatile compounds that may be employed. At certain temperature and pressure levels, the precursors decompose on the substrate surface and chemically react to form a thin solid layer. This makes it possible to achieve the type of coating required in terms of thickness, composition, and microstructure as a result of high control over deposition parameters such as temperature, pressure, and gas flow rates [59–61]. Perhaps the major strength of this technique lies in its potential to deliver conformal films on various structures. Since the precursors for the layers are gases, they are capable of reaching many small forms and contours and provide an even layering even with a difficult substrate. This feature proves especially useful when coating components with intricate and irregular shapes, for example, implantable and prosthetic devices where uniform coating thickness is critical for best performance. CVD is also versatile with regard to the material which it can deposit such as metals, oxides, nitrides, and carbides. This flexibility in the process results in the ability to create nanocomposite coatings with specific characteristics suited for specific medical uses. For example, using precursors that produce materials that possess high biocompatibility, including titanium oxide or hydroxyapatite (HAp), biocompatible coatings can be created. These coatings are important because they enhance the interaction of the implant with the tissue and avoid rejection by the body system.

In addition, CVD also enables the functional groups or bioactive agents to be introduced simultaneously into the coatings during the deposition process. Thus, the direct introduction of specific precursors or additives improves the ability of the coatings to fight infection-related complications of medical devices [62]. This capability is especially beneficial in industries like catheters or some surgical instruments where a patient's risk of getting infections should be reduced. However, there are disadvantages associated with CVD that need to be controlled during the practical application of the technology to produce medical coatings. As one of the critical factors affecting the technology, great concerns have been raised regarding the cost of the equipment and materials needed in the CVD process. Increased production cost is an implication of the need for specialized chamber and control systems to be used in the fabrication process, unlike other methods. Furthermore, achieving a similar type of coatings for different substrates can be difficult, and thereby, process control and optimization become very critical. The rate of deposition is also a factor in the CVD technique. The idea of getting the right balance between deposition rate and the quality of the formed layer is the key to leading a variety of applications. High deposition rates could result in poor coating characteristics; on the other hand, slow rates may be very unproductive. Scientists are always engaging in finding ways of increasing deposition rate while at the same time achieving the best quality film to be used in medical devices.

8.3.3 Electrospinning

Electrospinning is a novel fabrication process that offers the production of nanofibres and nanocoating having huge surface area and porosity [63, 64]. This method entails the use of a high-voltage electric field to draw a polymer solution into thin fibers and deposit it on a substrate. Electrospinning is especially beneficial when used to produce nanocoatings that have improved mechanical, biocompatibility and functionality characteristics which are ideal for use in the medical field. The growth of nanofiber mats by the electrospinning process starts with the preparation of a polymer solution which should be made by dissolving the polymer in solvent. It is then loaded into a syringe containing a spinneret that is connected to a source of high voltage electric field. When the voltage is applied the electrostatic forces between the charged polymer molecules overcome the surface tension of the solution and a fine jet of the polymer solution is formed. When the jet approaches the grounded collector, the solvent evaporates and forms continuous nanofiber fabric. However, one of the significant benefits that come with the use of electrospun nanofibers is the creation of materials with a high surface area/height ratio to the biological systems [65–67]. This high surface area is desirable in uses such as drug delivery where increased surface area contacts can improve cell on-tube attachment, mitosis, and drug assimilation. The fact that the fiber diameter and the morphology of coatings can be produced by using the electrospinning technique can be altered to suit various medical applications. Unlike other techniques for increasing surface area, electrospinning also facilitates the addition of bioactive agents into the nanofibers during the molding phase. Any drugs, growth factors or antimicrobial agents can be incorporated within the polymer solution in a liquid state and as we precipitate the solution to form nanofibers it maintains itself in the form of a reservoir in the nanofibers thus permitting it for a controlled release and targeted delivery. This capability makes it especially useful in tasks such as wound healing and tissue engineering since the controlled release of therapeutic agents will improve the healing process and overall patients' quality of life.

Electrospinning also permits the fabrication of structures with a hierarchical organization, providing the opportunity to develop nanocoatings that resemble the natural ECM [68–70]. The researchers say that the ability to manipulate the orientation and order of the nanofibers enables them to develop coatings that mimic the interactions of cells with natural tissue. This biomimetic strategy is critical to improving biocompatibility and supporting tissue incorporation in applications, including orthopedic implants and tissue engineering scaffolds. Furthermore, the electrospinning process can be coupled with other methods effectively to augment the characteristics of the developed nanocoatings. For instance, electrospun nanofibers may be further surface-modified through approaches like dip-coating or layer-by-layer deposition. This combination can lead to the synthesis of multitalented surface coatings with improvements in mechanical features, antibacterial features, and controlled drug release mechanisms. Due to the capacity of electrospinning for the generation of multilayered coatings, such a technique is very useful for the further proliferation of medical devices. However, the electrospinning technique also poses some difficulties that scientists have to solve. A major issue that must be addressed is the convergence and control of the electrospinning process in large-scale production. Despite the ability of electrospinning to manufacture a small quantity of nanofibers, the technique can hardly meet the large-scale requirement of an industry [71, 72]. Scientists are looking for ways to improve the productivity of this process and invent methods for creating a constant-mode electrospinning process to overcome this problem. Further, the use of polymer and solvent significantly influences the properties of the electrospun nanofibers. Easy availability and control over the selection of appropriate materials, which are biocompatible, biodegradable, and capable of electrospinning, is critical when used for health purposes. Current work includes prospects of new polymers and their blends to improve electrospinning characteristics and the types of materials that can be used in biomedical coatings.

8.4 Applications of nanocoatings in medical devices

Nanocoatings have opened a new era in the domain of medical devices coming up with new features and improvements in biocompatibility, mechanical characteristics, and applicability. As a result of adopting nanotechnology into the creation of medical devices, better patient conditions, lower incidences of infections, and longer durations of equipment have been achieved. The most popular applications of nanocoatings that are described in the medical devices industry are implants, instruments for surgery, and diagnostic equipment [73–75]. The different categories of medical devices also see specific advantages arising from the properties and functionality given by nanocoatings. Of all the branches where nanocoatings have made the greatest advancement, implants are among the most notable. The application of nanocoatings in implantation acts as a significant improvement in how implants work and are incorporated into the body. Regular materials used in implants including metallic ions like titanium as well as stainless steel have some issues of biocompatibility and corrosion. Nonetheless, nanocoatings can enhance these factors quite a lot. For example, nanostructured HAp coating on titanium implants results in increased bioactivity associated with improved osseointegration. This enhanced osseointegration is cardinal for orthopedic implants since it contributes to the increased stability and service duration of the device in the host tissues [76–79]. In addition, using nanocoatings, antimicrobial characteristics can be provided to implants, solving the problem of device-associated infection. These microorganisms can attach themselves to the implant surface and the various effects of their presence include failure in implants or the need for replacement by surgery. The nanocoating layer consists of other materials including a material with silver nanoparticle incorporation or other antimicrobial agents that prevent bacteria from attaching to the surface of the implant. These bactericidal layers do more than prevent infection, but also reduce the frequency of long-term antibiotic usage, which is upsetting to the overall health of the patient. The use of nanocoatings in implants, therefore, provides better safety to patients and enhances their associated surgeries. Nanocoatings can not only make biocompatibility and antimicrobial coatings but also improve the mechanical characteristics of implants [80, 81]. The addition of nanomaterials into coatings can improve the surface properties of implants such as wear resistance and toughness. Nanocomposite type of coatings, for example, where polymers and ceramics in the form of nanoparticles, are even more effective mechanically while at the same time flexible and strong. This is specifically important in load-bearing implants like the hip and knee joint prostheses, which mainly rely on their mechanical properties to achieve long-term fixation. The increase in mechanical properties together with better biological characteristics in nanocoated implants indicates that this research area can be a viable pathway toward improving orthopedic and dental procedures. In addition, nanoscopic layers can be specifically fashioned to meet the demands of clinical use. For example, surface roughness and hydrophilicity can be tuned by the selection of nanomaterial and by the methods of fabrication.

Such changes also affect protein adsorption, cell attachment and the subsequent cellular behaviors. Thus, the engineering of the implant surface can always lead to an improved response of the body tissue to the implant, accelerating its integration into the tissue, which provides a means of minimizing the occurrence of complications [82, 83]. The prospects of such being a reality are even better since one can engineer implants at the nanometer level, meaning that implants can be made distinct in relation to the patient's requirements.

Interestingly, the surgical instruments industry is harnessing nanocoatings to improve task efficacy and decrease risk. Surgical instruments are essential in operations that require accuracy and minimal injury to the patient and mainly rely on the fabricating materials. Many conventional metal tools share certain problems including corrosion, wearing out, and even the challenge of owning acute edges. The factors stated above could be alleviated by using nanocoatings since they would yield better surface characteristics required in surgical applications. For example, nanocoatings are presented that can offer enhanced corrosion protection thereby decreasing the probability of instrument degradation during sterilization procedures. Stainless steel instruments coated with nanostructured oxide layers are considered to possess enhanced corrosion and staining performances, and this is crucial in the preservation of the instruments' functionality in the long run. This not only helps to elongate the life of the brand's surgical tools but also guarantees that they will not harm patients. In addition, some of the nanocoatings possess antimicrobial characteristics that help to avoid bacterial accumulation on instruments, thereby, decreasing surgical site infections. In addition to enhancing the overall performance of cutting edges on surgical instruments, it is also factual that nanocoatings can considerably enhance the sharpness and durability of these cutting edges. Recoating or using some nanomaterials such as diamond-like carbon and titanium nitride (TiN) improves the hardness and the wear resistance of cutting tools [84, 85]. These coatings enable acuter angles to be used which do not blunt or dull as quickly reducing trauma to the surrounding tissue. Better cutting can translate to better surgery, less time required to heal and fewer complications to patients. In addition, the use of nanocoatings on surgical instruments enhances the engineering of other features that are not previously covered, including multiplexing functionality. For example, the surfaces of surgical instruments can be treated to have additional nanostructures that will act as an anti-reflective coating coupled with visibility characteristics during operations [86, 87]. This is most advantageous in bar service, especially in doing laparoscopic surgeries which are characterized by poor visibility all in all. Anti-glare and high transmission nanocoatings provide the surgeon better visibility of the operating table thus increasing the probability of a successful and fast surgery. That is why diagnostics tools are also concerned with the help of nanocoatings; implants, surgical instruments as well as diagnostic tools are some of the areas that are in focus at present. The need for fast and reliable diagnostic tools has been on the rise recently primarily associated with the diagnostics of infectious diseases as well as modern trends of individualized approach. Nanocoatings can enhance the performance of diagnostic devices by improving sensitivity, specificity, and ease of use. Another well-recognized application is based on biosensors, devices for the detection of biological analytes and providing real-time information on health state [88–90]. Another advantage of nanocoatings is to enhance the sensitivity of biosensors because of the availability of more area to the target molecules. Examples of these PITs include the incorporation of the sensor surface with gold or silver nanoparticles or an increase in the electrochemical or optical signal generated during the detection process. Being sensitive is important to help in the detection of diseases hence performing early diagnoses for necessary treatment. Furthermore, using nanocoating, the capture of biomolecules, including antibodies or enzymes, on the sensor surface is possible so that the expected sensor stability and repeatability of the measurements would be achieved. Imaging techniques are the other common application area of nanocoating in diagnostic tools and equipment. Such forms can be applied in imaging carrying contrast-enhancing tissue and organ visualization. For instance, super-paramagnetic nanoparticles may enhance magnetic resonance imaging contrast and make tumor and other abnormality detection easier. Consequently, quantum dots, which are semiconductor nanoparticles, can be employed in fluorescence imaging to produce well-delineated images of different structures at the cellular level. The degree of freedom in adjusting the optical properties of nanomaterials through coating makes it possible to create better imaging agents aimed at improving the diagnostic capability of equipment. In an experiment, Pirușcă et al. [91] involved the use of magnetite (Fe_3O_4) nanoparticles as the base material for coatings. These Fe_3O_4-based coatings have been enhanced with the incorporation of two potent antimicrobial agents: usnic acid (UA) and ceftriaxone (CEF). Usnic acid, a natural compound with broad-spectrum antibacterial properties, and ceftriaxone, a well-known antibiotic, are both encapsulated within the Fe_3O_4 nanoparticles. This functionalization, facilitated by sodium lauryl sulfate (SLS) as a stabilizing agent, enhances the nanoparticles' polarity, dispersibility, and structural integrity. Additionally, SLS contributes to the improved anti-biofilm capabilities of the nanoparticles by helping them more effectively interact with bacterial cells. The functionalized nanoparticles, which include Fe_3O_4@SLS, Fe_3O_4@SLS/UA, and Fe_3O_4@SLS/CEF, were synthesized using the co-precipitation method. A variety of analytical techniques were employed to fully characterize these nanoparticles, ensuring their suitability for medical applications. X-ray diffraction (XRD) was used to analyze the crystal structure of the particles, while transmission electron microscopy (TEM) and scanning electron microscopy (SEM) provided insights into their size, shape, and surface morphology. Selected area electron diffraction (SAED) confirmed the crystallinity of the nanoparticles, and Fourier-transform infrared spectroscopy (FTIR) helped identify the functional groups involved in the nanoparticles' composition. Finally, thermogravimetric analysis was conducted to assess the thermal stability of the coatings, which is crucial for their performance in real-world medical environments. Once characterized, the nanoparticles were used to fabricate nanostructured coatings via the matrix-assisted pulsed laser evaporation technique. This method allowed for the precise deposition of the nanoparticles onto surfaces, creating uniform coatings suitable for medical devices. To ensure that these coatings were not only effective in combating bacterial biofilms but also safe for use in the body, their biocompatibility was rigorously tested. Experiments measuring cell viability, lactate dehydrogenase release, and nitric oxide production in cultured murine pre-osteoblasts confirmed that the coatings did not induce significant cytotoxicity. Additionally, the morphology of the actin cytoskeleton in these cells remained unchanged, further indicating that the coatings supported healthy cell growth, adhesion, and migration. In parallel, the coatings' ability to inhibit biofilm formation was tested against two common bacterial pathogens: *Staphylococcus aureus* and *Pseudomonas aeruginosa*. These bacteria are frequently involved in biofilm-related infections and represent models for gram-positive and gram-negative species, respectively. The results showed that the Fe_3O_4-based coatings were effective at reducing biofilm growth at both 24 and 48 h, suggesting that they could provide prolonged protection against infection (Figure 8.2). This is particularly important in medical environments where devices remain in place for extended periods, offering bacteria the opportunity to form biofilms.

The issue of bacterial biofilm formation on medical devices is a significant concern in healthcare, as biofilms are resistant to standard cleaning procedures and antibiotic treatments, often leading to severe infections. Traditional methods to prevent biofilm formation involve the use of coatings that release antimicrobial agents like silver ions (Ag^+), known for their potent antibacterial properties. Nano-silver (nAg) has been particularly effective due to its ability to combat a wide range of bacteria. However, one of the primary limitations of nAg coatings is their tendency to degrade prematurely in harsh environments, especially in diseased tissues. This rapid dissolution can lead to the clearance of silver nanoparticles before they can provide long-lasting protection. As a result, researchers have been exploring more durable solutions, such as immobilizing nAg directly onto the surface of medical devices to ensure sustained antimicrobial activity. In an effort to address this challenge, a study was conducted by Riau et al. [92] to test the efficacy of nano-silver and nano-HAp (nHAp) immobilized on poly(methyl methacrylate) (PMMA), a polymer commonly used in medical devices, such as corneal prostheses. The immobilization process aimed to create a surface that could resist bacterial biofilm formation while remaining biocompatible with human cells. Three different formulations were developed, each with varying ratios of nano-HAp to nano-silver: a high silver content (75:25 HAp), an intermediate silver content (95:5 HAp), and a low silver content (99:1 HAp). These formulations were carefully analyzed for their ability to prevent biofilm formation and their biocompatibility with human stromal cells, the primary cells found in the corneal stroma. The results revealed that the high silver content formulation (75:25 HAp) was highly effective in preventing biofilm formation, particularly against two common pathogens associated with medical device infections: *P. aeruginosa* and *S. aureus*. However, despite its strong antibacterial properties, the 75:25 HAp formulations was found to be cytotoxic, negatively impacting the viability of human stromal cells. This raised concerns about its safety for medical applications, par-

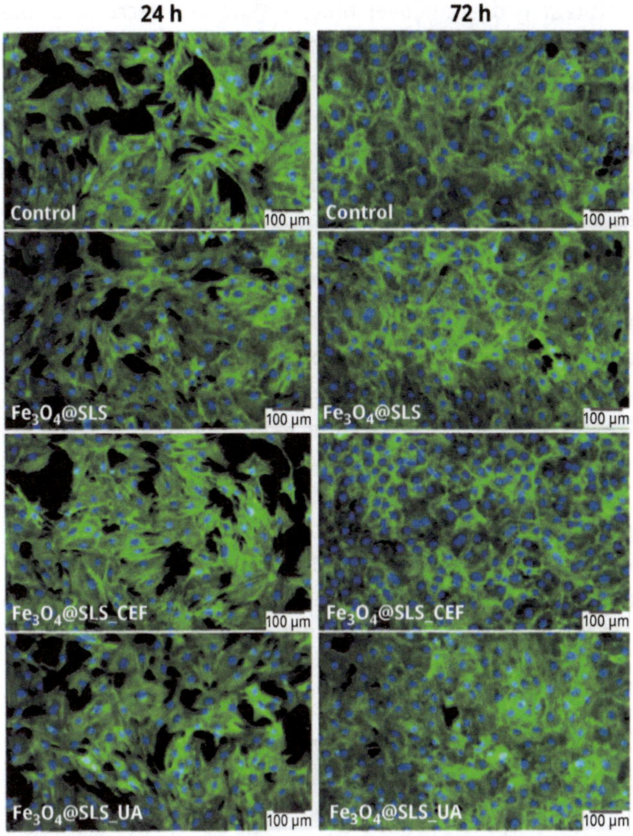

Figure 8.2: Actin cytoskeleton organization of MC3T3-E1 murine cells after 24 and 72 h of incubation with Fe_3O_4@SLS, Fe_3O_4@SLS/CEF, and Fe_3O_4@SLS/UA coatings. F-actin (green) was stained with phalloidin-FITC, and nuclei (blue) with DAPI. Magnification: 20×. Adapted from Ref. [91] under CCBY 4.0.

ticularly in sensitive tissues such as the eye. In contrast, the intermediate silver content formulation (95:5 HAp) struck an ideal balance, exhibiting strong antimicrobial effects against both bacterial species while maintaining excellent biocompatibility. This formulation prevented biofilm formation without causing harm to human cells, making it the most promising candidate for further development. The 95:5 Hap coating also demonstrated an additional benefit by eliminating non-adherent bacteria in the surrounding culture media, offering a dual mechanism of protection. The illustration in Figure 8.3 highlights two distinct methods for functionalizing hard polymers with antimicrobial silver nanoparticles (nAg). In traditional approaches, as depicted in part (A), most studies have focused on incorporating silver nanoparticles or silver salts into a soft, biodegradable polymeric coating, which is then applied over a hard polymer or substrate. While this method provides antimicrobial protection, the biodegradable nature of the polymer means that the coating may degrade over time, leading to a decrease in the availability of silver ions and, thus, reduced long-term antimicrobial efficacy.

In contrast, part (B) of the figure introduces a more durable method: the direct immobilization of silver nanoparticles onto the surface of the hard polymer. This innovative approach eliminates the need for a biodegradable layer, ensuring that the silver nanoparticles remain intact and functional for an extended period, as long as the hard polymer itself remains stable. Furthermore, the antimicrobial properties of the silver nanoparticles can be enhanced by combining them with HAp nanoparticles, which provide additional structural support for cells. This dual-functionalization approach not only offers sustained antimicrobial protection through the slow release of silver ions but also promotes cell adhesion and biocompatibility. The cross-sectional SEM image of a PMMA sheet coated with both silver and HAp nanoparticles is also shown. The thickness of the coating was measured to be approximately 32 ± 3 µm. At high magnification (30,000×), individual nanoparticles were clearly visible, as shown in the SEM image with a blue border, further illustrating the fine structural details and effective distribution of the nanoparticles on the polymer surface. This method holds promise for providing long-lasting antimicrobial protection for medical devices, reducing the risk of biofilm formation and infection over extended periods. However, despite the initial success of the 95:5 HAp formulation, long-term tests highlighted a potential drawback. After a 28-day incubation in artificial tear fluid, it was observed that the release of silver ions from the coated surface decreased by approximately 40%. This reduction in Ag^+ release had a noticeable impact on the coating's ability to inhibit *S. aureus*, a gram-positive bacterium. Interestingly, the antimicrobial efficacy against *P. aeruginosa*, a gram-negative bacterium, remained unaffected. This discrepancy is likely due to the differences in the bacterial cell walls of gram-positive and gram-negative organisms, which can influence their susceptibility to silver ions. The findings of this study underscore the importance of silver ion release kinetics in the long-term prevention of biofilm formation. While surface-immobilized nano-silver coatings may offer effective short-term protection against bacterial infections, particularly in the perioperative and early postoperative periods, their antimicrobial activity may diminish over time, especially against gram-positive bacteria. This suggests that while such coatings can be valuable in preventing initial infections, they may be less effective in the long-term management of biofilm-related infections on medical devices.

Dwivedi et al. [93] introduced a novel approach involving a layer-by-layer nanocoating technique that integrates vancomycin-loaded niosomes. Niosomes, which are vesicular drug delivery systems, are composed of nonionic surfactants and are particularly effective for encapsulating hydrophilic and lipophilic drugs. In this study, the niosomes were specifically formulated using the thin-film hydration method, a technique known for its ability to produce uniform vesicle sizes with optimal characteristics. The resulting niosomes had an average diameter of 340.5 ± 2.95 nm, which is suitable for enhancing the interaction with bacterial cells while minimizing potential

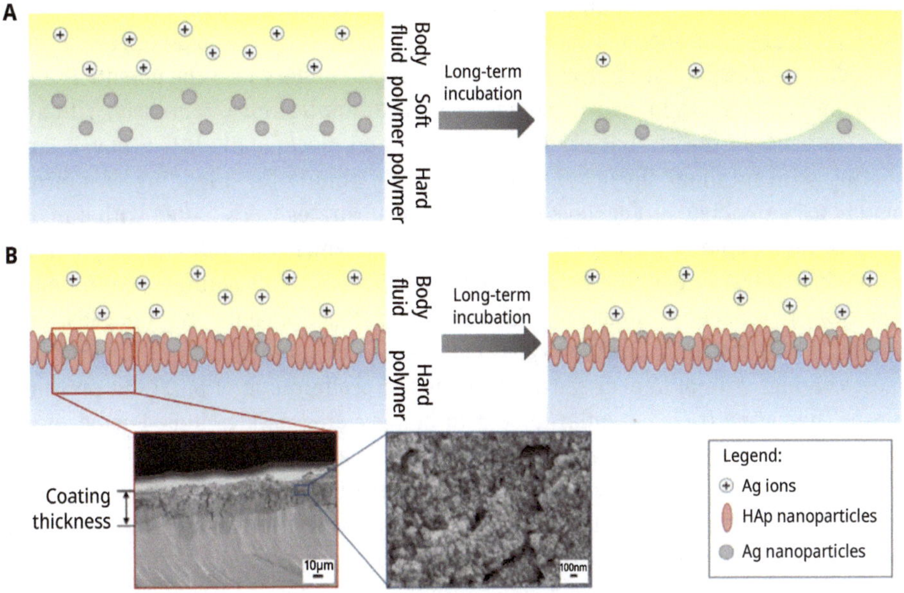

Figure 8.3: Cross-sectional SEM of PMMA coated with silver and hydroxyapatite nanoparticles, showing a coating thickness of 32 ± 3 μm and nanoparticle distribution at 30,000× magnification. Adapted from Ref. [92] under CCBY 4.0.

cytotoxic effects. Key physicochemical properties of the niosomes were thoroughly assessed, including their zeta potential and encapsulation efficiency (EE). The zeta potential, measured at 45.4 ± 0.77 mV, indicates good stability of the niosome dispersion, as higher zeta potentials often correlate with reduced aggregation. The EE of 50.47 ± 3.66% suggests that a significant portion of the vancomycin was successfully loaded into the niosomes, making them effective carriers for the antibiotic. This formulation is crucial as it allows for the targeted delivery of vancomycin directly to the site of infection. To apply this innovative coating to orthopedic implants, the dip-coating technique was employed. This method facilitates the even application of the niosome-based coating onto the implant surface, which was subsequently characterized morphologically using field-emission SEM (FE-SEM). This characterization confirmed the uniformity and integrity of the coating, which is essential for its functionality. The study also investigated the drug release profile from the coated bone plates compared to control plates lacking the niosome treatment. The results indicated that the bone plates coated with vancomycin-loaded niosomes achieved a significantly higher accumulation of vancomycin, leading to a prolonged release of the antibiotic over duration of 2 weeks. This sustained release is particularly advantageous for preventing bacterial colonization and biofilm formation, critical factors in the success of orthopedic implants. Furthermore, the niosome-coated bone plates exhibited impressive antibacterial activity over an extended period, effectively combating bacterial strains commonly associated with implant infections, such as Staphylococcus aureus. Importantly, this antimicrobial activity was achieved without any cytotoxic effects on normal cells, specifically L929 fibroblast cells. This aspect underscores the potential safety of using vancomycin-loaded niosomes in clinical applications. Similarly, Tîlmaciu et al. focused on the antibacterial effectiveness and biocompatibility of novel nanocoating with minimal silver content, specifically aiming for a loading of approximately 0.65 nmol cm^2. To achieve this, titanium substrates were coated with a monolayer of mercaptododecylphosphonic acid (MDPA), which was subsequently reacted with silver nitrate ($AgNO_3$). The resulting MDPA/$AgNO_3$ nanocoatings were found to significantly reduce the adhesion and biofilm formation of common pathogens, such as *Escherichia coli* and *Staphylococcus epidermidis*, when tested in vitro. Importantly, these coatings did not impede the attachment and proliferation of MC3T3-E1 preosteoblasts, indicating their potential for use in bone-related applications. Furthermore, the osteogenic differentiation of both MC3T3 cells and murine mesenchymal stem cells remained unaffected by the presence of these nanocoatings, highlighting their compatibility with bone-forming cells. The researchers also assessed the impact of sterilization via ethylene oxide on the properties of the nanocoatings. They found that this sterilization method did not compromise the antibacterial activity or biocompatibility of the coatings, making them suitable for clinical use. Following subcutaneous implantation in murine models, the MDPA/$AgNO_3$ nanocoatings exhibited robust antibacterial properties and maintained excellent biocompatibility, even after exposure to *S. epidermidis* postoperatively. These findings underscore the effectiveness of utilizing subnanomolar quantities of silver in coating strategies, which can be strategically positioned at the surface of metallic or ceramic medical devices. Such an approach not only helps prevent bacterial adhesion and biofilm formation but also ensures that the biocompatibility of the devices is preserved. This research contributes valuable insights into developing safer and more effective antimicrobial coatings for medical applications, ultimately enhancing patient outcomes and reducing the burden of device-related infections.

Additionally, nanocoatings can go toward designing diagnostic devices that can be used based on the patient's needs, they can be single-use and can be carried around easily. These devices are especially useful in resource-constrained areas since access to the laboratory facilities may be a limitation. By enhancing the interaction between the sample and the detection elements nanocoated surfaces make tests fast and reproducible [94–98]. For example, microfluidic devices with channel walls coated by nanomaterials increase the flow rates of samples, decrease the chances of blockages, and improve the corresponding testing procedures. Nanocoatings also enhance the use of combined diagnostic tools due to their aspects of using multiple biomarkers in a single nanocoated diagnostic tool. Thus, diverse nanomaterials and coating methods allow scientists to design sensors capable of detecting numerous biological molecules within the same analysis. This capability becomes especially valuable when handling the problem of personalized medicine where a patient's biomarker readings need to be taken into account. In this case, nanocoatings act as tools that improve the multi-functionality of diagnostic equipment. Besides the application of nanocoatings in medical devices, there are more concerns related to the improvement of comprehensive health and patient care. The application of superior nanocoatings can go a long way in preventing patient harm, containing expenditure on medical care, and boosting the overall well-being of individuals. Nanocoatings have the potential to reduce the chances of infections, increase the lifespan of devices and facilitate quicker diagnosis – all of which make it a major leap in this field. There are also growing research and development activities as well as cooperation between various actors: researchers, engineers, clinicians, and regulatory authorities, which have led to the successful application of nanocoatings in medical technologies. Systematic advancements in nanomaterials and fabrication methods are crucial in broadening application areas and to the complexity inherent when nanocoatings are used in the healthcare industry. Together, we will strive to improve next-generation medical devices to be more safe, efficacious, and responsive to patients' needs.

8.5 Biocompatibility and safety considerations

Functional coatings always serve the need to address biocompatibility and safety issues when developing medical devices especially those with nanocoatings. However, as applications of nanotechnology in medicine progress, it becomes important to establish how such nanomaterials respond to biological systems. The guidelines and protocols governing the utilization of industrial nanocoatings on medical devices are very many and ever-evolving from independent or perhaps a consortium of agencies depending on the country or continent in question [99–101]. In the United States, the Food and Drug Administration (FDA) is the government seat that has the primary responsibility of regulating any medical device that has or may have elements of nanomaterials. In its opportunity and risk assessment of nanotechnology, the FDA has developed a comprehension of compelling issues as well as the goals of nanotechnology; the practical nativeness of nanomaterials in FDA-regulated products and premarket noticed Submissions submission of products containing nanomaterials. Described documents state that FDA requires information on the properties of nanomaterials, and their application, as well as potential hazards of nanomaterial utilization in medical products manufacturing. In Europe, it is equally challenging with the EMA and the European Commission which give guidelines for medical devices that contain nanotechnology. In the case of devices containing nanomaterials, the Medical Device Regulation (MDR) and In Vitro Diagnostic Regulation (IVDR) illustrate the modalities to demonstrate equivalent conformity with the status of safety and performance of devices [102, 103]. The regulations promulgated require manufacturers to comprehensively evaluate the risks posed by nanomaterials based on their physicochemical characteristics, biological effects, and their pathways of entry into the environment and organism. One component in the legal arrangements is biocompatibility testing. Biocompatibility in its simplest form means the capacity of a material to perform as intended within a biological system without provoking inflammation or other unfavorable reactions from the surrounding environment. This is because before a medical device can be recommended for use, especially one that is to have direct contact with body fluids or tissue, compatibility at the "biological level" must be determined. ISO has set up a sequence of rules for testing and implementing biocompatibility: the most commonly used ISO part 1:2009 and ISO 10993 part 2:2016. This standard provides a step-by-step approach to evaluation, from the outer layers/covering of the device through in vitro and in vivo evaluation if the need arises.

The first step in deciding the level of biocompatibility of a certain material used in the device is to ascertain the chemical nature, physical properties and surface morphology of the materials for use in the body. This information is important in order to forecast the materials' behaviors in a biological context. Nanomaterials in general can be regarded as more problematic due to their size, and higher surface area can affect their biological response [104]. The biocompatibility part requires manufacturers to go through detailed chemical characterization and identification of any impurities or residuals. A car stops moving because you apply the brakes, and following material characterization, the subsequent tests that are prescribed in the evaluation of biocompatibility is a sequence of in vitro tests that seek to determine the biological response to such materials. Such tests often comprise cytotoxicity tests where the ability of the given material to cause cell death or inhibit cell division is examined. Some of the methods currently in use are the MTT assay, which measures the metabolic activity of the cells; and the hemolysis assay, which measures the ability of a material to hemolyze red blood cells. Other tests may also be performed such as genotoxicity tests, irritation tests, and sensitization tests depending on the type of use of the device and the kinds of materials used. When in vitro biocompatibility evaluation shows some signs of the material's compatibility with the living tissue in culture, more rigorous in vivo testing may be needed to determine the biological interactions affecting the organism. The most common tests for biocompatibility are in vivo tests would entail excluding the device or material under investigation in a test animal and comparing tissue reactions at different periods. Some factors considered during in vivo analysis include inflammation, tissue incorporation, and responses or otherwise to the implantation site. These studies are important in generating full details on the safety of the device before it is subjected to a human subject [105, 106].

Though regulatory rules and assessment frameworks are strict when it comes to nanomaterials in medical devices, the risks or toxicity of such structures continue to be debated and investigated. It has been observed that rather than the direct toxicity normally associated with the substance, a nanomaterial whose size is in the scale of one to hundred nanometers has remarkably distinct features, a high surface area, and size-dependent biological interactions that may result in three-dimensional toxicities. The first implication concerns the ability of the nanoparticles to cause inflammation or cytotoxicity after exposure to bio-tissues. Through activation of immune cells, the nanoparticles appear to cause inflammation and thereby the toxicities such as tissue injury or inflammation are manifested. This issue is particularly important for a number of implants and devices that interact with biological systems for a shorter or longer period. Another subject of worry is the ability of nanomaterials to cause oxidative stress. Oxidative stress may be defined as a state where the formation of free radicals known as reactive oxygen species (ROS) is greater than the ability of the body to eliminate them [107–109]. The metal nanoparticles produce ROS upon reacting with biomolecules and impart cellular damage which contributes to numerous diseases. It has been demonstrated that some nanoparticles can interact with cellular membranes and affect mitochondrial functionality, amplifying the shifts of oxidative stress and the level of cell mortality. This stresses the need to have a proper evaluation of the toxicological information of nanomaterials that are applied in medicine. Furthermore, the pathways by which people get in touch with nanomaterials also offer insight with regard to their safety. Nanoparticles can penetrate in human body through inhaled, ingested or dermal routes. The pathway of uptake plays a crucial role in stochiometry and phasing of nanoparticles across the tissues of the body and the overall prospect for toxicity. For instance, nanoparticles inhaled may get to the lungs and blood streams with implications for respiratory and cardiovascular health. Likewise, the nanoparticles that come in contact with skin or mucous membrane surface may produce different biological effects rather than the nanoparticles that are placed in tissues. The safe use and potential hazards of nanomaterials are dependent on their pharmacokinetics and biodistribution profiles.

As pointed out earlier, the degradation and long-term stability of nanomaterials is another important factor that has been commonly used while evaluating them. Nanoparticles were also found to exhibit all-scale-different behavior in biological environments as compared to what is observed in vitro or during production. Changes that are recognized include particle size, particle shape, and surface charge, which may alter the way in which nanoparticles interact with cells and tissues when they come in contact with biological fluids. Similarly, the possibility of toxic elements leaching out of nanocoating must be properly evaluated since this side effect can create other hazards to a patient's life. The proliferation of nanotech in medical devices is ever-revealing gaps in knowledge and regulating agencies are now addressing the issue of requiring an Integrative risk assessment of nanomaterials present in disease-fighting instruments. Such characterization also emphasizes the investigation of biocompatibility additionally, nanotoxicity studies that focus on nanomaterial characteristics and interactions. The great emphasis placed on preventing adverse effects caused by nanomaterials introduces the necessity to adapt unified testing methods and protocols that will serve as essential prerequisites to safety testing. The safety of nanomaterials depends on the useful cooperation of regulatory agencies, industries and researchers. This can be used to foster the exchange of data, identification of benchmarking standards, and distillation of standard operating procedures for the assessment of nanocoating in biomedical devices. Further, studies have been carried out to understand the dynamics of how nanoparticles interact with biological systems any information acquired concerning toxicity will form a rich database for future evaluation [110–112]. Subsequently, there remain a number of concerns regarding the development of nanopharmaceuticals, which are yet to be addressed in the broader field of nanotechnology and subsequently, the regulatory policies that are involved in this area will also have to be developed appropriately as the field becomes more complicated. The use of nanomaterial in health-related commodities and products will enhance the patient's quality of life if the principle of risk management is followed when using nanomaterial in implant technology. Each of the evaluation parameters; biocompatibility, toxicity considerations, and compliance with a strict regulatory structure will be crucial to the achievement of the expected gains from nanotechnology without compromising the health of the patients.

8.6 Challenges and future perspectives

There are several issues associated with the introduction of nanocoating that need to be addressed for medical devices to be safe, effective, and properly registered. Probably the most critical issue is the definition and categorization of nanomaterials and subsequent standard setting. The physicochemical characteristics of nanomaterials; including size, shape, surface chemistry as well as propensity to aggregate can strongly determine their bioreactivity and thereby define their safety and performance in any medical application [113]. The current safety regulation of these nanomaterials may not have comprehensive loopholes of complexities that are likely to result in variations in the testing parameters and the final results. The key goal of this paper is to provide detailed recommendations on the further development of nanocoating characterization standards as well as compare the efficiency and biocompatibility of different types of nanocoatings for medical applications. It will facilitate the reduction of the differences in the existing regulations, and build confidence with manufacturers, regulators, and other health workers. One of the most worrisome risks is the unpredictable toxicological effects that may be traced to nanomaterials. Although the scientific literature is replete with data on nanoparticles, the authors stress that the picture of biological interactions and possible negative consequences has not been painted in full. Based on the mechanisms of toxicity that are related to nanoparticles, such as oxidative stress, inflammation, and immune response activation, it is necessary to develop deeper insight into their biological activity. Subsequent research is needed to better understand these mechanisms, especially as they apply in more elaborate biological contexts where factors other than size and shape might contribute to the toxicity of nanoparticles. In addition, information about the long-term effects of nanocoatings is still lacking, which becomes a serious concern in the case of nanocoating use for the long-enduring devices implanted in the human body.

Looking at the future of neutron scattering, advances in the design of new nanomaterials with improved characteristics remain a promise and a threat. Exploitation of bio-inspired and biomimetic nanocoatings might result in enhancing the effectiveness of various types of medical devices. However, new materials require an assessment to check whether they will cause other risks that are not foreseen. Increased predictions are made on the concept of bespoke nanocoating – applying a new material for a direct patient's need; this, however, will present a challenge to nanotechnology and the identification of how various materials interact with diverse biological environments. Nanotechnology will also require the extending and reforming of the corresponding regulations. With nanocoatings expanding their usage in medical applications, including medical devices, there is a need for improved techniques for testing and regulating nanomaterials. Industry, academia and regulatory groups will have to work hand in hand to set up standards and protocols for testing. Such an approach may enhance the exchange of information and material that will enhance the efficiency of the regulations while keeping the patient's safety in mind even as the nation encourages the development of new products [114, 115]. Moreover, the opinion of the public regarding nanotechnology in medicine is a key factor that determines its use. This is a major problem because fear of nanomaterial safety may cause consumers and even care professionals to avoid the use and adoption of nanotechnology in medical applications. As a result of the discovery of nanocoating, there is a need to educate people about the pros and cons of using nanocoating. Patients, clinicians, and policymakers should be involved in the participation of nanotechnology in medical applications to enhance the acceptance and understanding of it. Education of the public, authorities, and research community about the procedure, procedures followed and the assessments conducted will be central to reversing the current perception of nanotechnology.

8.7 Conclusion

The specific characteristics of nanomaterials including large surface area and the ability to tailor their mechanical properties as well as the possibility of carrying out multiple functions within them make nanomaterials ideal for use in implants, surgical instruments and diagnostics. However, this technology holds much potential in the future of data mining but it also has problems that have to be solved to harness its potential. Currently, it is questionable that nanocoatings are adequately regulated for the determination of the appropriate application where health, safety, and environmental concerns are taken into consideration related to nanostructured materials' properties and behaviors. It is evident that the need for the increased standardization of characterization methods is crucial for the assessment of nanocoatings. Through the specification of acceptable protocols for the testing and evaluation of these materials, the stakeholders will enhance the process of regulation and make efforts toward developing safer and more efficient medical devices. However, the knowledge of the mechanisms of biocompatibility and toxicity of nanomaterials is still developing. This approach will remain critical for assessing the potential toxicity of therapeutics and the necessary patient precautions as well as for setting future regulatory guidelines that may be expected from the mapping of toxicity mechanisms such as oxidative stress and immune response activation.

Furthermore, it is expected that nanotechnology applications to medicine will advance in the future with the potential of creating bio-inspired nanocoating and innovative approaches to personalized medicine. The future innovation of creating nanomaterials will possibly be tormented by a process of specific specifications of treatments for patients to improve the result of treatment. However, such potential has to be achieved taking into account the importance of highly strict safety assessments as well as the need for openness in the regulatory sphere. In addition, the awareness that the public has of nanotechnology also directly affects its acceptance and implementation in the medical field. It is therefore necessary to explain how nanocoating works, their advantages, and disadvantages, and the legal policies governing the technology in order to foster the acceptance of nanotechnology in the healthcare industry. Through public participation, and having people discuss nanotechnology and its effects on their lives, we are able to overcome negative impressions about concerns and embrace this branch of technology in medicine. In conclusion, the experience of nanocoating in medical devices just started and many difficulties should be viewed as prospects in the future. But by working together with other researchers, industrialists, and other relevant authorities, then those challenges can easily be handled and thus paving the way for the full utilization of nanotechnology that will lead to the development of better medical devices for improved patient care all over the world.

References

[1] Erdonmez, D., Mosayyebi, S., Erkan, K., Salimi, K., Nagizade, N., Saglam, N., et al. (2014). Nanofabrication and characterization of PVA–organofiller/Ag nanocoatings on pMAD plasmids. *Applied Surface Science* [Internet], *318*, 127–131. Available from: https://www.sciencedirect.com/science/article/pii/S0169433214002918

[2] Roest, R., Latella, B. A., Heness, G., & Ben-Nissan, B. (2011). Adhesion of sol–gel derived hydroxyapatite nanocoatings on anodised pure titanium and titanium (Ti6Al4V) alloy substrates. *Surface and Coatings Technology* [Internet], *205*(11), 3520–3529. Available from: https://www.sciencedirect.com/science/article/pii/S0257897210012806

[3] Pavoor, P. V., Gearing, B. P., Muratoglu, O., Cohen, R. E., & Bellare, A. (2006). Wear reduction of orthopaedic bearing surfaces using polyelectrolyte multilayer nanocoatings. *Biomaterials* [Internet],

27(8), 1527–1533. Available from: https://www.sciencedirect.com/science/article/pii/S0142961205007878

[4] Graziani, G., Ghezzi, D., Boi, M., Baldini, N., Sassoni, E., Cappelletti, M., et al. (2024). Ionized jet deposition of silver nanostructured coatings: Assessment of chemico-physical and biological behavior for application in orthopedics. *Biomaterials Advances* [Internet], *159*, 213815. Available from: https://www.sciencedirect.com/science/article/pii/S277295082400058X

[5] Zhan, J., Li, L., Yao, L., Cao, Z., Lou, W., Zhang, J., et al. (2024). Evaluation of sustained drug release performance and osteoinduction of magnetron-sputtered tantalum-coated titanium dioxide nanotubes. *RSC Advances* [Internet], *14*(6), 3698–3711. Available from: https://www.sciencedirect.com/science/article/pii/S2046206924002961

[6] Singh, H., & Kaur, K. (2023). Role of nanotechnology in research fields: Medical sciences, military & tribology- A review on recent advancements, grand challenges and perspectives. *Materials Today: Proceedings* [Internet]. Available from: https://www.sciencedirect.com/science/article/pii/S2214785323005783

[7] Patiyananuwat, N., Safarzadeh, M., Chartpitak, T., Riansuwan, K., Ritshima, W., & Nasongkla, N. (2024). In vitro and in vivo evaluation of the layer-by-layer vancomycin with poly(ε-caprolactone) nanosphere-coated Schanz pins for prolonged release. *Journal of Drug Delivery Science and Technology* [Internet], *100*, 106029. Available from: https://www.sciencedirect.com/science/article/pii/S1773224724006981

[8] Tan, Y., Ding, Y., Liu, S., Liu, P., & Cai, K. (2024). Titanium-based substrate modified with nanoenzyme for accelerating the repair of bone defect. *Colloids Surfaces B: Biointerfaces* [Internet], *234*, 113737. Available from: https://www.sciencedirect.com/science/article/pii/S0927776523006318

[9] Li, X., Sun, L., Zhang, P., & Wang, Y. (2021). Novel approaches to combat medical device-associated biofilms. *Coatings*, *11*(3), 1–31.

[10] Dey, N., Vickram, S., Thanigaivel, S., Kamatchi, C., Subbaiya, R., Karmegam, N., et al. (2022). Graphene materials: Armor against nosocomial infections and biofilm formation – A review. *Environmental Research* [Internet], *214*, 113867. Available from: https://www.sciencedirect.com/science/article/pii/S001393512201194X

[11] Valbuena, A., & Mateu, M. G. (2017). Kinetics of surface-driven self-assembly and fatigue-induced disassembly of a virus-based nanocoating. *Biophysical Journal* [Internet], *112*(4), 663–673. Available from: https://www.sciencedirect.com/science/article/pii/S0006349517300425

[12] Pituła, E., Janik, M., Sezemsky, P., Szymańska, K. P., Olszewski, M., Stranak, V., et al. (2023). Smartphone-based dynamic measurements of electro-optically modulated lossy-mode resonance and its biosensing applications. *Measurement* [Internet], *206*, 112349. Available from: https://www.sciencedirect.com/science/article/pii/S0263224122015457

[13] Graziani, G., Barbaro, K., Fadeeva, I. V., Ghezzi, D., Fosca, M., Sassoni, E., et al. (2021). Ionized jet deposition of antimicrobial and stem cell friendly silver-substituted tricalcium phosphate nanocoatings on titanium alloy. *Bioactive Materials* [Internet], *6*(8), 2629–2642. Available from: https://www.sciencedirect.com/science/article/pii/S2452199X20303418

[14] Kocgozlu, L., Mutschler, A., Tallet, L., Calligaro, C., Knopf-Marques, H., Lebaudy, E., et al. (2024). Cationic homopolypeptides: A versatile tool to design multifunctional antimicrobial nanocoatings. *Materials Today Bio* [Internet], *28*, 101168. Available from: https://www.sciencedirect.com/science/article/pii/S2590006424002291

[15] Huo, D., Liu, T., Huang, K., Que, C., Jiang, S., Yang, Y., et al. (2024). AgBiS2@CQDs/Ti nanocomposite coatings for combating implant-associated infections by photodynamic /photothermal therapy. *Biomaterials Advances* [Internet], *158*, 213763. Available from: https://www.sciencedirect.com/science/article/pii/S2772950824000062

[16] Karagkiozaki, V., Karagiannidis, P. G., Gioti, M., Kavatzikidou, P., Georgiou, D., Georgaraki, E., et al. (2013). Bioelectronics meets nanomedicine for cardiovascular implants: PEDOT-based nanocoatings

for tissue regeneration. *Biochimica et Biophysica Acta – General Subjects* [Internet], *1830*(9), 4294–4304. Available from: https://www.sciencedirect.com/science/article/pii/S0304416512003807
[17] Tawakal, M. S., Abdelghany Metwally, A. M., El-Wassefy, N. A., Tawfik, M. A., & Shamaa, M. S. (2023). Static friction, surface roughness, and antibacterial activity of orthodontic brackets coated with silver and silver chitosan nanoparticles. *Journal of the World Federation of Orthodontists* [Internet], *12*(6), 260–268. Available from: https://www.sciencedirect.com/science/article/pii/S221244382300067X
[18] Attia, N. F., Zakria, A. M., Nour, M. A., Abd El-Ghany, N. A., & Elashery, S. E. A. (2023). Rational strategy for construction of multifunctional coatings for achieving high fire safety, antibacterial, UV protection and electrical conductivity functions of textile fabrics. *Materials Today Sustainability* [Internet], *23*, 100450. Available from: https://www.sciencedirect.com/science/article/pii/S2589234723001367
[19] Giri J, T. S., & Saravanan, R. (2024). Optimizing laser cladding powder injection parameters to shape bioactive glass nano-coated zirconium oxide for biomedical application. *Heliyon* [Internet], *10*(3), e25277. Available from: https://www.sciencedirect.com/science/article/pii/S2405844024013082
[20] Sen, A., Kurian, M. S., Narayanan, D. P., Abraham, A., Thomas, T. S., Jayalakshmi, P. S., et al. (2024). Bio-nanocomposites: Innovative solutions for addressing issues in health, agriculture, energy and environmental domains. *Nano-Structures and Nano-Objects* [Internet], *39*, 101270. Available from: https://www.sciencedirect.com/science/article/pii/S2352507X24001811
[21] Horprasertkij, K., Dwivedi, A., Riansuwan, K., Kiratisin, P., & Nasongkla, N. (2019). Spray coating of dual antibiotic-loaded nanospheres on orthopedic implant for prolonged release and enhanced antibacterial activity. *Journal of Drug Delivery Science and Technology* [Internet], *53*, 101102. Available from: https://www.sciencedirect.com/science/article/pii/S1773224719300656
[22] Tsiapla, A. R., Bakola, V., Karagkiozaki, V., Pavlidou, E., & Logothetidis, S. (2019). Biodegradable electrosprayed NPs as drug carriers for optimal treatment of orthopaedic infections. *Materials Today: Proceedings* [Internet], *19*, 110–116. Available from: https://www.sciencedirect.com/science/article/pii/S2214785319328603
[23] Mansoori, A., Ahmad, S., Sonia, Vashishath, M., & Kumar, D. (2022). Low-cost inkjet-printed humidity sensor using nanoporous surface on coated paper. *Sensors and Actuators B: Chemical* [Internet], *370*, 132389. Available from: https://www.sciencedirect.com/science/article/pii/S0925400522010310
[24] Wang, P., Gong, J. Y., Li, Y., Tian, X. B., Xie, R., Ju, X. J., et al. (2024). Superamphophilic antimicrobial membrane engineered with zinc-coordinated metallo-polymeric framework nanocoating. *Journal of Membrane Science* [Internet], *702*, 122779. Available from: https://www.sciencedirect.com/science/article/pii/S0376738824003739
[25] Nakao, S., Ishihara, T., Tsujimura, T., Hata, Y., Higashino, N., Kusuda, M., et al. (2024). Two-year real world clinical outcomes after intravascular imaging device guided percutaneous coronary intervention with ultrathin-strut biodegradable-polymer sirolimus-eluting stent. *International Journal of Cardiology* [Internet], *399*, 131686. Available from: https://www.sciencedirect.com/science/article/pii/S0167527323018442
[26] Calabrese, G., De Luca, G., Franco, D., Morganti, D., Rizzo, M. G., Bonavita, A., et al. (2023). Structural and antibacterial studies of novel ZnO and ZnxMn(1−x)O nanostructured titanium scaffolds for biomedical applications. *Biomaterials Advances* [Internet], *145*, 213193. Available from: https://www.sciencedirect.com/science/article/pii/S2772950822004708
[27] Ul Haq, I., & Krukiewicz, K. (2023). Antimicrobial approaches for medical implants coating to prevent implants associated infections: Insights to develop durable antimicrobial implants. *Applied Surface Science Advances* [Internet], *18*, 100532. Available from: https://www.sciencedirect.com/science/article/pii/S2666523923001666

[28] Dwivedi, A., Mazumder, A., & Nasongkla, N. (2018). Layer-by-layer nanocoating of antibacterial niosome on orthopedic implant. *International Journal of Pharmaceutics* [Internet], *547*(1), 235–243. Available from: https://www.sciencedirect.com/science/article/pii/S0378517318303892

[29] Poyyamozhi, N., Kumar, S. S., Renugadevi, P., & Prabhakar, P. (2024). Experimental investigation of nano-coating effects on extended surfaces for enhanced thermal management of heat sinks. *International Journal of Thermal Sciences* [Internet], *201*, 109049. Available from: https://www.sciencedirect.com/science/article/pii/S1290072924001716

[30] Su, C., Huang, X., Zhang, L., Zhang, Y., Yu, Z., Chen, C., et al. (2023). Robust superhydrophobic wearable piezoelectric nanogenerators for self-powered body motion sensors. *Nano Energy* [Internet], *107*, 108095. Available from: https://www.sciencedirect.com/science/article/pii/S2211285522011739

[31] Fischer, N. G., Kobe, A. C., Dai, J., He, J., Wang, H., Pizarek, J. A., et al. (2022). Tapping basement membrane motifs: Oral junctional epithelium for surface-mediated soft tissue attachment to prevent failure of percutaneous devices. *Acta Biomaterialia* [Internet], *141*, 70–88. Available from: https://www.sciencedirect.com/science/article/pii/S1742706121008448

[32] Moro, G., Chiavaioli, F., Liberi, S., Zubiate, P., Del Villar, I., Angelini, A., et al. (2021). (INVITED) Nanocoated fiber label-free biosensing for perfluorooctanoic acid detection by lossy mode resonance. *Results of the Optional Practical Training* [Internet], *5*, 100123. Available from: https://www.sciencedirect.com/science/article/pii/S2666950121000717

[33] Tîlmaciu, C. M., Mathieu, M., Lavigne, J. P., Toupet, K., Guerrero, G., Ponche, A., et al. (2015). In vitro and in vivo characterization of antibacterial activity and biocompatibility: A study on silver-containing phosphonate monolayers on titanium. *Acta Biomaterialia* [Internet], *15*, 266–277. Available from: https://www.sciencedirect.com/science/article/pii/S1742706114005935

[34] Bijli, M. K., Verma, P., & Singh, A. P. (2024). A systematic review on the potency of swarm intelligent nanorobots in the medical field. *Swarm and Evolutionary Computation* [Internet], *86*, 101524. Available from: https://www.sciencedirect.com/science/article/pii/S2210650224000579

[35] Branecky, M., Aboualigaledari, N., & Cech, V. (2020). Plasma nanotechnology for controlling chemical and physical properties of organosilicon nanocoatings. *Materials Today Communications* [Internet], *24*, 101234. Available from: https://www.sciencedirect.com/science/article/pii/S235249282030502X

[36] Merkl, P., Aschtgen, M. S., Henriques-Normark, B., & Sotiriou, G. A. (2021). Biofilm interfacial acidity evaluation by pH-Responsive luminescent nanoparticle films. *Biosensors and Bioelectronics* [Internet], *171*, 112732. Available from: https://www.sciencedirect.com/science/article/pii/S095656632030720X

[37] Maillard, L., Vochelet, F., Peycher, P., Ayari, A., Barra, N., Billé, J., et al. (2020). MAPT (Mono Antiplatelet Therapy) as regular regimen after COBRA PzF™ NanoCoated Coronary Stent (NCS) implantation. *Cardiovascular Revascularization Medicine* [Internet], *21*(6), 785–789. Available from: https://www.sciencedirect.com/science/article/pii/S1553838919306566

[38] Nizami, M. Z. I., Campéon, B. D. L., & Nishina, Y. (2022). Electrodeposition of hydroxyapatite and graphene oxide improves the bioactivity of medical grade stainless steel. *Materials Today Sustainability* [Internet], *19*, 100193. Available from: https://www.sciencedirect.com/science/article/pii/S2589234722000859

[39] Wei, X., Cai, B., Chen, K., Cheng, L., Zhu, Y., Wang, Z., et al. (2019). Enhanced isolation and release of fetal nucleated red blood cells using multifunctional nanoparticle-based microfluidic device for non-invasive prenatal diagnostics. *Sensors and Actuators B: Chemical* [Internet], *281*, 131–138. Available from: https://www.sciencedirect.com/science/article/pii/S0925400518317908

[40] Karacan, I., Ben-Nissan, B., Wang, H. A., Juritza, A., Swain, M. V., Müller, W. H., et al. (2019). Mechanical testing of antimicrobial biocomposite coating on metallic medical implants as drug delivery system. *Materials Science and Engineering: C* [Internet], *104*, 109757. Available from: https://www.sciencedirect.com/science/article/pii/S0928493119301602

[41] Wang, X., Hu, T., Hu, B., Liu, Y., Wang, Y., He, Y., et al. (2022). Imparting reusable and SARS-CoV-2 inhibition properties to standard masks through metal-organic nanocoatings. *Journal of Hazardous Materials* [Internet], *431*, 128441. Available from: https://www.sciencedirect.com/science/article/pii/S0304389422002291

[42] Cutlip, D. E., Jauhar, R., Meraj, P., Garratt, K. N., Novack, V., Novack, L., et al. (2022). Five-year clinical outcomes of the COBRA Polyzene F NanoCoated Coronary Stent system. *Cardiovascular Revascularization Medicine* [Internet], *41*, 76–80. Available from: https://www.sciencedirect.com/science/article/pii/S1553838921008319

[43] Chandna, S., Gogde, K., Paul, S., & Bhaumik, J. (2024). Lignin-based sustainable antifungal gel nanocoatings for disinfecting biomedical devices††Electronic supplementary information (ESI) available: Instruments and methods, materials and chemicals, and tables. *RSC Sustainability* [Internet], *2*(8), 2348–2356. Available from: https://doi.org/10.1039/d4su00180j, https://www.sciencedirect.com/science/article/pii/S275381252400168X

[44] Kister, F., Specht, O., Warkentin, M., Geis-Gerstorfer, J., & Rupp, F. (2017). Peri-implantitis cleaning instrumentation influences the integrity of photoactive nanocoatings. *Dental Materials* [Internet], *33*(2), e69–78. Available from: https://www.sciencedirect.com/science/article/pii/S0109564116305346

[45] Zhou, Q., Wu, T., Bai, Z., Hong, G., Bian, J., Xie, H., et al. (2022). A silane-based coupling strategy for enhancing the mechanical properties of proanthocyanidin nanocoatings on Ti dental implants. *Applied Surface Science* [Internet], *602*, 154400. Available from: https://www.sciencedirect.com/science/article/pii/S016943322201933X

[46] Mehta, P., Al-Kinani, A. A., Arshad, M. S., Singh, N., van der Merwe, S. M., Chang, M. W., et al. (2019). Engineering and development of chitosan-based nanocoatings for ocular contact lenses. *Journal Pharmaceutical Sciences* [Internet], *108*(4), 1540–1551. Available from: https://www.sciencedirect.com/science/article/pii/S0022354918307846

[47] Girbas, M. G., Riedel, T., Riedelová, Z., Wolf, M., Schlensak, C., & Avci-Adali, M. (2024). Comparison of the hemocompatibility of neurovascular flow diverters with anti-thrombogenic coatings. *Journal of Science: Advanced Materials and Devices* [Internet], *9*(1), 100666. Available from: https://www.sciencedirect.com/science/article/pii/S2468217923001351

[48] Pattnaik, P. K., Mishra, S. R., Anwar Bég, O., Khan, U. F., & Umavathi, J. C. (2022). Axisymmetric radiative titanium dioxide magnetic nanofluid flow on a stretching cylinder with homogeneous/heterogeneous reactions in Darcy-Forchheimer porous media: Intelligent nanocoating simulation. *Materials Science and Engineering: B* [Internet], *277*, 115589. Available from: https://www.sciencedirect.com/science/article/pii/S0921510721005420

[49] Kwiatkowska, A., Wierzchowski, K., Lipko, A., Grzeczkowicz, A., Antosiak-Iwańska, M., Drabik, M., et al. (2024). Albumin-derived perfluorodecalin-based hybrid systems with developed effective surfaces aimed at biomedical application. *Colloids and Surfaces A: Physicochemical and Engineering Aspects* [Internet], *694*, 134157. Available from: https://www.sciencedirect.com/science/article/pii/S0927775724010185

[50] Wulandari, E., Bilimoria, K., Krasowska, M., Al-Bataineh, S., Beattie, D., Gillam, T., et al. (2023). Nanoscale iodophoric poly(vinyl amide) coatings for the complexation and release of iodine for antimicrobial surfaces. *Applied Surface Science* [Internet], *641*, 158422. Available from: https://www.sciencedirect.com/science/article/pii/S0169433223021025

[51] Priyanka, S., Arvind Bharani, R. S., & John, A. (2023). Biocompatible green technology principles for the fabrication of food packaging material with noteworthy mechanical and antimicrobial properties– A sustainable developmental goal towards the effective, safe food preservation strategy. *Chemosphere* [Internet], *336*, 139240. Available from: https://www.sciencedirect.com/science/article/pii/S0045653523015072

[52] Wongsuwan, N., Dwivedi, A., Tancharoen, S., & Nasongkla, N. (2020). Development of dental implant coating with minocycline-loaded niosome for antibacterial application. *Journal of Drug Delivery Science and Technology* [Internet], *56*, 101555. Available from: https://www.sciencedirect.com/science/article/pii/S1773224719315849

[53] Ismail, A. A., Alzubi, F., Al-Hajji, L. A., Alseidi, M., Ahmad, S., & Alduweesh, A. (2023). Self-cleaning coatings for minimizing the impact of dust precipitation on the power production of solar cells utilizing mesoporous TiO_2/SiO_2 and ZnO/SiO_2 films. *Ceramics International* [Internet], *49*(14, Part A), 22788–22796. Available from: https://www.sciencedirect.com/science/article/pii/S0272884223010489

[54] Agarwalla, S. V., Ellepola, K., Silikas, N., Castro Neto, A. H., Seneviratne, C. J., & Rosa, V. (2021). Persistent inhibition of *Candida albicans* biofilm and hyphae growth on titanium by graphene nanocoating. *Dental Materials* [Internet], *37*(2), 370–377. Available from: https://www.sciencedirect.com/science/article/pii/S0109564120303481

[55] Van Duong, T., Nguyen, H. T., Wang, F., Wang, M., Narwankar, P. K., & Taylor, L. S. (2022). Surface nanocoating of high drug-loading spray-dried amorphous solid dispersions by atomic layer coating: Excellent physical stability under accelerated storage conditions for two years. *International Journal of Pharmaceutics* [Internet], *620*, 121747. Available from: https://www.sciencedirect.com/science/article/pii/S0378517322003027

[56] Kurniawan, F. H., Chinavinijkul, P., & Nasongkla, N. (2021). Hydrophobic and antibacterial bed sheet using ZnO nanoparticles: A large-scale technique. *Journal of Drug Delivery Science and Technology* [Internet], *62*, 102339. Available from: https://www.sciencedirect.com/science/article/pii/S1773224721000204

[57] Karagkiozaki, V., Logothetidis, S., Kalfagiannis, N., Lousinian, S., & Giannoglou, G. (2009). Atomic force microscopy probing platelet activation behavior on titanium nitride nanocoatings for biomedical applications. *Nanomedicine Nanotechnology, Biology and Medicine* [Internet], *5*(1), 64–72. Available from: https://www.sciencedirect.com/science/article/pii/S1549963408001202

[58] Baghdan, E., Raschpichler, M., Lutfi, W., Pinnapireddy, S. R., Pourasghar, M., Schäfer, J., et al. (2019). Nano spray dried antibacterial coatings for dental implants. *European Journal of Pharmaceutics and Biopharmaceutics* [Internet], *139*, 59–67. Available from: https://www.sciencedirect.com/science/article/pii/S0939641119301043

[59] Xu, Y., Hou, L., Zhao, H., Bi, S., Zhu, L., & Lu, Y. (2019). Alumina sandpaper-supported nickel nanocoatings and its excellent application in non-enzymatic glucose sensing. *Applied Surface Science* [Internet], *463*, 1028–1036. Available from: https://www.sciencedirect.com/science/article/pii/S0169433218323912

[60] Merugu, R., Sharma, S., & Rastogi, M. (2023). An overview of nanotechnological applications in biomedical engineering. *Materials Today: Proceedings* [Internet], *76*, 412–417. Available from: https://www.sciencedirect.com/science/article/pii/S2214785322073965

[61] Felipe, B. H. S., Cabral, R. L. B., Ladchumananandasivam, R., Zille, A., Kim, S., Fechine, P. B. A., et al. (2022). Nanocoating on cotton fabric with nitrogen-doped graphene quantum dots/titanium dioxide/PVA: An erythemal UV protection and photoluminescent finishing. *Jouranal of Materials Research and Technology* [Internet], *18*, 2435–2450. Available from: https://www.sciencedirect.com/science/article/pii/S2238785422003854

[62] Kopernik, M., & Milenin, A. (2012). Two-scale finite element model of multilayer blood chamber of POLVAD_EXT. *Archives of Civil and Mechanical Engineering* [Internet], *12*(2), 178–185. Available from: https://www.sciencedirect.com/science/article/pii/S1644966512000325

[63] Agarwalla, S. V., Ellepola, K., Sorokin, V., Ihsan, M., Silikas, N., Neto, A. H. C., et al. (2022). Antimicrobial-free graphene nanocoating decreases fungal yeast-to-hyphal switching and maturation of cross-kingdom biofilms containing clinical and antibiotic-resistant bacteria.

Biomaterials and Biosystems [Internet], *8*, 100069. Available from: https://www.sciencedirect.com/science/article/pii/S2666534422000319

[64] Li, Y., Miao, Y., Yang, L., Wang, G., Fu, M., Wang, Y., et al. (2023). Malate-based polyester chemically shielded metal-phenolic networks coated artificial hair fibers with long-lasting antimicrobial and anti-inflammatory performance. *Chemical Engineering Journal* [Internet], *455*, 140572. Available from: https://www.sciencedirect.com/science/article/pii/S1385894722060521

[65] Wang, D. K., Rahimi, M., & Filgueira, C. S. (2021). Nanotechnology applications for cardiovascular disease treatment: Current and future perspectives. *Nanomedicine Nanotechnology, Biology and Medicine* [Internet], *34*, 102387. Available from: https://www.sciencedirect.com/science/article/pii/S1549963421000307

[66] Park, S., Park, J., Heo, J., Lee, S. E., Shin, J. W., Chang, M., et al. (2018). Polysaccharide-based superhydrophilic coatings with antibacterial and anti-inflammatory agent-delivering capabilities for ophthalmic applications. *Journal of Industrial and Engineering Chemistry* [Internet], *68*, 229–237. Available from: https://www.sciencedirect.com/science/article/pii/S1226086X18304337

[67] Fatma, B., Andrabi, S. M., Gupta, S., Verma, V., Kumar, A., Pitsalidis, C., et al. (2023). Biocompatible, breathable and degradable microbial cellulose based triboelectric nanogenerator for wearable transient electronics. *Nano Energy* [Internet], *114*, 108628. Available from: https://www.sciencedirect.com/science/article/pii/S2211285523004652

[68] Yuan, B., Xue, M., Zhao, Y., Guo, Q., Zheng, G., Xu, Z., et al. (2023). A self-degradable "nanoarmor" coating of medical implant potentiates bone fracture healing. *Nano Today* [Internet], *52*, 101959. Available from: https://www.sciencedirect.com/science/article/pii/S1748013223002086

[69] Andreeva, T., Drieschner, T., Golovko, D. S., Lorenz, A., Rebner, K., & Krastev, R. (2023). Process validation and a new method for quality control of ultrathin polyelectrolyte multilayer coatings. *Colloids and Surfaces A: Physicochemical and Engineering Aspects* [Internet], *676*, 132157. Available from: https://www.sciencedirect.com/science/article/pii/S0927775723012414

[70] Hou, L., Zhao, H., Bi, S., Xu, Y., & Lu, Y. (2017). Ultrasensitive and highly selective sandpaper-supported copper framework for non-enzymatic glucose sensor. *Electrochimica Acta* [Internet], *248*, 281–291. Available from: https://www.sciencedirect.com/science/article/pii/S0013468617315608

[71] Dai, D., Zhou, D., He, L., Wang, C., & Zhang, C. (2022). Graphene oxide nanocoating for enhanced corrosion resistance, wear resistance and antibacterial activity of nickel-titanium shape memory alloy. *Surface and Coatings Technology* [Internet], *431*, 128012. Available from: https://www.sciencedirect.com/science/article/pii/S0257897221011865

[72] Marqués-Marchán, J., Fernandes, M. M., Lanceros-Méndez, S., & Asenjo, A. (2023). Magneto-mechanically induced antimicrobial properties of cone-like shaped surfaces. *Applied Materials Today* [Internet], *34*, 101900. Available from: https://www.sciencedirect.com/science/article/pii/S2352940723001701

[73] Wang, Y., Chen, L., Hou, R., Bai, L., & Guan, S. (2023). Rapamycin-loaded nanocoating to specially modulate smooth muscle cells on ZE21B alloy for vascular stent application. *Applied Surface Science* [Internet], *615*, 156410. Available from: https://www.sciencedirect.com/science/article/pii/S0169433223000867

[74] Janjić, K., Valentova, A., Arellano, S., Unterhuber, A., Krause, A., Oberoi, G., et al. (2024). The impact of print orientation and graphene nanoplatelets on biaxial flexural strength and cytotoxicity of a 3D printable resin for occlusal splints. *Dental Materials* [Internet]. Available from: https://www.sciencedirect.com/science/article/pii/S010956412400232X

[75] Ben-Nissan, B., Choi, A. H., & Bendavid, A. (2013). Mechanical properties of inorganic biomedical thin films and their corresponding testing methods. *Surface and Coatings Technology* [Internet], *233*, 39–48. Available from: https://www.sciencedirect.com/science/article/pii/S0257897212011553

[76] Manikandan, V., & Min, S. C. (2023). Roles of polysaccharides-based nanomaterials in food preservation and extension of shelf-life of food products: A review. *International Journal of Biological*

Macromolecules [Internet], *252*, 126381. Available from: https://www.sciencedirect.com/science/article/pii/S0141813023032774

[77] Lu, P., & Ai, J. (2023). Fluorescence probes and their sensing applications in nanomaterials system. *Talanta Open* [Internet], *8*, 100248. Available from: https://www.sciencedirect.com/science/article/pii/S2666831923000681

[78] Islam, M. R., Parimalam, M., Sumdani, M. G., Taher, M. A., Asyadi, F., & Yenn, T. W. (2020). Rheological and antimicrobial properties of epoxy-based hybrid nanocoatings. *Polymer Testing* [Internet], *81*, 106202. Available from: https://www.sciencedirect.com/science/article/pii/S0142941819310438

[79] Shen, W., Cai, K., Yang, Z., Yan, Y., Yang, W., & Liu, P. (2012). Improved endothelialization of NiTi alloy by VEGF functionalized nanocoating. *Colloids Surfaces B: Biointerfaces* [Internet], *94*, 347–353. Available from: https://www.sciencedirect.com/science/article/pii/S092777651200118X

[80] Nadeem, Q. U. A., Gill, R., Cheung, D. W. F., & Shchukin, D. G. (2021). Ensnaring animate-inanimate toxins using ZnO–MgO @ Elastomer thin films with self-cleaning and regenerative properties. *Environmental Technology and Innovation* [Internet], *24*, 102011. Available from: https://www.sciencedirect.com/science/article/pii/S2352186421006593

[81] Kohse, M., Witzdam, L., Jakob, F., Boes, A., Mescheder, H., Day, R., et al. (2024). Bioinspired active hemocompatible coating systems for mechanical circulatory support devices: When engineering meets nano and molecular technology. *Procedia CIRP* [Internet], *125*, 60–65. Available from: https://www.sciencedirect.com/science/article/pii/S2212827124003652

[82] Karabulut, G., Beköz Üllen, N., Karakuş, S., & Toruntay, C. (2023). Enhancing antibacterial and anticorrosive properties of 316L stainless steel with nanocoating of copper oxide nanoparticles. *Materials Chemistry and Physics* [Internet], *308*, 128265. Available from: https://www.sciencedirect.com/science/article/pii/S0254058423009732

[83] Tagad, H. D., Brito, J., Marin, A., Buckley, C., Wang, H., Sun, J., et al. (2023). 4-methylumbelliferone-functionalized polyphosphazene and its assembly into biocompatible fluorinated nanocoatings with selective antiproliferative activity. *Biomacromolecules* [Internet], *24*(5), 2278–2290. Available from: https://www.sciencedirect.com/science/article/pii/S1525779723001381

[84] Gurzawska, K., Svava, R., Yihua, Y., Haugshøj, K. B., Dirscherl, K., Levery, S. B., et al. (2014). Osteoblastic response to pectin nanocoating on titanium surfaces. *Materials Science and Engineering: C* [Internet], *43*, 117–125. Available from: https://www.sciencedirect.com/science/article/pii/S0928493114003907

[85] Arshad, I., Kanwal, A., Zafar, I., Unar, A., Mouada, H., Razia, I. T., et al. (2024). Multifunctional role of nanoparticles for the diagnosis and therapeutics of cardiovascular diseases. *Environmental Research* [Internet], *242*, 117795. Available from: https://www.sciencedirect.com/science/article/pii/S0013935123025999

[86] Tsiapla, A. R., Karagkiozaki, V., Pappa, F., Bakola, V., Choli-Papadopoulou, T., Moutsios, I., et al. (2017). Drug delivery nanoplatform for orthopaedic-associated infections. *Materials Today: Proceedings* [Internet], *4*(7, Part 1), 6880–6888. Available from: https://www.sciencedirect.com/science/article/pii/S2214785317313056

[87] Zubiate, P., Urrutia, A., Zamarreño, C. R., Egea-Urra, J., Fernández-Irigoyen, J., Giannetti, A., et al. (2019). Fiber-based early diagnosis of venous thromboembolic disease by label-free D-dimer detection. *Biosensors and Bioelectronics: X* [Internet], *2*, 100026. Available from: https://www.sciencedirect.com/science/article/pii/S2590137019300160

[88] He, L., Dai, D., Xie, L., Chen, Y., & Zhang, C. (2021). Biological effects, applications and strategies of nanomodification of dental metal surfaces. *Materials and Design* [Internet], *207*, 109890. Available from: https://www.sciencedirect.com/science/article/pii/S0264127521004433

[89] Bangera, A. E., & Appaiah, K. (2020). Tunable wettability and prediction of total surface energy for nanocoatings: An empirical model. *Surface and Coatings Technology* [Internet], *383*, 125277. Available from: https://www.sciencedirect.com/science/article/pii/S0257897219312678

[90] Xu, Y., Li, Y., Chen, T., Dong, C., Zhang, K., & Bao, X. (2024). A short review of medical-grade stainless steel: Corrosion resistance and novel techniques. *Journal of Materials Research and Technology* [Internet], *29*, 2788–2798. Available from: https://www.sciencedirect.com/science/article/pii/S2238785424002400

[91] Piruşcă, I. A., Balaure, P. C., Grumezescu, V., Irimiciuc, S. A., Oprea, O. C., Bîrcă, A. C., et al. (2024). New Fe3O4-based coatings with enhanced anti-biofilm activity for medical devices. *Antibiotics*, *13*(7).

[92] Riau, A. K., Aung, T. T., Setiawan, M., Yang, L., Yam, G. H. F., Beuerman, R. W., et al. (2019). Surface immobilization of nano-silver on polymeric medical devices to prevent bacterial biofilm formation. *Pathogens*, *8*(3), 1–17.

[93] Dwivedi, A., Mazumder, A., & Nasongkla, N. (2018). Layer-by-layer nanocoating of antibacterial niosome on orthopedic implant. *International Journal of Pharmaceutics* [Internet], *547*(1–2), 235–243. Available from: https://doi.org/10.1016/j.ijpharm.2018.05.075

[94] Thakur, A., Dagdag, O., Berisha, A., Ebenso, E. E., Kumar, A., Sharma, S., et al. (2024). Mechanistic insights into the corrosion inhibition of mild steel by eco-benign *Asphodelus tenuifolius* aerial extract in acidic environment: Electrochemical and computational analysis. *Surface and Coatings Technology* [Internet], *480*, 130568. Available from: https://www.sciencedirect.com/science/article/pii/S0257897224001981

[95] Sharma, D., Thakur, A., Sharma, M. K., Bhardwaj, A., Kumar, D., Kumar, A., et al. (2025). Electrochemicaland computational insightsinto the utilization of an N-heteroaromatic containing compound2-(4-Methoxy-phenyl)-5-naphthalen-2-yl-[1,3,4]oxadiazole as a promising anticorrosive agent for mild steel in corrosive medium. *Journal of Molecular Structure* [Internet], *1321*, 140055. Available from: https://www.sciencedirect.com/science/article/pii/S002228602402564X

[96] Thakur, A., Kumar, A., & Singh, A. (2024). Adsorptive removal of heavy metals, dyes, and pharmaceuticals: Carbon-based nanomaterials in focus. *Carbone New York* [Internet], *217*, 118621. Available from: https://www.sciencedirect.com/science/article/pii/S0008622323008667

[97] Thakur, A., & Kumar, A. (2024). Emerging paradigms into bioremediation approaches for nuclear contaminant removal: From challenge to solution. *Chemosphere* [Internet], *352*, 141369. Available from: https://www.sciencedirect.com/science/article/pii/S0045653524002625

[98] Bensalah, J., Thakur, A., & Kumar, A. (2024). Investigating the adsorption processes of polymer resins for the removal of micropollutants: A comprehensive review in the field of environmental remediation. *Environmental Research* [Internet], *254*, 119128. Available from: https://www.sciencedirect.com/science/article/pii/S0013935124010338

[99] Pizarek, J. A., Fischer, N. G., & Aparicio, C. (2023). Immunomodulatory IL-23 receptor antagonist peptide nanocoatings for implant soft tissue healing. *Dental Materials* [Internet], *39*(2), 204–216. Available from: https://www.sciencedirect.com/science/article/pii/S0109564123000192

[100] Magovac, E., Budimir, A., Jordanov, I., Bischof, S., & Grunlan, J. C. (2021). Antibacterial cotton from novel phytic acid-based multilayer nanocoating. *Green Materials* [Internet], *10*(1), 35–40. Available from: https://www.sciencedirect.com/science/article/pii/S204912392100002X

[101] Hoyo, J., Ivanova, K., Guaus, E., & Tzanov, T. (2019). Multifunctional ZnO NPs-chitosan-gallic acid hybrid nanocoating to overcome contact lenses associated conditions and discomfort. *Journal of Colloid and Interface Science* [Internet], *543*, 114–121. Available from: https://www.sciencedirect.com/science/article/pii/S0021979719302176

[102] Zhang, X., Yang, Y., Sun, S., Wu, Q., Wan, N., Pan, D., et al. (2016). Multifunctional ZrF4 nanocoating for improving lithium storage performances in layered Li[Li0.2Ni0.17Co0.07Mn0.56]O2. *Solid State Ionics* [Internet], *284*, 7–13. Available from: https://www.sciencedirect.com/science/article/pii/S0167273815301247

[103] Cutlip, D. E., Garratt, K. N., Novack, V., Barakat, M., Meraj, P., Maillard, L., et al. (2017). 9-month clinical and angiographic outcomes of the COBRA Polyzene-F NanoCoated Coronary Stent system. *JACC: Cardiovascular Interventions* [Internet], *10*(2), 160–167. Available from: https://www.sciencedirect.com/science/article/pii/S1936879816319100

[104] Dziekan, Z., Pituła, E., Kwietniewski, N., Stonio, B., Janik, M., Śmiarowski, T., et al. (2022). Performance of nanoimprinted and nanocoated optical label-free biosensor – Nanocoating properties perspective. *Optics and Lasers in Engineering* [Internet], *153*, 107009. Available from: https://www.sciencedirect.com/science/article/pii/S0143816622000641

[105] Yang, J., Qian, H., Wang, J., Ju, P., Lou, Y., Li, G., et al. (2021). Mechanically durable antibacterial nanocoatings based on zwitter ionic copolymers containing dopamine segments. *Journal of Material Science Technology* [Internet], *89*, 233–241. Available from: https://www.sciencedirect.com/science/article/pii/S100503022030949X

[106] Shen, Y., Zhang, W., Xie, Y., Li, A., Wang, X., Chen, X., et al. (2021). Surface modification to enhance cell migration on biomaterials and its combination with 3D structural design of occluders to improve interventional treatment of heart diseases. *Biomaterials* [Internet], *279*, 121208. Available from: https://www.sciencedirect.com/science/article/pii/S0142961221005652

[107] Sharma, D., Thakur, A., Sharma, M. K., Kumar, A., Jakhar, K., Kumar, S., et al. (2023). A convenient synthesis, electrochemical profiling, and morphological studies of a pyridine-based 1,3,4-oxadiazole hybrid: A promising study for corrosion mitigation of mild steel in strongly acidic environment. *Inorganic Chemistry Communications* [Internet], *158*(P1), 111554. Available from: https://doi.org/10.1016/j.inoche.2023.111554

[108] Kaya, S., Lgaz, H., Thakkur, A., Kumar, A., Özbakır Işın, D., Karakuş, N., & Ben Ahmed, S. (2023). Molecular insights into the corrosion inhibition mechanism of omeprazole and tinidazole: A theoretical investigation. *Molecular Simulation* [Internet], *49*(17), 1632–1646. Available from: https://doi.org/10.1080/08927022.2023.2256888

[109] Sharma, D., Thakur, A., Sharma, M. K., Jakhar, K., Kumar, A., Sharma, A. K., et al. (2023). Synthesis, electrochemical, morphological, computational and corrosion inhibition studies of 3-(5-Naphthalen-2-yl -[1,3,4]oxadiazol-2-yl)-pyridine against Mild Steel in 1 M HCl. *Asian Journal of Chemistry*, *35*(5), 1079–1088.

[110] Pattnaik, P. K., Mishra, S. R., Anwar Bég, O., Khan, U. F., & Umavathi, J. C. (2022). Axisymmetric radiative titanium dioxide magnetic nanofluid flow on a stretching cylinder with homogeneous/heterogeneous reactions in Darcy-Forchheimer porous media: Intelligent nanocoating simulation. *Materials Science and Engineering: B*, *277*(November 2020).

[111] Sultana, A., Zare, M., Luo, H., & Ramakrishna, S. (2021). Surface engineering strategies to enhance the in situ performance of medical devices including atomic scale engineering. *International Journal of Molecular Sciences*, *22*(21).

[112] Karabulut, G., Beköz Üllen, N., Karakuş, S., & Toruntay, C. (2023). Enhancing antibacterial and anticorrosive properties of 316L stainless steel with nanocoating of copper oxide nanoparticles. *Materials Chemistry and Physics*, *308*(July).

[113] Subhashree, S., Senthil Kumar, P., et al. (2022). New analytical strategies amplified with carbon-based nanomaterial for sensing food pollutants. *Chemosphere* [Internet], *295*, 133847. Available from: https://www.sciencedirect.com/science/article/pii/S004565352200340X

[114] Chandna, S., Thakur, N. S., Kaur, R., & Bhaumik, J. (2020). Lignin–bimetallic nanoconjugate doped pH-responsive hydrogels for laser-assisted antimicrobial photodynamic therapy. *Biomacromolecules* [Internet], *21*(8), 3216–3230. Available from: https://www.sciencedirect.com/science/article/pii/S152577972100684X

[115] Kunčák, J., Forinová, M., Pilipenco, A., Procházka, V., Horák, P., Sycheva, S. D., et al. (2024). Automating data classification for label-free point-of-care biosensing in real complex samples. *Sensors and Actuators A: Physical* [Internet], *374*, 115501. Available from: https://www.sciencedirect.com/science/article/pii/S0924424724004953

Dileep Pathote, Ranjit Barua, Sachin Latiyann, Priyanka Dash, and Sudipto Datta*

Chapter 9 Surface modification of bone screws, reconstruction surgeries

Abstract: Bone screws are integral components in orthopedic reconstruction surgeries, providing mechanical stability and facilitating the healing of fractured or reconstructed bone. However, poor osseointegration, implant loosening, and susceptibility to infection often limit their success. This study investigates advanced surface modification techniques to enhance the functional performance of bone screws. Methods such as plasma spraying, chemical etching, anodization, and surface coating with bioactive materials like hydroxyapatite, titanium dioxide, and antimicrobial agents were evaluated. The impact of these modifications on biocompatibility, osseointegration, and mechanical stability was analyzed using in vitro and in vivo models. Results demonstrated that surface coatings significantly enhanced the bioactivity of bone screws, promoting faster and stronger integration with bone tissue. Additionally, antimicrobial surface treatments reduced the incidence of peri-implant infections without compromising cellular viability. Finite element analysis confirmed the mechanical integrity of modified screws under physiological loading conditions. This study underscores the potential of surface engineering to address critical challenges in orthopedic implantology, paving the way for improved outcomes in reconstruction surgeries. Further research into long-term clinical applications and patient-specific customization is recommended.

Keywords: Bone screw, implant surgeries, surface modifications, biomedical applications

9.1 Introduction

Bone screws serve an essential role in orthopedic and maxillofacial operations by providing support during bone healing and restoration. They are often made of stainless steel, titanium alloys, and biodegradable polymers. While these materials have good mechanical qualities, they are not biologically active, which can restrict osseointegra-

***Corresponding author: Sudipto Datta**, Department of Materials Engineering, Indian Institute of Science, C.V. Raman Avenue, Bangalore 560012, Karnataka, India, e-mail: sudiptodatta1990@gmail.com
Dileep Pathote, Sachin Latiyann, Priyanka Dash, Department of Materials Engineering, Indian Institute of Science, C.V. Raman Avenue, Bangalore 560012, Karnataka, India
Ranjit Barua, Mechanical Engineering Department, Omdayal Group of Institutions, Howrah 711316, West Bengal, India

https://doi.org/10.1515/9783111571423-009

tion and long-term success. Surface modification of bone screws seeks to solve these constraints by modifying their surface properties while maintaining their bulk mechanical integrity. Coordinated control of several biological processes that promote tissue growth, homeostasis, and damage healing is a key component of bone physiology [1]. Due in part to the many and complex inputs that contribute to the control of its niche, bone regenerating processes can be extremely difficult to mimic. These include the tissue's complex makeup and its many related pathways, which include things like the soluble microenvironment of bone, the composition and renewal of extracellular matrix (ECM) insoluble proteins and glycoproteins, interactions between cells and between cells and the ECM, mechanical stimulation of cells, and the function of microRNAs.

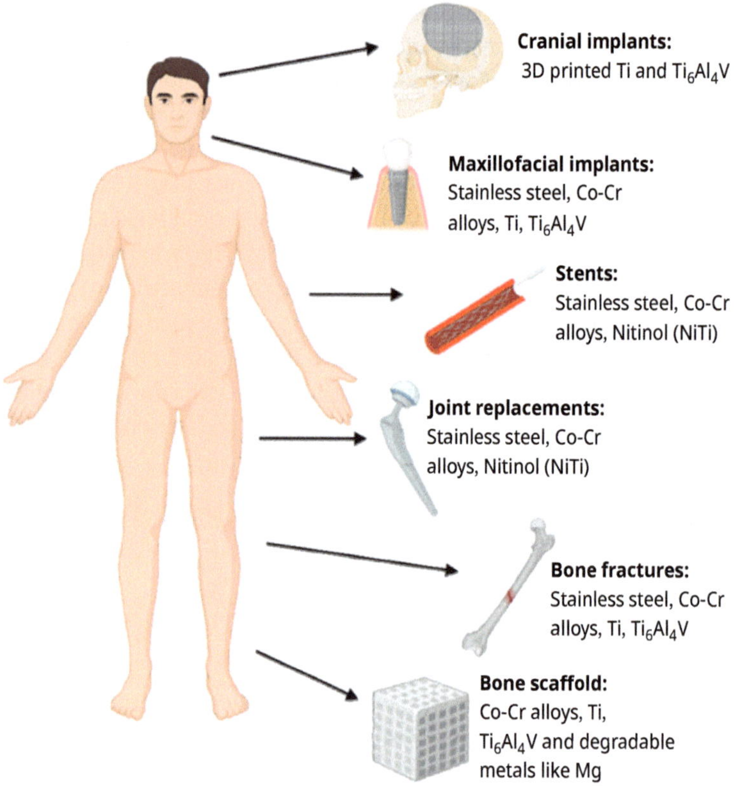

Figure 9.1: Types of metal implants that are used in human anatomy [2].

Improved surface features can significantly impact cell adhesion, bone development, and microbial colonization resistance. This chapter discusses numerous surface modification tactics, including coating technologies, surface texturing, and chemical treatments, emphasizing their use in reconstruction surgery. Numerous biomimetic and/or bioinspired bone regeneration techniques have been developed as a result of the understanding of each intervenient's unique and coordinated influence on bone tissue physiology as well as their interrelated activities (Figure 9.1). In actuality, the creation of novel, efficient, and appropriate therapies for bone-related injuries is desperately needed.

Throughout an adult's (i.e., non-embryonic) existence, bone grows, models, and remodels both longitudinally and radially. During childhood and adolescence, both longitudinal and radial development take place. In contrast, adults are less likely to undergo bone modeling, an anabolic process that involves the deposition of new bone in response to mechanical or physiological stimuli, than remodeling. Bone modeling involves the independent action of osteoblasts and osteoclasts, i.e., the activity of both cells may not be coupled anatomically or temporally, in contrast to bone remodeling, where osteoclasts and osteoblasts work sequentially in the same bone remodeling unit. Adult humans undergo constant remodeling to create and preserve intricate and useful skeletal structures [1]. The human body's structural stability is maintained by this mechanism, which helps offset the impacts of growing bone fragility over time.

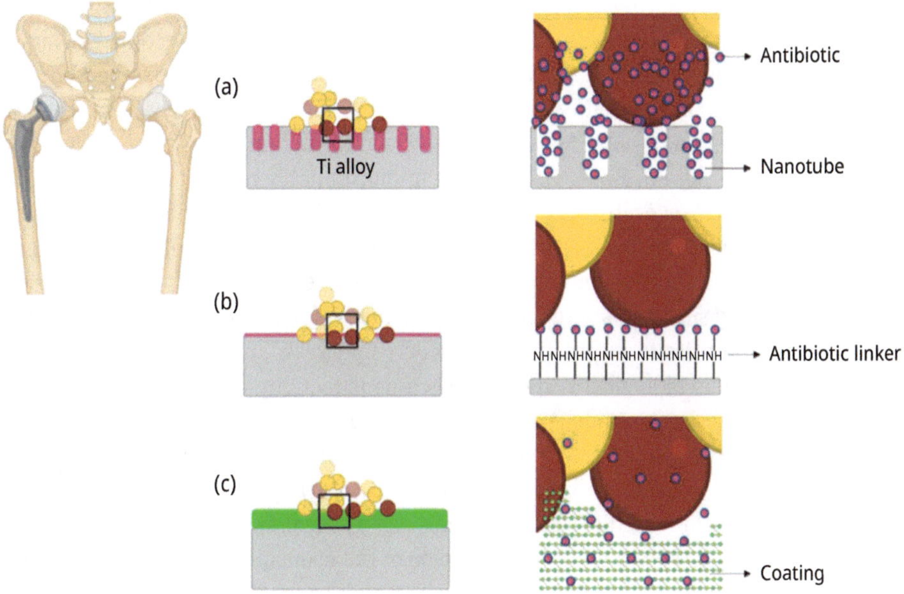

Figure 9.2: Numerous localized antibiotic treatment techniques: (a) nanotubes packed with antibiotics, (b) Ti alloy is covalently bonded to an antibiotic, and (c) coating with antibiotics. Yellow suggests a live bacterium. Red indicates dead bacteria [3].

The main source of stability is the lack of bone-implant movement. Depending on the implant's properties, the quantity and quality of bone, and the implantation methods, mechanical locking takes place between the implant and bone [4, 5]. Research has shown that applying orthodontic pressures to mini-screws requires adequate main stability [6]. For the spine to remain stable and to help heal from a variety of spinal conditions, spinal fixation devices are necessary. However, the majority of orthopedic implants are composed of metal, which makes them more rigid than bone, prone to the pull-out effect, and susceptible to failure over time, necessitating revision treatments. In this regard, clinical practice uses reinforcing techniques, each with their own set of dangers and limits, including changing the screw trajectory, changing the screw form, injecting bone cement, and using hybrid constructions.

The intended healing surrounding the screw cannot occur if primary stability is not attained, which leads to early screw loss [7, 8]. As the bone structure develops around the miniscrew throughout the healing process, secondary stability is the support that is given [9]. The effective establishment of secondary stability is influenced by the level of primary stability that is created following miniscrew insertion [1, 10]. For implanted devices in biomedical applications, such as orthopedic stents, dental implants, cardiovascular stents, cranial fixations, and surgical suture materials, medical-grade stainless steel (MSS) is one of the most popular material. Due to its extensive corrosion resistance, biocompatibility, and high-temperature resistance, MSS is widely utilized in biomedical, particularly in the fields of medicine, dentistry, medicinal chemistry, and pharmaceuticals [11]. MSS is made with a layer of protection. However, it releases toxic metal ions like nickel, chromium, and others when it is exposed to a corrosive environment of bodily fluids (chloride, proteins, and amino acids) and low oxygen concentrations in bodily fluids prevent the formation of a stable oxide layer on corroded MSS [12–14]. Even while MSS is unquestionably important in biomedical applications today, several researches have been done to address its drawbacks by functionalizing or changing its surface by utilizing a variety of chemicals, biomaterials, and methods. One efficient way to modify bioactive ingredients to satisfy therapeutic needs is by surface modification. One such technique is coating, which not only improves implant-tissue interactions and corrosion resistance but also influences medication release, biological responses (such as bioactivity and cytocompatibility), and implant-tissue interactions [15]. Orthopedic implants, particularly those for bone fixation (bone plates, screws, wires, and microplates), dental implants, spinal fixations, catheters, and cardiovascular (terminal and stent) applications, were typically made using MSS [16]. Vascularized bone flaps are unusual among mandibular reconstruction techniques because they allow repair of the oromandibular complex despite the recipient bed being frequently damaged by previous irradiation and salivary contamination. Vascularized bone grafts, as opposed to non-vascularized ones, can eventually tolerate the loading stresses brought on by mastication and can still repair the nearby native mandible. In most patients, the fibular, iliac, and scapular donor sites give enough bone supply for dental implants, which is crucial for complete oral rehabilitation. Additionally, each of these three composite flap donor locations yields soft tissue components that can be used as a source of tissue for extraoral covering or intraoral lining. Which of the numerous donor sites will offer the best source for reconstruction depends on inherent variances in bone stock, soft tissue quality, capacity for sensory reinnervation, and pedicle shape. Although the most common kind of flap, the fibula osteocutaneous flap, permits osteotomy, and implant implantation, several adjustments may be made to strengthen the oral sphincter, lower the risk of postoperative wound infection, and lessen donor site morbidity. The enhancements include a pedicled myoosseous flap with a free skin flap, a fibula flap with a segmental soleus muscle to augment submandibular dead space, and double free flaps with the tensor fascia lata for composite oromandibular defect restoration.

To create therapeutic approaches and biological replacements that preserve, repair, or enhance biological processes, tissue engineering is an interdisciplinary discipline that blends the concepts of engineering, material science, and biology. Adequate regeneration can be induced by growth factors, smad proteins, such as bone morphogenic proteins, and platelet-rich plasma osteoprogenitor cells generated on a recipient bed. Autologous cells from patients can be used to create functional tissue thanks to the tissue engineering approach. Free bone transplants and tissue engineering methods are among the treatment alternatives available in benign conditions. Microvascular bone flaps are the only clinical alternative available for malignancies, particularly for irradiated individuals with substantial mandibular abnormalities.

9.2 Materials commonly used for bone screws

9.2.1 Metallic materials

Stainless steel provides excellent strength and corrosion resistance but lacks bioactivity and can cause unfavorable tissue reactions owing to ion release. Titanium and its alloys have good biocompatibility, a high strength-to-weight ratio, and superior corrosion resistance. However, poor wear resistance and bioactivity necessitate surface changes. Magnesium alloys are biodegradable and have mechanical qualities similar to natural bone. However, their rapid deterioration can result in localized alkalinity and hydrogen evolution. Infections linked to implants are among the most unpleasant issues, frequently resulting in infected non-union, elevated morbidity, and much poorer chronic infection outcomes. Harmful bacteria can colonize, multiply, and develop biofilms on implant surfaces in wounds [17]. An extracellular polysaccharide biofilm is formed when an implant is introduced because the implant surface is covered by an adaptive layer made of host-derived adhesions, which encourages planktonic bacterial adhesion. The formation of a biofilm makes it simple for bacteria to evade the immune system and medications [18]. Usually, these infections necessitate implant removal, prolonged systemic antibiotic treatment, or many debridement procedures. In addition to raising recurrence and failure rates, more procedures and treatments also raise healthcare expenses [19]. Strategies aimed at preventing infection and biofilm formation during surgery and the early postoperative period may be more effective improvements that can totally avoid implant-related infections be-

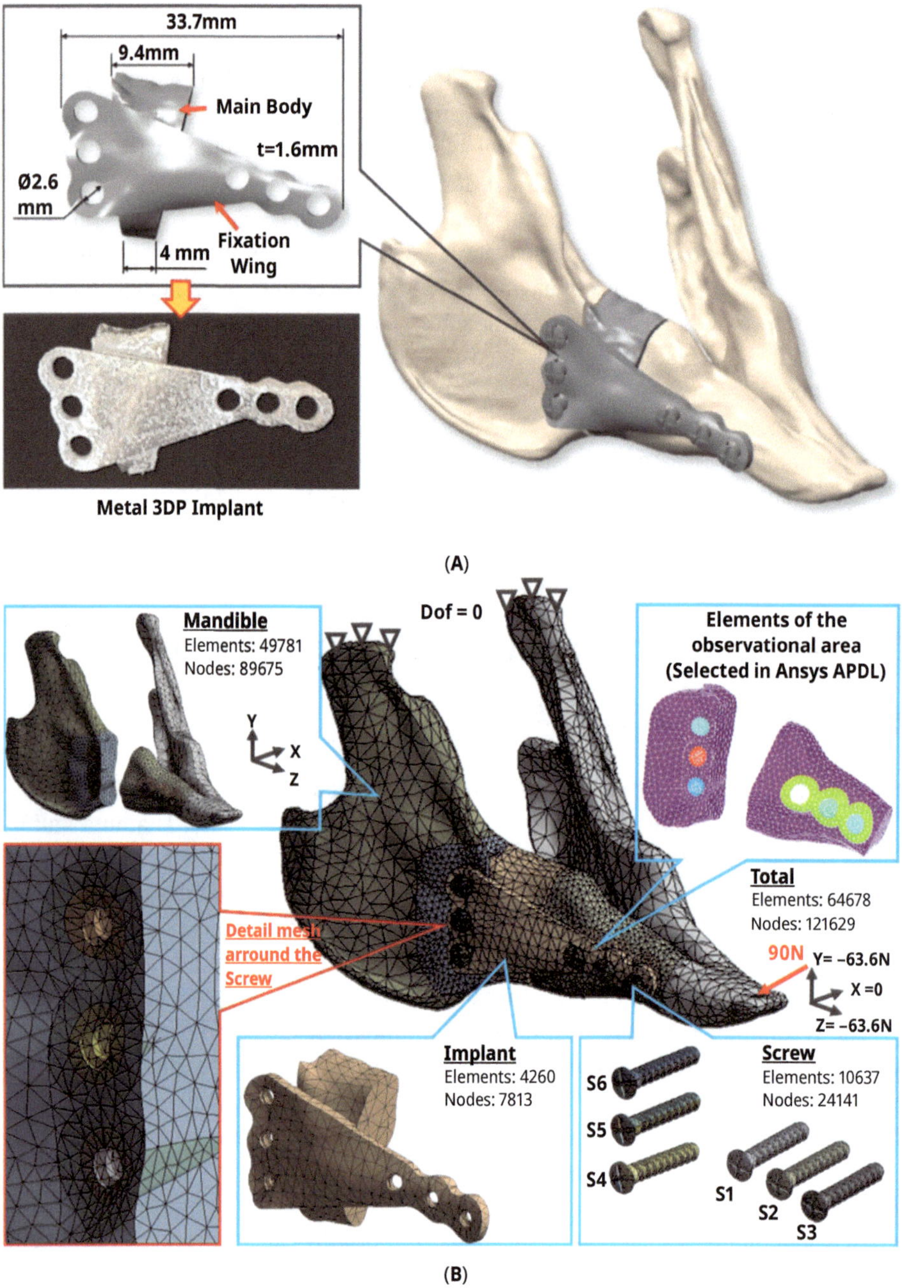

Figure 9.3: Finite element (FE) mesh model of the simulated mandible with reconstruction implant and fixation screw (S4, S5, and S6 at distal side and S1, S2, and S3 at mesial side) followed by a detailed mesh pattern surrounding the fixation screws. (A) Solid model of rabbit mandible with unilateral defect at right posterior teeth area and placed with a specific reconstruction implants [22].

cause of the difficulties in treating these illnesses. The use of anti-infective biomaterials as supplemental measures to stop implant infections and stop microbes from developing biofilms has grown.

In general, local antibiotic administration methods were employed, such as coating antimicrobial molecules on implant surfaces, injecting biomaterials, and using degradable sponges [20]. To enhance adhesion of antimicrobial molecules or coating materials to implants, these materials may need to be prefabricated due to their varying antimicrobial degradation and elution profiles. In this situation, antimicrobial coatings have a lot of promise. The effectiveness of cinnamon oil and CH as bio-adhesives to prevent biofilms has been shown in a study intended to ascertain the impact of coating on MSS implant surfaces [21]. Another research found that a particulate Ag coating applied to MSS prevented bacterial infection and was non-toxic to host cells [22]. Using spray pyrolysis, Ag-substituted HA-functionalized multiwalled carbon nanotubes (Ag-HA-f-MWCNTs) on MSS implants were discovered to be corrosion-resistant and antibacterial [23]. Ag ions implanted on MSS by plasma immersion ion implantation demonstrated enhanced bactericidal activity and decreased biofilm in different research [24]. For the preservation of the body's structure and the protection of its organs, a healthy bone system is essential. A system that ensures its continued functioning [23] by triggering healing processes upon fracture and preventing the development of scar tissues is crucial for this. In humans, fractures are the most frequent large-organ traumatic injuries. Bone fracture healing is a postnatal regenerative process that replicates many of the ontological processes of embryonic skeletal development, as was previously highlighted. Even while the injured skeletal organ is often returned to its pre-injury cellular composition, structure, and biomechanical function by fracture repair procedures, severe fractures will not heal properly [25].

9.2.2 Biodegradable polymers

Biodegradable polymers, such as polylactic acid (PLA) and polyglycolic acid (PGA), are increasingly being used for bone screws. These materials decay in the body, removing the need for surgery. However, their mechanical properties are inferior to metals, and changes are needed to improve their strength and surface features. In the medical field, biodegradable polymers are commonly used in surgical implant devices [26], wound healing products [27, 28], and medication delivery systems [29, 30]. Nowadays, biopolymeric drug delivery methods are being developed with great attention, particularly in controlled delivery [31]. Furthermore, biodegradable capsules can be used to control the way drugs are delivered to people [32]. The high permeability of buccal mucosa makes it a suitable target for drug administration and biodegradable polymers are specifically utilized to create innovative formulations [33]. In this sense, medication administration via buccal channels, biopolymers, and shielding is quick, safe, and effective [34]. Similar to this, bioresorbable non-wovens can be used in place of human tissue repair in wound healing [35], and basic sutures, staples, or meshes are available [36]. In contrast, it is worthwhile to disclose the use of biologically resorbable scaffolds for tissue engineering [37]. Moreover, biodegradable polymers come in a variety of forms, are affordable, and are renewable [38]. Notably, because of their bioactive properties, which promote cell growth and regeneration potential and offer antibacterial conditions and immunomodulation, biodegradable polymers are regarded as an outstanding choice for wound healing [39]. Furthermore, because biodegradable polymers have a large capacity for water absorption, they are likely to be used in wound treatment. These polymers may release medications at the site of harm, which makes them suitable for healing purposes. Many biopolymers can be used in traditional commodity applications because of their superior film-forming properties [40]. Furthermore, non-woven biodegradable polymers have applications in agriculture, filtration, personal hygiene, and protective clothing [41]. Though it hasn't been fully applied in many applications, this finding marked a significant turning point in the growth of research on biodegradable polymers. Due to the release kinetics of these polymer particles and the concerns about biopolymers' compatibility with tissue, biodegradable polymers are particularly exploited in drug delivery applications and are highly stimulating. More significantly, the depletion characteristics of the polymer's drug release dictate the drug release. Additionally, after these biopolymers have completed their anticipated function, they decompose and are eliminated from the human body. Essentially, biopolymers need to degrade at a pace that is similar to that of healing and regeneration. It has been demonstrated that the mechanical properties of biodegradable polymers are both desired and tenable [42]. The biodegradable polymers must not trigger immunological responses and must have biocompatibility properties. Furthermore, the broken-down components need to be easily eliminated, metabolized, and non-toxic. For the produced polymers to survive the production process, they must be simple to synthesize and maintain a high degree of durability [43, 44]. As biological applications maintain their stronger entity impacts, they have been chosen as a focal area for the usage of biodegradable polymers. Applications for biodegradable polymers in medicine are becoming more widespread, especially in the field of medical implants. More significantly, biopolymers may be utilized in tissue regeneration applications and as surgical sutures, providing exceptional strength and endurance. Furthermore, the biopolymers used as sutures may be quickly extracted or allowed to naturally dissolve from the body. For example, because of their exceptional flexibility and dependable knot stability, poly(glycolic acid) (PGA), poly(L-lactic acid) (PLA), and their composite materials are frequently used as sutures [45, 46]. It is microbes or enzymes that break down the biopolymers into low-molecular-weight molecules. Heteroatoms in biopolymers like PCL and PLA, amide-rich biopolymers, polyurethanes, and the majority of natural biopolymers may be susceptible to hydrolytic cleavage of the amide and ester groups [47]. Biodegradable polymers may be divided into three categories based on the source of their raw materials, which include sugar, starch, cellulose, and fossil oil. These categories are: (1) synthetic,

(2) natural, and (3) microorganism-built biodegradable polymers [48]. While synthetic polymers are derived from petroleum, natural biopolymers are produced from natural resources. The most often investigated aliphatic polyesters and typical natural macromolecules were the majority of what we mentioned. The use of biopolymers is expanding quickly, and the global economy is worth billions of dollars each year. Biodegradable polymers have been used in a wide range of industries, including agriculture, packaging, medicine, and the automobile sector [49]. Furthermore, biodegradable polymers that are suitable for packaging are still being given more thought than those that are utilized for other purposes. According to projections, 41% of plastics are used in packaging, with food items accounting for over half of that proportion [50]. Because biodegradable polymers absorb water more readily and are less soluble in it, they can be used as absorbent options in horticultural, agricultural, medical, and biological applications. In particular, the production of artificial tissue is one area where biodegradable polymers have been utilized when plastics are not suitable. Biomaterial properties including environmental responsiveness, biocompatibility, and biodegradable candidates that are sensitive to physicochemical and thermal changes as well as pH fluctuations may be required for these applications [51]. In general, biopolymers are less processable, chemically resistant, and have poorer mechanical and thermal properties (tensile strength and brittleness) than synthetic polymers. In particular, they can be strengthened with fillers that significantly improve these properties, making them suitable for particular applications. The use of biopolymers as polymeric vectors to deliver genetic or medicinal cargo has been investigated. It is possible to transfect cells without the need for active guiding by using several biocompatible natural polymers, such as chitosan, dextran, and hydroxyapatite (HA). This allows for cell- or tissue-specific transfection. When dextran sulfate binds to receptors on sinusoidal endothelial cells, it can accumulate in the liver more readily than HA [52]. Because chitosan may bind with cilia, improving nanoparticles' retention over the surface, covering it with poly(lactic-co-glycolic acid) (PLGA) materials increases the antiepileptic efficacy by intranasal distribution [53].

9.2.3 Composite materials

Composite bone screws combine polymers and ceramics, leveraging the strength of ceramics and the degradability of polymers. Surface modifications can optimize their bioactivity and degradation rate.

9.3 Surface modification techniques

9.3.1 Coating technologies

Coatings made of HA mimic the mineral composition of bone. Increases osteoblast adhesion, which improves osseointegration. Applied by electrophoretic deposition, sol-gel methods, or plasma spraying. Antibacterial coatings: use copper, zinc, or silver to stop the growth of biofilms. Antibiotic-loaded polymeric coatings give medications locally. Biodegradable polymer coatings: protect implants from early-stage bacterial colonization by acting as temporary barriers and medicine delivery systems.

9.3.2 Roughening and texturing the surface

Acid etching and sandblasting: make micro rough surfaces to improve cell adhesion and mechanical interlocking and used a lot for titanium screws. Cell alignment and differentiation are guided by the exact surface patterns created by laser texturing. It enables surface topography to be customized at the micro and nanoscale.

9.3.3 Surface modification techniques and their benefits

- **Silane coatings:** Enhance surface hydrophilicity, promoting protein adsorption and improved cell attachment.
- **Bioactive molecules:** Functionalizing surfaces with peptides, growth factors, or proteins strengthens cellular interactions and supports bone regeneration.
- **Ionic implantation:** Introduces ions like calcium or phosphate to stimulate osteogenic activity.

9.4 Clinical implications of surface modifications

9.4.1 Enhanced osseointegration

Surface modifications, such as HA coatings and micro-texturing, promote osteoblast activity, improving bone-screw integration and reducing micromotion. As the morphological and physiological features of native human tissue have been further clarified, the area of bone regeneration has benefitted. The form and content of adult tissues, as well as naturally occurring bone precursors, have influenced treatments aimed at repairing bone injury. By controlling biochemical signaling by imitating the endochondral ossification pathway or by providing stem cells with biophysical cues that mimic the mechanical characteristics of the unmineralized precursor of bone (the osteoid), some of the most intriguing and promising methods have made it possible to simulate bone development phenomena.

The construction of novel scaffolds and biomaterials based on the architecture of bones has shown great promise when holistic techniques incorporating multiscale characteristics and properties are used [54]. By combining top-down and bottom-up approaches, it is now possible to fully regulate the scaffolds' characteristics and enhance the complexity of the materials that are developed. Well-accepted critical features, such as the type and spatial-temporal distribution of constructs' properties, the suitability of their mechanical properties for the site of implantation, and the appropriate triggering of the immune response, must be taken into account regardless of the manufacturing strategies employed to create implantable biomaterials. Other facets of bone biology and its comprehension as a source of reliable knowledge to develop innovative regeneration techniques have recently drawn interest. It has been determined that one of the most effective ways to influence stem cell fate and encourage bone regeneration is to exert control over the insoluble portion of bone, which includes ECM proteins, glycoproteins, cellular structures, and even immobilized growth factors that interact with cellular membranes. Studies focusing on the utilization of glycans in biomaterials as instruments to modify cell behavior are still uncommon, especially despite their pervasiveness in human ECMs [55]. The broader knowledge, characterization, and use of soluble substances (like growth factors) and nanostructures that include soluble cues (like EVs) generated by cells themselves are undoubtedly significant trends that have a high effect as adaptable instruments to guide cell differentiation. Their use as readily available, naturally occurring nanoparticles incorporated into biomaterial matrices might be a useful aid in upcoming regeneration methods, especially for EVs.

9.4.2 Lowered infection risk

Antibacterial coatings are crucial in mitigating infection risks, which primarily cause implant failure. Controlled drug-release systems are particularly beneficial for high-risk patients.

9.4.3 Improved mechanical stability

Techniques like anodization and sandblasting enhance the pull-out strength of bone screws, reducing the likelihood of implant loosening.

9.4.4 Evolution of implants

The history of implantology traces back to the Mayan era when volcanic glass shaped like a tooth was used to replace lost teeth. In 1809, Italian dentist Maggiolo introduced gold as the first implant metal, employing an immediate implantation technique similar to the Mayans, where the implant root conformed to the tooth socket. Over time, materials like platinum, iridium, lead, rubber, and porcelain entered the field [56]. By 1934, Hans Abel experimented with ferrous alloys. The modern dental implant design began with Strock in 1983, using cobalt-molybdenum alloys. Currently, titanium and its alloys dominate dental implantology due to their biocompatibility, resistance to oral fluids, and low elastic modulus, with pure titanium being the most preferred material. The aesthetic demands led to the introduction of non-metal materials like ceramics. The first ceramic, aluminum oxide, was widely used for its osseointegration properties but was later discontinued due to inadequate mechanical strength. Zirconium ceramics emerged as a superior alternative, but challenges in material performance persist [57].

9.4.5 Methods to determine osseointegration

Osseointegration, a critical aspect of implant success, is assessed through invasive and non-invasive methods. Non-invasive techniques like radiography detect density changes when demineralization exceeds 40%. Invasive methods like cutting torque analysis and removal torque tests disturb the bone-implant interface, making them destructive and irreversible [58, 59]. Alternative techniques, such as percussion tests, impact hammer methods, and resonance frequency analysis (RFA), utilize vibration and acoustic signals to evaluate stability. RFA is particularly effective, with Implant Stability Quotient (ISQ) values above 65 indicating good osseointegration [60].

9.4.6 Osseointegration: prime focus

The surface properties and chemistry of dental implants significantly influence osseointegration [61–65]. Research highlights that surface features promote osteoblast proliferation, accelerating osseointegration. Studies on Ti-Nb-Hf alloys suggest that these materials demonstrate properties comparable to pure titanium while offering enhanced biocompatibility [66–72]. Incorporating bioactive ions like calcium, zinc, and strontium into titanium alloys further improves bone formation [73].

9.4.7 Coating processes

Various coating techniques enhance bone regeneration at the titanium-implant interface [74]. These include physicochemical and biochemical methods, which aim to establish a physiological connection between the implant surface and surrounding bone tissue [75–77]. Methods like plasma spraying, sol-gel deposition, and biomimetic deposition are frequently employed to create bioactive layers. Coatings such as HA accelerate healing by mimicking bone tissue's organic and inorganic components. Additive and subtractive approaches modify implant surfaces, with additive techniques generally offering better surface characterization [78].

9.4.8 Latest research on coatings and future direction

The various localized antibiotic treatment methods are depicted in Figure 9.2: (a) antibiotic-filled nanotubes, (b) an antibiotic is covalently bound to a Ti alloy, and (c) applying an antibacterial coating. A living bacteria is shown by yellow. Dead bacteria are indicated by red. An FE (finite element) mesh model of a simulated mandible with reconstruction implant and fixation screws (S1, S2, and S3 at the mesial side and S4, S5, and S6 at the distal side) is shown in Figure 9.3. A detailed mesh pattern around the fixation screws is then displayed. Figure 9.3(A) shows a solid model of a rabbit mandible attached to a particular reconstructive implant that has a unilateral deficiency at the right posterior tooth region [22]. Recent advancements in coating technology focus on embedding biomolecules like growth factors and bioactive proteins to expedite healing. Since medical gadgets are becoming essential components of contemporary healthcare, their use is steadily growing. Implanted medical devices (IMDs), such as heart valves, stents, vascular grafts, and replacement hips, are presently utilized all over the world to improve the quality of life for a large number of patients. To enhance the quality of life for patients, non-implanted, short-term medical devices such as contact lenses, orthopedic fixation screws, and catheters are also introduced [79]. In this way, the need for medical devices has led to the development of a sizable industry as well as a substantial corpus of research in materials science and medical devices, including clinical investigations and device design [80, 81]. Device-related or implant-associated infections can affect any IMD, from temporary, early introduced, and retrieved devices (like urinary catheters and endotracheal tubes) to more permanently surgically implanted devices (like cardiac valves, embolic coils, vascular grafts, hip, knee, and shoulder joints, pacemakers, and coronary stents) [60]. Complex interactions between the pathogens, the biomaterial, and the host immune response are all part of the infection of IMDs. The infection component raises the danger of tissue devitalization, repeated surgeries throughout the implant period, and distant infections in other body areas, extended hospital stays, and operation intervals. The medical community is always looking for efficient ways to prevent and treat implant-related infections since they continue to be a problem, with few improvements in infection rates over several decades despite some tried-and-true treatments. One intriguing approach to intervention is to alter the surface of IMDs to prevent microbial adhesion. It is possible to develop coatings that have improved antibacterial qualities in addition to strong biological activity by combining surface modifications with antimicrobial agents [61].

Studies reveal that acid-etched zirconia and bioactive glass coatings enhance cell proliferation and osseointegration. Techniques such as plasma nitriding and fluoride treatment further improve mechanical strength, corrosion resistance, and integration efficiency. Emerging trends include incorporating bioactive drugs and antibacterial coatings to enhance implant performance. Despite advancements, challenges persist, particularly in ensuring the durability of coatings and achieving optimal osseointegration. Innovations in materials and surface treatment methods aim to overcome these barriers and enhance clinical outcomes, ultimately reducing healing times and improving patient satisfaction.

9.5 Conclusion

Surface modification of bone screws has emerged as a game-changing approach in orthopedic and reconstructive surgeries. By refining surface properties, these modifications effectively tackle critical issues such as inadequate osseointegration, infection susceptibility, and mechanical instability. Progress in nanotechnology, biomaterials, and additive manufacturing is anticipated to revolutionize these innovations further, enhancing clinical outcomes and improving patients' quality of life. Over the past few decades, a significant amount of research has been devoted to the creation of biopolymers. In general, the development of biodegradable polymers and their composites has been receiving a lot of attention, although it is still in its early stages. It is impossible to resist the need for new types of candidates from potential producers of biodegradable polymers. Specifically, despite the urgent need for a vast array of sustainable goods, their manufacturing is lowered compared to traditional synthetic polymers due to their poor performance characteristics and expensive research costs. The creation of contemporary polymeric materials and procedures that are acknowledged to adhere to the welfare of typical habitats has improved as a result of renewed ecological rules aimed at allaying environmentalist fears. To fully use the ecological, social, and industrial advantages, highly active biodegradable polymer products must be developed. The nano-safety of their composite and the biological safety of created biopolymers remain negligible. This new perspective places a strong emphasis on anticipating, evaluating, and illustrating potential issues with the use of sophisticated polymers. The performance of bio-derived materials, the cost of manufacturing, and the management of primary resources must be the main challenges facing biodegradable polymers in the future. The production of bio-derived monomers and bio-derived polymers from renewable resources will also face commercial manufacturing challenges. Due to inexperience with new technology and the evaluation of stock/demand balance, building industrial plants can be challenging, and increase with time. The long-term viability of biodegradable polymers still depends on several factors, even though new types of bio-derived polymers are being created on an industrial scale. Feedstock rivalry is expected to occur as the world's need for food and energy increases gradually. People's daily lives involve externally applied stimulation of their bones, which is recognized to be essential for bone remodeling. The formation of high-quality neonates' skeletal growth has recently been emphasized as dependent on appropriate fetal mobility [79], and mechanical forces in utero have been proposed as potentially having long-term effects.

The mechanical strength of bones is attributed to their inorganic HA crystals, which range from 60 to 100 nm in length and align along the bone's longitudinal axis [73]. Collagen fibrils form a triple-helical structure that is subsequently mineralized by apatite crystals. These mineralized collagen fibrils, known as lamellae, arrange concentrically around the Haversian canal to create osteons. The ECM of bone primarily comprises three cell types: osteoblasts (responsible for bone formation), osteoclasts (involved in bone resorption), and osteocytes (which maintain the function and structure of differentiated osteoblasts).

References

[1] Clarke, B. (2008). Normal bone anatomy and physiology. *Clinical Journal of the American Society of Nephrology: CJASN*, *3*(Suppl 3), S131–S139. https://doi.org/10.2215/CJN.04151206

[2] Benčina, M., Resnik, M., Starič, P., & Junkar, I. (2021). Use of plasma technologies for antibacterial surface properties of metals. *Molecules (Basel, Switzerland)*, *26*(5), 1418. https://doi.org/10.3390/molecules26051418

[3] Esteban, J., Vallet-Regí, M., & Aguilera-Correa, J. J. (2021). Antibiotics-and heavy metals-based titanium alloy surface modifications for local prosthetic joint infections. *Antibiotics*, *10*(10), 1270.

[4] Langdahl, B., Ferrari, S., & Dempster, D. W. (2016). Bone modeling and remodeling: Potential as therapeutic targets for the treatment of osteoporosis. *Therapeutic Advances in Musculoskeletal Disease*, *8*(6), 225–235. https://doi.org/10.1177/1759720X16670154

[5] Chowdhury, S., Jain, N., Waidi, Y. O., Barua, R., Das, S., Prasad, A., & Datta, S. (2024). Biological Smart Biomaterials: Materials for Biomedical Applications. In: Kumar, A., Kumar, A., Kumar, A. (eds) *Applications of Biotribology in Biomedical Systems* (pp. 313–325). Cham: Springer Nature Switzerland.

[6] Molly, L. (2006). Bone density and primary stability in implant therapy. *Clinical Oral Implants Research*, *17*(S2), 124–135.

[7] Javed, F., & Romanos, G. E. (2010). The role of primary stability for successful immediate loading of dental implants. A literature review. *Journal of Dentistry*, *38*(8), 612–620. https://doi.org/10.1016/j.jdent.2010.05.013

[8] Wilmes, B., Rademacher, C., Olthoff, G., & Drescher, D. (2006). Parameters affecting primary stability of orthodontic mini-implants. *Journal of Orofacial Orthopedics = Fortschritte der Kieferorthopadie:*

Organ/official Journal Deutsche Gesellschaft Fur Kieferorthopadie, *67*(3), 162–174. https://doi.org/10.1007/s00056-006-0611-z

[9] Trisi, P., De Benedittis, S., Perfetti, G., & Berardi, D. (2011). Primary stability, insertion torque and bone density of cylindric implant ad modum Branemark: Is there a relationship? An in vitro study. *Clinical Oral Implants Research*, *22*(5), 567–570. https://doi.org/10.1111/j.1600-0501.2010.02036.x

[10] Song, Y. Y., Cha, J. Y., & Hwang, C. J. (2007). Mechanical characteristics of various orthodontic mini-screws in relation to artificial cortical bone thickness. *The Angle Orthodontist*, *77*(6), 979–985. https://doi.org/10.2319/090606-363.1

[11] Motoyoshi, M., Yoshida, T., Ono, A., & Shimizu, N. (2007). Effect of cortical bone thickness and implant placement torque on stability of orthodontic mini-implants. *International Journal of Oral & Maxillofacial Implants*, *22*(5).

[12] Cheng, S. J., Tseng, I. Y., Lee, J. J., & Kok, S. H. (2004). A prospective study of the risk factors associated with failure of mini-implants used for orthodontic anchorage. *International Journal of Oral & Maxillofacial Implants*, *19*(1).

[13] Miyawaki, S., Koyama, I., Inoue, M., Mishima, K., Sugahara, T., & Takano-Yamamoto, T. (2003). Factors associated with the stability of titanium screws placed in the posterior region for orthodontic anchorage. *American Journal of Orthodontics and Dentofacial Orthopedics*, *124*(4), 373–378.

[14] Raghavendra, S., Wood, M. C., & Taylor, T. D. (2005). Early wound healing around endosseous implants: A review of the literature. *International Journal of Oral & Maxillofacial Implants*, *20*(3).

[15] Resnik, M., Benčina, M., Levičnik, E., Rawat, N., Iglič, A., & Junkar, I. (2020). Strategies for improving antimicrobial properties of stainless steel. *Materials (Basel, Switzerland)*, *13*(13), 2944. https://doi.org/10.3390/ma13132944

[16] Williams, R. L., Brown, S. A., & Merritt, K. (1988). Electrochemical studies on the influence of proteins on the corrosion of implant alloys. *Biomaterials*, *9*(2), 181–186. https://doi.org/10.1016/0142-9612(88)90119-6

[17] Williams, D. F. (2008). On the mechanisms of biocompatibility. *Biomaterials*, *29*(20), 2941–2953. https://doi.org/10.1016/j.biomaterials.2008.04.023

[18] Barua, R. & Datta, S. (2025). Revolutionizing Nerve Repair: The Transformative Role of Nanoparticles in Peripheral Nerve Regeneration. In D. Erisen, K. Uludag, & N. Ahmad (Eds.), Nanomaterials and the Nervous System (pp. 275–300). IGI Global Scientific Publishing. https://doi.org/10.4018/979-8-3693-3065-4.

[19] Prasad, A., Datta, S., De, S., Singh, P., Mahto, B. (2024). Bioresorbable Composite for Orthopedics and Drug Delivery Applications. In: Kumar, A., Kumar, A., Kumar, A. (eds) Applications of Biotribology in Biomedical Systems. Springer, Cham. https://doi.org/10.1007/978-3-031-58327-8_12

[20] Niinomi, M., Nakai, M., & Hieda, J. (2012). Development of new metallic alloys for biomedical applications. *Acta Biomaterialia*, *8*(11), 3888–3903.

[21] Costerton, J. W., Stewart, P. S., & Greenberg, E. P. (1999). Bacterial biofilms: A common cause of persistent infections. *Science (New York, N.Y.)*, *284*(5418), 1318–1322. https://doi.org/10.1126/science.284.5418.1318

[22] Wu, L.-J., Hsieh, K.-H., & Lin, C.-L. (2020). Integrating finite element death technique and bone remodeling theory to predict screw loosening affected by radiation treatment after mandibular reconstruction surgery. *Diagnostics*, *10*(10), 844. https://doi.org/10.3390/diagnostics10100844

[23] Costerton, J. W. (2005). Biofilm theory can guide the treatment of device-related orthopaedic infections. *Clinical Orthopaedics and Related Research (1976-2007)*, *437*, 7–11.

[24] Gristina, A. G., & Costerton, J. W. (1985). Bacterial adherence to biomaterials and tissue. The significance of its role in clinical sepsis. *Journal of Bone & Joint Surgery*, *67*(2), 264–273.

[25] Jennings, J. A., Carpenter, D. P., Troxel, K. S., Beenken, K. E., Smeltzer, M. S., Courtney, H. S., & Haggard, W. O. (2015). Novel antibiotic-loaded point-of-care implant coating inhibits biofilm. *Clinical Orthopaedics and Related Research®*, *473*(7), 2270–2282.

[26] Magetsari, R., Dewo, P., Saputro, B. K., & Lanodiyu, Z. (2014). Cinnamon oil and chitosan coating on orthopaedic implant surface for prevention of *Staphylococcus epidermidis* biofilm formation. *Malaysian Orthopaedic Journal*, *8*(3), 11.

[27] DeVasConCellos, P., Bose, S., Beyenal, H., Bandyopadhyay, A., & Zirkle, L. G. (2012). Antimicrobial particulate silver coatings on stainless steel implants for fracture management. *Materials Science and Engineering: C*, *32*(5), 1112–1120.

[28] Sivaraj, D., & Vijayalakshmi, K. (2019). Enhanced antibacterial and corrosion resistance properties of Ag substituted hydroxyapatite/functionalized multiwall carbon nanotube nanocomposite coating on 316L stainless steel for biomedical application. *Ultrasonics Sonochemistry*, *59*, 104730. https://doi.org/10.1016/j.ultsonch.2019.104730

[29] Einhorn, T. A., & Gerstenfeld, L. C. (2015). Fracture healing: Mechanisms and interventions. *Nature Reviews. Rheumatology*, *11*(1), 45–54. https://doi.org/10.1038/nrrheum.2014.164

[30] Meyer-Kobbe, V., Doll, K., Stiesch, M., Schwestka-Polly, R., & Demling, A. (2018). Comparison of intraoral biofilm reduction on silver-coated and silver ion-implanted stainless steel bracket material: Biofilm reduction on silver ion-implanted bracket material. *Journal of Orofacial Orthopedics*, *80*(1), 32.

[31] Van de Velde, K., & Kiekens, P. (2002). Biopolymers: Overview of several properties and consequences on their applications. *Polymer Testing*, *21*(4), 433–442.

[32] Barua, R. (2024). Innovations in Minimally Invasive Surgery: The Rise of Smart Flexible Surgical Robots. In M. Garcia & R. de Almeida (Eds.), Emerging Technologies for Health Literacy and Medical Practice (pp. 110–131). IGI Global Scientific Publishing. https://doi.org/10.4018/979-8-3693-1214-8.ch006

[33] Kurowiak, J., Klekiel, T., & Będziński, R. (2023). Biodegradable Polymers in Biomedical Applications: A Review-Developments, Perspectives and Future Challenges. International journal of molecular sciences, 24(23), 16952. https://doi.org/10.3390/ijms242316952

[34] Barua, R., Das, S., Datta, S., Datta, P., & Chowdhury, A. R. (2022). Analysis of surgical needle insertion modeling and viscoelastic tissue material interaction for minimally invasive surgery (MIS). *Materials Today: Proceedings*, *57*, 259–264.

[35] Datta, S. & Barua, R. (2024). 3D Printing in Modern Healthcare: An Overview of Materials, Methods, Applications, and Challenges. In M. Garcia & R. de Almeida (Eds.), Emerging Technologies for Health Literacy and Medical Practice (pp. 132-152). IGI Global Scientific Publishing. https://doi.org/10.4018/979-8-3693-1214-8.ch007

[36] Barua, R., Giria, H., Datta, S., Roy Chowdhury, A., & Datta, P. (2020). Force modeling to develop a novel method for fabrication of hollow channels inside a gel structure. *Proceedings of the Institution of Mechanical Engineers, Part H: Journal of Engineering in Medicine*, *234*(2), 223–231.

[37] Râpă, M. A. R. I. A., Popa, M. E., Cinelli, P., Lazzeri, A., Burnichi, R., Mitelut, A., & Grosu, E. (2011). Biodegradable alternative to plastics for agriculture application. *Romanian Biotechnological Letters*, *16*(6), 59–64.

[38] Prasad, A. (2022). Biomaterial-Based Nanofibers Scaffolds in Tissue Engineering Application. In: Jana, S., Jana, S. (eds) Functional Biomaterials. Springer, Singapore. https://doi.org/10.1007/978-981-16-7152-4_9

[39] Waidi, Y. O., Jain, N., Chowdhury, S., Barua, R., Das, S., Prasad, A., & Datta, S. (2024). Challenges and Perspective of Manufacturing Techniques in Biomedical Applications. In: Kumar, A., Kumar, A., Kumar, A. (eds) Applications of Biotribology in Biomedical Systems. Springer, Cham. https://doi.org/10.1007/978-3-031-58327-8_14

[40] Jain, N., Waidi, Y. O., Chowdhury, S., Barua, R., Das, S., Prasad, A., & Datta, S. (2024). Medical Devices Tribology. In: Kumar, A., Kumar, A., Kumar, A. (eds) *Applications of Biotribology in Biomedical Systems* (pp. 235–250). Cham: Springer Nature Switzerland.

[41] Barua, R. & Datta, S. (2024). The Ongoing Advancements in Surgical Robotics: Transforming the Landscape of Surgery. In A. Khang (Ed.), AI-Driven Innovations in Digital Healthcare: Emerging

Trends, Challenges, and Applications (pp. 288–303). IGI Global Scientific Publishing. https://doi.org/10.4018/979-8-3693-3218-4.ch015

[42] Penning, J. P., Dijkstra, H., & Pennings, A. J. (1993). Preparation and properties of absorbable fibres from L-lactide copolymers. *Polymer*, *34*(5), 942–951.

[43] Jain, N., Barua, R., Waidi, Y. O., & Datta, S. (2024). Nanowires for Bio-Sensing Applications. In S. Kulkarni, B. Kharisov, A. Haghi, & V. Srivastava (Eds.), Next Generation Materials for Sustainable Engineering (pp. 205–219). IGI Global Scientific Publishing. https://doi.org/10.4018/979-8-3693-1306-0.

[44] Prasad, A., Datta, S., Kumar, A., & Gupta, M. (2023). Introduction to next-generation materials for biomedical applications. *Advanced Materials and Manufacturing Techniques for Biomedical Applications*, 1–24.

[45] Sahana, T. G., & Rekha, P. D. (2018). Biopolymers: Applications in wound healing and skin tissue engineering. *Molecular Biology Reports*, *45*(6), 2857–2867. https://doi.org/10.1007/s11033-018-4296-3

[46] Asokan, P., Firdoous, M., & Sonal, W. (2012). Properties and potential of bio fibres, bio binders, and bio composites. *Reviews on Advanced Materials Science*, *30*(3), 254–261.

[47] Barua, R. (2024). Exploring Artificial Intelligence in Evolving Healthcare Environments: A Comprehensive Analysis. In I. Shah & Q. Sial (Eds.), Advances in Computational Intelligence for the Healthcare Industry 4.0 (pp. 123–138). IGI Global Scientific Publishing. https://doi.org/10.4018/979-8-3693-2333-5.ch006

[48] Laycock, B., Nikolić, M., Colwell, J. M., Gauthier, E., Halley, P., Bottle, S., & George, G. (2017). Lifetime prediction of biodegradable polymers. *Progress in Polymer Science*, *71*, 144–189.

[49] Chandra, R. U. S. T. G. I., & Rustgi, R. (1998). Biodegradable polymers. *Progress in Polymer Science*, *23*(7), 1273–1335.

[50] Barua, R., Bhowmik, S., Dey, A., Das, S., & Datta, S. (2022, September). Analysis of Robotically Controlled Percutaneous Needle Insertion into Ex Vivo Kidney Tissue for Minimally Invasive Percutaneous Nephrolithotomy (PCNL) Surgery. In *International Conference on Emergent Converging Technologies and Biomedical Systems* (pp. 249–257). Singapore: Springer Nature Singapore.

[51] Hirvikorpi, T., Vähä-Nissi, M., Harlin, A., Salomäki, M., Areva, S., Korhonen, J. T., & Karppinen, M. (2011). Enhanced water vapor barrier properties for biopolymer films by polyelectrolyte multilayer and atomic layer deposited Al_2O_3 double-coating. *Applied Surface Science*, *257*(22), 9451–9454.

[52] Tian, H., Tang, Z., Zhuang, X., Chen, X., & Jing, X. (2012). Biodegradable synthetic polymers: Preparation, functionalization and biomedical application. *Progress in Polymer Science*, *37*(2), 237–280.

[53] Yin, G. Z., & Yang, X. M. (2020). Biodegradable polymers: A cure for the planet, but a long way to go. *Journal of Polymer Research*, *27*(2), 38.

[54] Gopi, S., Amalraj, A., & Thomas, S. (2016). Effective drug delivery system of biopolymers based on nanomaterials and hydrogels-a review. *Drug Diethylstilbestrol*, *5*(129), 2169–0138.

[55] Andreopoulos, A. G., & Theophanides, T. (1994). Degradable plastics: A smart approach to various applications. *Journal of Elastomers & Plastics*, *26*(4), 308–326.

[56] Petersen, K., Nielsen, P. V., Bertelsen, G., Lawther, M., Olsen, M. B., Nilsson, N. H., & Mortensen, G. (1999). Potential of biobased materials for food packaging. *Trends in Food Science & Technology*, *10*(2), 52–68.

[57] Oliveira, S. M., Reis, R. L., & Mano, J. F. (2015). Towards the design of 3D multiscale instructive tissue engineering constructs: Current approaches and trends. *Biotechnology Advances*, *33*(6 Pt 1), 842–855. https://doi.org/10.1016/j.biotechadv.2015.05.007

[58] Sgambato, A., Russo, L., Montesi, M., Panseri, S., Marcacci, M., Caravà, E., Raspanti, M., & Cipolla, L. (2016). Different sialoside epitopes on collagen film surfaces direct mesenchymal stem cell fate. *ACS Applied Materials & Interfaces*, *8*(24), 14952–14957. https://doi.org/10.1021/acsami.5b08270

[59] Barua, R. (2024). An Investigation of AI Techniques for Detecting Kidney Stones in CT Scan Images Through Advanced Image Processing. In A. Sharma, N. Chanderwal, S. Tyagi, P. Upadhyay, & A. Tyagi (Eds.), Enhancing Medical Imaging with Emerging Technologies (pp. 133–150). IGI Global Scientific Publishing. https://doi.org/10.4018/979-8-3693-5261-8.ch008
[60] Guerrero-Cázares, H., Tzeng, S. Y., Young, N. P., Abutaleb, A. O., Quiñones-Hinojosa, A., & Green, J. J. (2014). Biodegradable polymeric nanoparticles show high efficacy and specificity at DNA delivery to human glioblastoma in vitro and in vivo. *ACS Nano*, *8*(5), 5141–5153.
[61] Kaur, S., Manhas, P., Swami, A., Bhandari, R., Sharma, K. K., Jain, R., . . . Wangoo, N. (2018). Bioengineered PLGA-chitosan nanoparticles for brain targeted intranasal delivery of antiepileptic TRH analogues. *Chemical Engineering Journal*, *346*, 630–639.
[62] Chen, Q., & Thouas, G. A. (2015). Metallic implant biomaterials. *Materials Science and Engineering: R*, *87*, 1–57.
[63] Smith, J. R., & Lamprou, D. A. (2014). Polymer coatings for biomedical applications: a review. Transactions of the IMF, 92(1), 9–19. https://doi.org/10.1179/0020296713Z.000000000157
[64] Ahangar, P., Cooke, M. E., Weber, M. H., & Rosenzweig, D. H. (2018). Current biomedical applications of 3D printing and additive manufacturing. *Applied Sciences*, 9(8), 1713. https://doi.org/10.3390/app9081713
[65] Vallet-Regí, M. (2000). Bioactive ceramics: State of the art. *Journal of Biomedical Materials Research Part A*, 53(4), 321–340.
[66] Sidambe, A. T. (2014). Biocompatibility of advanced manufactured titanium implants – A review. *Materials*, 7(12), 8168–8188.
[67] Zhao, L., et al. (2009). Antibacterial coatings on titanium implants. *Journal of Biomedical Materials Research Part B: Applied Biomaterials*, *91*(2), 470–480.
[68] Amini, A. R., Laurencin, C. T., & Nukavarapu, S. P. (2012). Bone tissue engineering: Recent advances and challenges. *Critical Reviews in Biomedical Engineering*, *40*(5), 363–408. https://doi.org/10.1615/CritRevBiomedEng.v40.i5
[69] Prasad, A., Bhasney, S. M., Prasannavenkadesan, V., Sankar, M. R., & Katiyar, V. (2023). Polylactic acid reinforced with nano-hydroxyapatite bioabsorbable cortical screws for bone fracture treatment. *Journal of Polymer Research*, *30*(5), 177.
[70] Clarke, B. (2008). Normal bone anatomy and physiology. *Clinical Journal of the American Society of Nephrology*, *3*(Suppl 3), S131–S139. https://doi.org/10.2215/CJN.04151206
[71] Einhorn, T. A. (1998). The cell and molecular biology of fracture healing. *Clinical Orthopaedics and Related Research*, *355*(Suppl), S7–S21. https://doi.org/10.1097/00003086-199810001-00003
[72] Rho, J. Y., Kuhn-Spearing, L., & Zioupos, P. (1998). Mechanical properties and the hierarchical structure of bone. *Medical Engineering & Physics*, *20*(2), 92–102. https://doi.org/10.1016/S1350-4533(98)00007-1
[73] Spiro, D. (1983). Collagen fibril diameters in bone. *Journal of Ultrastructure Research*, *83*(1), 44–54. https://doi.org/10.1016/S0022-5320(83)90160-9
[74] Li, Y., Zhang, X., & Wang, L. (2021). Surface modification of titanium implants for improved osseointegration. *Journal of Materials Science: Materials in Medicine*, *32*(4), 58. https://doi.org/10.1007/s10856-021-06475-7
[75] Parikh, A. M., & Patel, K. K. (2020). Antimicrobial coatings on orthopedic implants: A review. *International Journal of Infection Control*, *16*(3), 345–357. https://doi.org/10.1080/13489393.2020.1743841
[76] Rupp, M. E., & Langhorst, A. (2019). Bioactive coatings for orthopedic implants: Enhancing osteointegration and antibacterial performance. *Journal of Bone and Joint Surgery*, *101*(2), 120–127. https://doi.org/10.2106/JBJS.18.00360

[77] Zhang, Y., Zhang, F., & Du, S. (2018). The effects of surface modification on the mechanical properties and biological performance of titanium-based bone implants. *Biomaterials*, *166*, 95–110. https://doi.org/10.1016/j.biomaterials.2018.02.038
[78] Prasad, A., Bhasney, S. M., Prasannavenkadesan, V., Sankar, M. R., & Katiyar, V. (2023). Nano-hydroxyapatite reinforced polylactic acid bioabsorbable cancellous screws for bone fracture fixations. *Journal of Applied Polymer Science*, *140*(43), e54577.
[79] Gupta, R., & Singhal, R. (2019). Advances in surface modification of metallic implants for improved orthopedic applications. *Journal of Orthopaedic Research*, *37*(3), 839–849. https://doi.org/10.1002/jor.24024
[80] Wu, C., & Hu, H. (2017). Surface functionalization of bone screws for enhanced osteointegration and reduced infection risk. *Surface and Coatings Technology*, *314*, 105–113. https://doi.org/10.1016/j.surfcoat.2017.02.015
[81] Targeting pathogenic biofilms: newly developed superhydrophobic coating favors a host-compatible microbial profile on the titanium surface. *ACS Applied Materials & Interfaces*, *12*(9), 10118–10129. https://doi.org/10.1021/acsami.9b22741

Ranjit Barua*, Arbind Prasad, and Siddharth Mohan Bhasney

Chapter 10
Emerging application of modern additively manufactured medical implants

Abstract: Using the method of additive manufacturing (AM) or 3D printing allows us to make better medical implants. Using AM technologies to make medical implants works better because this technology creates exact customized shapes at high precision. This chapter explains how AM creates medical implants and describes its important traits including material choices along with manufacturing processes and outcomes. This chapter reviews specific examples and shares predictions about better implants that match better with body tissues.

Keywords: Additive manufacturing, medical implants, 3D printing, biocompatibility, titanium implants, patient-specific design, biomedical applications, advanced materials, customization, healthcare innovation

10.1 Introduction

Healthcare uses advanced manufacturing technology called additive manufacturing (AM) to create medical implants [1, 2]. Compared to standard manufacturing techniques, AM allows us to create parts through stacked layers to produce unique personalized items [3]. Medical implants require customized solutions for patient-specific anatomy which AM delivers effectively [4, 5]. Medical implants receive unmatched custom details now that 3D printing technology adds material instead of taking away from it [6]. This modern production approach shows remarkable results in healthcare because it can create implants exactly matched to patient needs and complex designs that boost treatment results (Figure 10.1) [7]. AM technology helps make precise personalized implants that fit patient anatomy and supports medical implant progress in every healthcare area including orthopedics and dentistry [8, 9]. The process of creating physical components using computer-aided design and building things layer-by-layer is known as additive manufacturing, or 3D printing in a non-technical context

***Corresponding author: Ranjit Barua**, Mechanical Engineering Department, Omdayal Group of Institutions, Howrah 711316, West Bengal, India, e-mail: ranjitjgec007@gmail.com
Arbind Prasad, Mechanical Engineering Department, Katihar Engineering College (Under Department of Science, Technology and Technical Education, Government of Bihar), Katihar 854109, Bihar, India
Siddharth Mohan Bhasney, Machino Polymers Limited, Plot No. 2, Sector 33, Delhi–Jaipur Expressway, Phase 2, Info Technology Park, Sector 33, Gurugram 12200, Haryana, India

https://doi.org/10.1515/9783111571423-010

[10–13]. These processes are typically referred to as tool-less processes. Other methods, such laser forming or incremental sheet forming, also construct objects layer-by-layer, but they do it by adding the form rather than the material [14]. Even though these techniques have been applied similarly to the production of customized medical goods, for instance, they are not considered AM techniques.

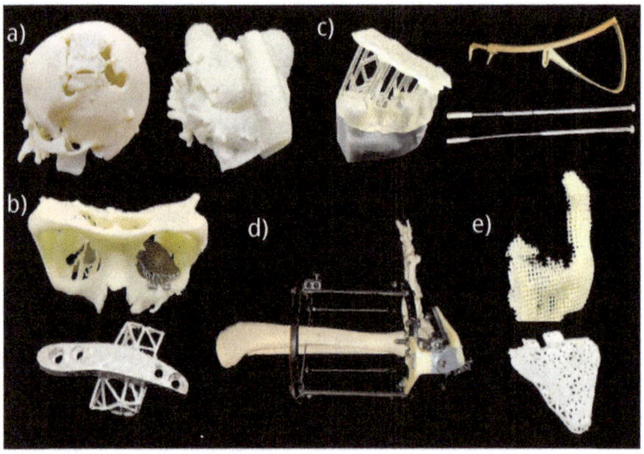

Figure 10.1: Medical models, implants, tools, instruments, and parts for medical devices, medical aids, supporting guides, splints, and prostheses, and biomanufacturing are among the aforementioned categories [15].

Nowadays, industries like the medical, transportation, aviation, and marine sectors, as well as industrial spare parts, are using and researching AM [16, 17]. One manufacturing technique that is said to offer little complexity or customization is AM [18]. In contrast to mass producing identical parts, this necessitates labeling and tracking the various components. However, AM provides a far more possibility for customization and complicated geometries than conventional production. Despite this, if the geometry is made for mass production and only the manufacturing cost is taken into account, AM is typically not less expensive when comparing costs [19]. Reiterating the full product design and examining the economics of the product lifetime would be sufficient [20]. At the moment, AM is growing quickly, and new competitors are constantly joining the market. Significant investments have been made in both internal development at major corporations in various sectors and in new businesses like Formlabs, Carbon, etc. Although the fundamental ideas behind the various AM techniques have not changed, there are now additional resources available to further these technologies, which will also lead to new opportunities in the medical field [21–23].

Since each patient is different in the medical arena, AM has a great chance of being used for individualized and customized medicinal applications [24]. Personalized implants, medical models, and saw guides are the most frequently used medical clinical applications [25]. AM is used in dentistry on drill guides, dental models, orthodontic equipment, and splints. Nonetheless, AM has also been investigated for the creation of synthetic organs and tissues [26]. Digitalization of medical imaging has a history in medicine, and it enables the reconstruction of three-dimensional models from human cadavers. Using computed tomography or other 3D scanning techniques to image or capture the patient's geometry is the first step in a typical workflow for customized medical equipment [27]. A 3D model of the patient's anatomy is then created by manipulating these data; this can already be an example of AM, such as a medical model. Additionally, patient-specific implants can be designed using the geometry and then additively fabricated. Post-processing, including polishing, is frequently required after production [28]. Clinical application and follow-up are the last steps once the medical gadget is prepared. Orthopedics benefits from AM technology because strong demand for custom implants requires this advanced production method in several medical fields including dentistry, cardiovascular, and other specialties [29–31]. Manufacturing medical implants by AM technology helps doctors create implants faster at lower costs while building strong bone connections and helpful features when used in these products. This chapter investigates the production methods, biological materials and current use cases of medical implants designed through AM. Main features of AM manufacturing and its improvements over traditional processes are explained in this start. This section reviews distinct AM fabrication techniques SLM, EBM, and FDM with their metal, ceramic, and polymer implant production capabilities [32]. These materials – titanium alloys, cobalt-chromium alloys, and bioceramics – receive individual attention because they exhibit excellent compatibility with body tissue plus their needed strength traits for medical implant production.

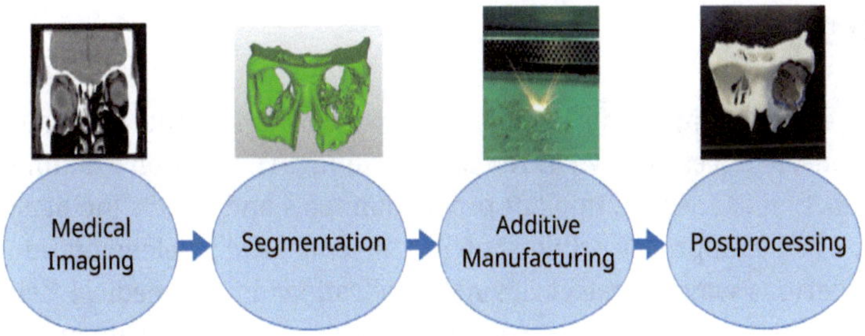

Figure 10.2: Common characteristics of medical model process flow [15].

AM technology now produces metal-tooth and bone replacements as well as cardiac equipment and biological tissue support structures for clinical practice. The introduction shows how patient-specific designs in production led to better medical results through examples of custom hip replacements and cranial plates. The chapter examines the production obstacles AM faces when making medical implants and explains how product regulation, available materials, post-production steps, and manufacturing costs impact performance. This chapter presents both the existing technology developments and identifies what trends will change the industry in the years ahead.

Multiple printing methods now let us embed bioactive materials in constructs and create hybrid implants using bio-printing technology while AI helps us design implants with better performance. This chapter explores how nanotechnology improves implant performance while helping to make AM more sustainable. This chapter explains the ways digital fabrication technology transforms medical implant practices both today and tomorrow. Researchers and healthcare experts across fields find the resources essential to develop AM solutions that help patients recover better.

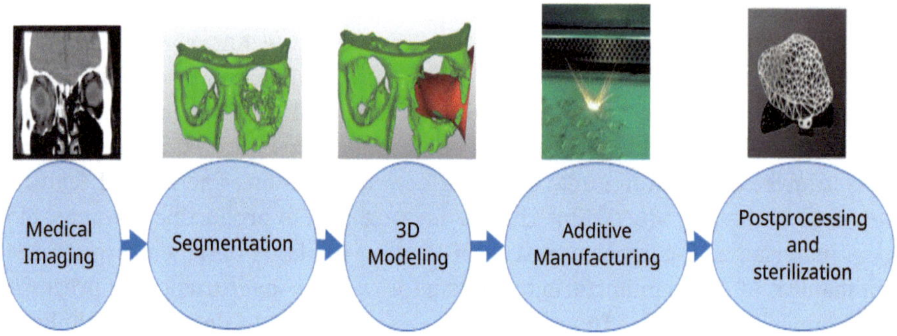

Figure 10.3: A typical workflow for implants made using additive manufacturing [15].

10.2 Additive manufacturing techniques for medical implants

Manufacturers can create components layer-by-layer from digital file templates through the revolutionary process known as AM or 3D printing. This section examines how AM methods function as medical implant production tools and details the benefits and constraints of each approach while showing their medical implementation. Although there are various ways to categories AM applications in the medical field [33, 34], this study uses an application classes-based approach. "Models for preoperative planning, education and training," "inert implants," "tools, instruments and parts for medical devices," "medical aids, supportive guides, splints, and prostheses," and "bio-manufacturing" are the classes under which AM applications fall [34]. This can be changed to allow implants to be inert for a broader classification [35]. Postoperative and operative models could also be used for preoperative planning, teaching, and training under the umbrella term "clinical representations." Based on the anatomy of the patient, medical models can be utilized for pre- and postoperative planning and training, medical student education, and patient and family education [36, 37]. For instance, the geometry can be scaled up or down or only the interesting parts can be taken. It may be preferable for haptic response to be near the bone if models are being used for training, such as bone drilling. Although the cranio-maxillofacial region frequently uses medical models, there are also instances from other bone structures, such as the spine and pelvis, as well as from other limbs [37]. It may be advised to sterilize the models if they are used in operating rooms, but typically, there is a great deal of freedom in choosing the material, which further emphasizes the fact that these are among the most widely used applications. A typical process workflow for creating medical models is depicted in Figure 10.2, which begins with patient anatomy obtained through medical imaging techniques like computed tomography (CT), magnetic resonance imaging (MRI), or ultrasound. Next, segmentation algorithms are used to create 3D model geometry for AM [38, 39]. Post processing, like deleting the support structures, is frequently required after AM.

10.2.1 Selective laser melting (SLM)

SLM builds solid metal components through the focus of strong lasers that fuse metal powders [40]. For metal implants titanium and its alloys work best through this process because they show high strength along with good corrosion protection and compatibility with body tissues. SLM produces porous material designs to improve bone cell penetration [41, 42].

10.2.2 Electron beam melting (EBM)

Electron beam melting (EBM) uses an electron beam to heat metallic powders through vacuum processing. The production of titanium implants with exact dimensions works well with this technique [43]. EBM lets us make titanium-based implants with suitable pores for root-bone integration. The use of a vacuum environment helps protect implants from external contaminants which results in better quality products [44].

10.2.3 Binder jetting

In binder jetting technology, the machine laydowns layers of powder while spraying liquid binder. This method has smaller use for structural implants but shows growing potential to make bio-ceramic and polymer-based implants [45]. Material preparation steps after manufacturing are needed to deliver the required mechanical performance [46].

10.2.4 Fused deposition modeling (FDM)

Fused deposition modeling (FDM) deploys a heating system to push thermoplastic materials into thin layers as it prints. Though primarily seen in prototyping FDM technology now enables the production of polymer-based implants for medical use and helps develop biomedical parts. Biocompatible filaments enable medical implants to work better with patient tissue [47, 48].

10.2.5 Stereolithography (SLA)

SLA uses a laser light to create photopolymer resins in thin sections one above the other. SLA produces precise structures particularly well when used to make tissue engineering scaffold designs. New resin technology is helping SLA create strong implants for basic medical uses [49, 50].

Table 10.1: The additive manufacturing (AM) techniques for medical implants.

Technique	Process description	Materials	Applications	Advantages	Challenges
Selective laser melting (SLM)	Uses a laser to fully melt metal powders layer-by-layer	Titanium alloys, cobalt-chromium alloys, stainless steel	Orthopedic implants (e.g., hip, knee), craniofacial plates, dental implants	High resolution, dense and strong parts, patient-specific customization	Expensive equipment and raw materials, post-processing required
Electron beam melting (EBM)	Electron beam melts metal powders in a vacuum environment	Titanium alloys, cobalt-chromium alloys	Spinal cages, acetabular cups, craniofacial prostheses	Contamination-free, high fatigue resistance, no need for support structures in some geometries	Limited material choices, surface roughness
Fused deposition modeling (FDM)	Extrudes thermoplastic filaments layer-by-layer	PLA, PEEK, PCL, biocompatible polymers	Bioabsorbable implants, surgical models, drug delivery systems	Cost-effective, suitable for rapid prototyping, biodegradable implants	Limited to polymers, lower resolution and mechanical strength

Table 10.1 (continued)

Technique	Process description	Materials	Applications	Advantages	Challenges
Direct metal laser sintering (DMLS)	Laser selectively sinters metal powders to create dense parts	Stainless steel, titanium alloys, cobalt-chromium alloys	Dental prosthetics, stents, surgical tools	High precision, reduced material waste	Requires post-processing for thermal treatment and surface polishing
Binder jetting	Deposits binder onto powder layers, followed by sintering or infiltration	Titanium alloys, ceramics, polymers, composites	Porous scaffolds, cranial meshes	Cost-effective for large batches, suitable for porous structures	Requires post-processing for mechanical strength, limited for load-bearing applications
Stereolithography (SLA)	Cures liquid resin with UV light layer-by-layer	Biocompatible resins, hydrogels	Hearing aids, dental surgical guides, presurgical models	High resolution, smooth surface finish	Limited to polymers, lower mechanical and thermal resistance
Laser cladding	Deposits material in powder or wire form onto a substrate using a laser	Titanium, cobalt-chromium alloys, stainless steel	Repairing worn implants, fabricating large orthopedic devices	High deposition rate, ideal for repair	Lower resolution, less suitable for intricate components

The multiple AM methods help us design medical implants that precisely match the required dimensions and deliver specific material properties plus functionality. Table 10.1 lists basic technical elements of each method to help you easily see their medical implant fabrication capabilities. The selection of production method depends on both the kinds of materials available and what specific patients need. As researchers develop better materials and processing methods for AM, they make medical implants more practical through advanced techniques. AM is changing healthcare by enabling us to customize medical treatments and dental implants.

The potential of AM technologies to create effective surgical instruments that satisfy the growing demands of hospital operating rooms while having low production costs [51] and short manufacturing times [52] is one of their most important advantages. AM allows the construction of inguinal hernia surgical instruments without the need for traditional stainless steel implements [53]. In a recent study, space support for intraesophageal sutures was produced using 3DP imprint. Better intraesophageal resection was made possible by these outcomes, which resulted in increased intraesophageal suture insertion [54].

10.3 Materials for additively manufactured implants

10.3.1 Titanium and titanium alloys

Titanium leads as the primary choice for medical implant AM due to its remarkable combination of strength-to-weight properties and perfect compatibility with body tissues that resist corrosion. People widely choose Ti-6Al-4 V material for its use in medical implants for bones and teeth.

10.3.2 Cobalt-chromium alloys

Doctors use cobalt-chromium alloys for hip and knee implants because they resist wear and stay strong. Through AM we can make intricate shapes that boost how well the implant works.

10.3.3 Stainless steel

316L stainless steel is a common choice for making medical implants that need to stand in place for only a short time and surgical instruments. With AM users can build specialized tools directly according to their precise needs.

10.3.4 Bioceramics

AM technology uses hydroxyapatite and bioglass materials because these substances enhance bone restoration and connect with biological tissues.

10.3.5 Polymers

The printing of medical implants uses biocompatible polymers such as PEEK and PLA in FDM and SLA processes to create lightweight and flexible parts.

10.3.6 Magnesium alloys

Magnesium-based implants break down over time and work best for short-term uses. Manufacturers are working to produce better AM methods to slow down the fast corrosion of magnesium parts.

10.4 Applications of additively manufactured medical implants

AM through 3D printing systems now transforms medical implant production. AM technologies help design exact custom implants with defined physical attributes to enhance healthcare delivery. This study describes the different ways medical implants benefit medicine through the AM process in orthopedics, dentistry, heart treatments, maxillofacial procedures, and additional fields.

10.4.1 Orthopedic implants

Orthopedics stands out as the main beneficiary of modern AM. The creation of customized implants has made the production of joint replacement parts and medical equipment better for patients:

- **Joint replacements**: Customized hip and knee replacements made via SLM and EBM produce better fit and benefit patients after surgery. These implants use titanium alloy metals that keep well with body tissue while maintaining durability.
- **Spinal implants**: Porous spinal implants made through 3D printing increase bone growth and better connect to the body. The extra openings strengthen the implant which helps it stay in place more securely and make rejection problems less likely.
- **Fracture fixation devices**: Specialized plates and screws created by AM machines exactly match a patient's bone structure to help pieces fit together better during surgical repair. Patients recover faster from procedures when surgery teams spend less time making on-site adjustments because customized devices match their unique anatomy.

10.4.2 Dental implants

The dental field has made important progress because of AM technology. Dental implants, crowns, bridges, and orthodontic tools receive advanced precision thanks to AM. In order to replace lost or damaged tissue, implants are either directly or indi-

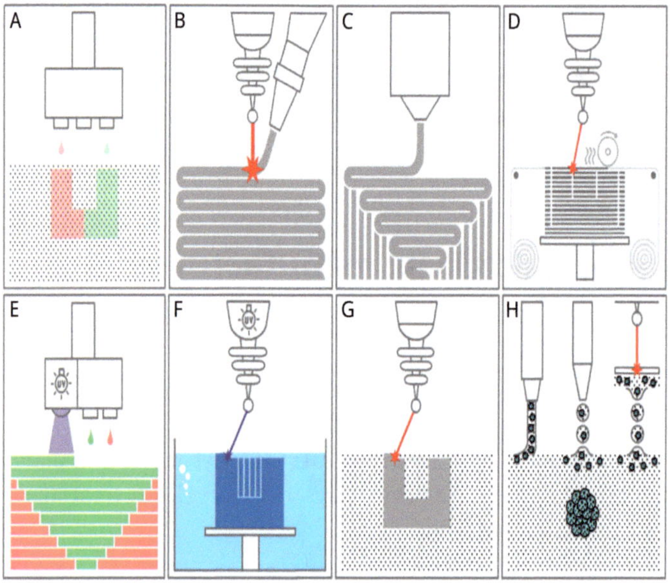

Figure 10.4: Different types of additive manufacturing (AM). (A) The procedure of binder jetting involves spraying a liquid binder over a specific area of powder, then adding more powder on top and repeating the process. (B) Directed energy deposition: metal is being extruded from a nozzle while being melted by a laser. (C) Material extrusion: the print bed rapidly cools molten thermoplastics once they are sprayed via a nozzle. (D) The process of sheet lamination involves cutting a sheet of material to size using a blade or laser, adding a second sheet of adhesive, and then using heat to fuse the layers together before cutting the second layer to form. (E) The process of material jetting (MJ) involves spraying a layer of liquid resin, curing it (usually with UV light), and then adding the next layer. The print bed lowers one layer at a time into a vat of liquid resin in stereolithography (F). The resin is then UV-cured, and the print bed keeps moving through it, layer by layer. (G) Powder bed fusion: a layer of powder, usually metal, is prepared on the print bed and then melted with a laser to produce a layer. The procedure is repeated once the print bed descends and the subsequent powder layer is applied on top of the hardened layer. As with material extrusion, (H) extrusion bioprinting (left) involves layer-by-layer extrusion of a bioink (a combination of cells and carrier material) from a nozzle or needle, followed by cross-linking or curing. Picoliter-sized drops of bioink are applied to the print bed and given time to solidify into fibres in inkjet bioprinting (centre). Small droplets of bioink are released from a surface into predetermined shapes using a laser pulse in laser-assisted bioprinting (right) [41].

rectly additively created [55]. Dental applications including crowns and bridges are also included in this class [49]. Approval procedures are lengthy, and the material must meet stringent requirements, including being tissue-compatible. Cell adhesion may be impacted by surface characteristics. Recent research has looked into embedding materials into implants, for instance, as a way to distribute drugs [56]. AM is an advantageous option for customized implants, and a standard procedure calls for capturing a patient's anatomy in a manner akin to that of medical models. In order to facilitate patient-specific fitting, this digital 3D model of the patient's anatomy is then utilized as a design reference [57]. The powder bed fusion method is used to create the majority of common implants, which necessitates several post processing procedures such heat treatments, polishing, and support machining. Sterilization of implants is required prior to clinical surgery. The usual process flow for additively manufactured implants is depicted in Figure 10.3, which begins with medical imaging and segmentation, moves on to 3D modeling of the implant, and then moves on to AM, postprocessing, and sterilization:

- **Implants and prosthetics**: DMLS produces strong dental implants that help replace lost teeth permanently. Dental implants made from titanium and cobalt-chromium materials deliver good patient matches over time [58].
- **Orthodontic appliances**: By using AM technology medical teams construct orthodontic devices and appliances that fit perfectly based on a patient's oral structure. This improves comfort and treatment efficacy.
- **Surgical guides**: The technology helps create precise implant placement guides for surgery that prevent mistakes and maintain correct implant position.

10.4.3 Cardiovascular applications

Cardiovascular implants are now created differently through the possibilities of AM technology:

- **Customized stents**: AM produces stents designed exactly for each patient using unique vessel geometry. These medical devices made from biocompatible materials help deliver precise support to damaged or blocked blood vessels.
- **Heart valves**: Research groups use bioprinting technology to make heart valves from materials that dissolve naturally into the body. Our new valves built using 3D printing resemble natural heart valve features and decrease rejection problems plus enhance patient health.
- **Vascular grafts**: Through AM methods engineers create customized vascular grafts that better match patients' blood vessels while lowering the risk of surgery problems.

10.4.4 Maxillofacial and craniofacial implants

Complex maxillofacial reconstruction work needs medical implants that match the exact shape of each patient's features. AM shows great value in helping these efforts:

- **Cranial plates**: AM technology creates custom-made implants that improve skull replacement. Custom-made implants through AM technology deliver both perfect fit and pleasing looks.

- **Facial reconstruction**: Doctors use AM to build bones that restore afflicted facial bones due to trauma, birth disorders, or tumors. These medical implants help both how the face functions and how it looks.
- **Mandibular reconstruction**: Titanium medical implants made for each patient at AM technology rebuilds jaw performance and cosmetic value following cancer extraction or trauma repair.

10.4.5 Bioabsorbable and drug-eluting implants

By using AM technology engineers created implants that absorb naturally into tissue while delivering medications to treat medical conditions:

- **Bioabsorbable scaffolds**: The body naturally breaks down implants constructed from PLA and PGA so patients won't need future operations to remove them. Doctors use these implants most often when treating children because their body structures change as they grow.
- **Drug-eluting implants**: New implants made through AM can send medicine doses at specific planned timelines. Special implants used for orthopedics can release medicine to protect patients from surgery side effects.

10.4.6 Presurgical planning and training

AM supports surgical planning as well as medical education without producing implants. Surgeons create better medical results by practicing operations with and planning procedures using 3D-printed anatomical models.

10.4.7 Emerging applications: bioprinting and tissue engineering

A new technology called bioprinting helps us make implants using living cells:

- **Tissue-engineered implants**: Through bio printing technologies, experts make tissue scaffolds that hold stem cells and let them transform into functional new tissues. Scientists test these developments for medical uses in cartilage repair, bone rebuilding, and organ transplants.
- **Organ-on-a-chip devices**: Although classified as different from implants these devices serve essential roles in drug screening and medical research. Scientists use bioprinting to produce small organs that work like natural human organs.

The benefits of fabricating medical implants through AM continue to multiply across various medical fields. Medical professionals now design and make better implants with AM technology by producing customized solutions that work with optimal efficiency for orthopedic and cardiovascular needs plus face surgeries. The future growth of medical treatments depends on material science and 3D printing solutions which will create a new way to design care around individual patients. Technology advances like artificial intelligence and machine learning will improve the accuracy and manufacturing capabilities of AM solutions to lead medical advancement.

10.4.8 Advantages of additive manufacturing for medical implants

3D printing technology under its full name AM has transformed the medical implant field with new methods to address medical problems. The technology lets us produce precision-made implants better than traditional methods could achieve. AM now plays a foundational role in medical progress because it builds unique patient-oriented solutions. The process of 3D printing has three primary steps. The initial stage involves creating a 3D CAD file of the item right before it is printed. The selection of the material with the necessary particular qualities comes next, after which the printing process itself begins. Materials used for printing include metals, polymers, glass, fabrics, ceramics, resins, sand, biomaterials, and even food. In order to get exact control over the printed items, the majority of these materials provide sufficient tolerances. To get the desired control during printing, however, extensive research and development efforts are still being undertaken since some materials, like glass, are not yet readily susceptible to precise control of production parameters [27]. Completing the last level necessitates matching the qualities of the functional materials with certain instrumentation skills. Usually, after an object is printed, it cannot be used until it has been cured, lacquered, sanded, or painted to get the desired finish [59].

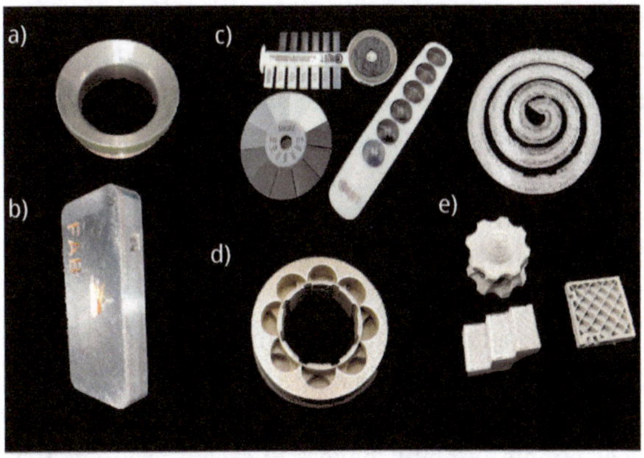

Figure 10.5: Prospects for AM in the dentistry and medical industries in the future. Repairing using directed energy deposition (a), sheet lamination for multi-material metal parts (b), material jetting for multi-material metal parts (c), binder jetting for metal parts (d), and material extrusion for metal and composite parts (e) [15].

The most valued strength of AM is its power to produce medical implants that match individual patient needs exactly. AM produces customized implants unlike bulk manufacturing which needs the same design for all patients. Medical implants adapt better to specific body parts because they use CT or MRI images that show each patient's anatomy. Personalized medical implants made through this technology decrease the risk of implant rejection while delivering faster patient outcomes.

The main benefit of AM technology lies in custom part design capabilities. Today's technology enables us to make complex shapes and detailed structures which standard manufacturing processes cannot reproduce. Through AM technology porous implants made for better bone integration at implant surfaces which leads to more steady outcomes. The optimization process lets us build strong and lightweight implants that use less material while maintaining full durability and making treatments more pleasant for patients. By using several materials in our printing technology we can now create items with special medical features like medication release and easy disposal which lets us make advanced medical products.

The way parts are built by AM system makes implant development reach the marketplace faster. In comparison to traditional methods AM skips all steps needed for prototyping and creating special production tools. Quick prototyping methods help test designs faster and improve innovation processes more rapidly. The system lets hospitals print implants when needed which means patients get their replacement parts faster. Our ability to quickly produce custom implants at times of medical emergency or for rare situations is most helpful.

AM technology saves both patients and medical providers money. Companies start their AM program by spending more at first yet they won't need the costly tooling and molding fixtures anymore to make implants. AM technologies build implants layer-by-layer so they need exactly the right number of materials to work which cuts down on waste. The production technology works well for small product orders including different implants made in limited quantities.

Medical implant biocompatibility improves through AM because it enables exact control of material choices and modifications. The materials used in manufacturing titanium and cobalt-chromium alloys alongside bioresorbable polymers enable exact adjustments for targeted biological and mechanical needs. The exact control of fabrication creates textured and porous surfaces which helps implants connect better with natural tissues so they work longer while increasing their chances of being accepted by patients.

Over standard implants made by AM behave better than those created through traditional production methods. When used in construction AM allows us to design better weight distribution which prevents parts from breaking down easily. Advanced manufacturing and material choices enable implant makers to create products that last longer while resisting damage better.

AM shows exceptional potential in combining intelligent functions within medical implants. Special implant designs deliver medication at controlled rates to eliminate the requirement of multiple treatments throughout patient recovery. Our technology enables implanted devices to contain monitoring sensors that simultaneously track stress levels, temperature variations, and healing advancement. Our new technology enables doctors to deliver better-individualized healthcare treatments.

Using digital manufacturing techniques to make parts brings several important benefits that include sustainability. With layer-by-layer manufacturing AM helps reduce material waste better than traditional machining that wastes raw materials. The energy needs for AM tools run more efficiently making these processes better for the planet. Both industries recognize that sustainability must become a fundamental part of medical care delivery.

With AM businesses can produce parts closer to their market needs and maintain smooth workflows. Implants can be created both locally within facilities and directly at their final destinations so shipping and storage overhead decrease. The finished product process cuts back on material waste and helps costs as it uses resources correctly for each patient's need.

AM leads medical research into creating and printing body tissues in new ways. The technology prints tissue scaffolds containing cells that support natural healing practices for damaged cartilage and bones while laying the groundwork for future transplantations. The technology improves regenerative medicine through its engineering capabilities which now help patients with advanced medical problems. Scientific teams now work to create complete functional implants that match specific patient needs through AM technology which will transform organ transplants.

Medical implant manufacturing with additive methods provides many advantages such as personalization, design adaptability, faster production timelines, reduced costs, and better product compatibility with the body. Our technology helps connect smart functions while making products more eco-friendly and easier to deliver. Medical technology keeps getting better so its applications can solve harder medical issues and help patients worldwide. Through its production of personalized medical implants, the technology beyond improving implants is reshaping healthcare's future.

10.5 Challenges in additive manufacturing of medical implants

The medical uses of AM have been reviewed extensively [60]. The majority of them concentrate on specific application areas, such orthopedics, cardiology, or surgical instruments. Others have a material focus, like metals for prosthodontics applications [61]. Various AM techniques (Figure 10.4) have been examined in some research, such as the use of material extrusion and binder jetting in the production of pharmaceuticals [62] or the use of material extrusion, binder jetting, powder bed fusion or VAT photopolymerization in the production of ceramics for dental purposes [63]. Additionally, a certain merely examined AM's broad applications [29] or the advancements throughout the whole production chain, including the design phase, for instance [40]. Application-focused evaluations exist that even classify those in some AM processes, as VAT photo-polymerization, powder bed fusion, and material extrusion [64], but they do not include the other processes. There are material-focused reviews in dentistry that classify various AM processes, including material extrusion, binder jetting, VAT photopolymerization, powder bed fusion, material jetting, and binder jetting [65–68]. Additionally, there are reviews that focus specifically on metallic implants in dentistry using the powder bed fusion procedure [31]. The drawbacks, however, are typically that the technology is presented using trade names, etc., or that the reviews do not link the applications to the specific AM processes. The used AM methods are also typically not specified at all. For statistics and systematic reviews, this is the largest challenge. Fused filament fabrication, fused deposition modeling, and filament freeform fabrication are some terminologies used to describe the material extrusion technique in the literature [69–73]. The reason these phrases are connected is that FDM was registered and trademarked by Stratasys, therefore other makers had to come up with a different term. Similar difficulties arise in other AM disciplines as well. However, most materials are adequately defined, and the most frequent ones include metals like titanium alloys, thermoplastics, and photopolymers:

- **Regulatory hurdles:** Meeting stringent regulatory standards for medical devices.
- **Material limitations:** Limited availability of biocompatible materials for AM.
- **Postprocessing:** Need for extensive post-processing to achieve desired properties.
- **Cost:** High initial investment for AM equipment and materials.
- **Surface finish:** Achieving the required surface roughness and texture for implants.

10.6 Future trends and innovations

It is highly recommended that future research on the industrial side of AM as well as medical and dental applications use a uniform language [74]. This would enable a more methodical comparison of how various AM techniques are used. Future and present advancements in AM materials, devices, and procedures would enable further applications in the dentistry and medical fields (Figure 10.5). There is a chance that some of the techniques, such sheet lamination, may never be employed again, or that new competitors will enter this sector. Nonetheless, metal sheet lamination offers intriguing opportunities for multi-metals. With the ability to create metal components, some procedures, such binder jetting and material jetting, have a great chance of becoming more extensively utilized in dentistry and medical applications. Both material extrusion for composite components and the repair of metallic parts may benefit from directed energy deposition. In multi-material components, material extrusion and material jetting will be possible. Aspects including printability, mechanical qualities, biocompatibility, degradation, and biomimicry would be the main emphasis of future bioprinting material development [34]. For precise deposition with the required spatial and temporal control, the material characteristics should be tailored to the particular AM operations. The biological activities of the constructions depend on the upkeep of 3D structures in order to resist or produce particular forces. One possible method for creating diverse tissues and organs with varying mechanical needs may be the development of innovative composite biomaterials with varied constituents to achieve reprogrammed mechanical characteristics and functioning. In order to satisfy the demands of various regeneration phases in vivo, the biomaterials should also degrade with spatial and temporal controllability. By printing micro-/nano-scale biomimetic features or adding biomimetic components, exciting progress has been achieved in accurately controlling the microstructural or physical microenvironments for cell development. One of the primary obstacles to printing cells or organs is creating new bio-inks that can support the development and proliferation of the implanted cells in addition to being printed to preserve certain tissue architectures. The functional complexity of the printed structures has increased along with manufacturing precision thanks to recent developments in the creation of different functional materials for innovative AM methods. In order to create complex structures or devices with multiple properties like electroconductibility, sensing, biocompatibility, and stimulus-responsiveness, advancements in material strategies and AM techniques should be prioritized concurrently. This will greatly expand the range of applications for the future of 3D printing [75, 76]:

- **Multi-material printing:** Development of implants with integrated functionalities such as drug delivery and antibacterial coatings.
- **Bio-printing:** Combining AM with tissue engineering to create hybrid implants.
- **Nanotechnology:** Integration of nanostructures for enhanced biocompatibility and mechanical performance.
- **Artificial intelligence:** AI-driven design optimization for patient-specific implants.
- **Sustainability:** Use of eco-friendly materials and energy-efficient processes in AM.

10.7 Conclusion

Despite a significant increase in recent years, most industrial companies still do not fully understand 3D printing of metal, ceramic, and polymer composite materials. The availability and compatibility of materials, the inconsistent mechanical performance of the printed 3D objects and components, and the lengthy printing process are some of the constraints that have slowed the fast market penetration. 3D printing techniques have shown remarkable development trends in the fields of aerospace, automotive, energy, and medical in recent years, despite these limitations. It has been demonstrated that scientists are investigating novel 3D printing technology-related materials and uses. Along with some of the benefits of 3D printing, such cost effectiveness, design flexibility, and manufacturing speed, there are also a number of drawbacks. These include the printing process's high energy consumption, issues with mass production, and the scarcity of printing ink. A 3D printer, however, is frequently a far superior manufacturing solution to traditional ones for short production runs and prototyping. The growth and expansion of 3D printing in recent years has been centered on both the size of yields and the improvement of printed item quality, opening up a wide range of new possibilities for the printing technique. Through the improvement of printing capabilities, three-dimensional printing is evolving from a speedy prototype method to a financially viable choice for mass production of materials, both in terms of final product size and complexity.

The medical industry uses AM to produce implants that now feature custom designs and advanced shapes with better device performance. AM technology now produces a wide range of medical tools including joint and tooth replacements along with heart equipment and artificial limbs. The ongoing development of materials science and manufacturing techniques plus updated regulations will lead to AM technologies fully transforming medical practices. AM brings together different medical fields to make new-generation implant products that heal patients better.

References

[1] Amaya-Rivas, J. L., Perero, B. S., Helguero, C. G., Hurel, J. L., Peralta, J. M., Flores, F. A., & Alvarado, J. D. (2024). Future trends of additive manufacturing in medical applications: An overview. *Heliyon, 10*(5), e26641. https://doi.org/10.1016/j.heliyon.2024.e26641

[2] Davoodi, E., Montazerian, H., Mirhakimi, A. S., Zhianmanesh, M., Ibhadode, O., Shahabad, S. I., Esmaeilizadeh, R., Sarikhani, E., Toorandaz, S., Sarabi, S. A., Nasiri, R., Zhu, Y., Kadkhodapour, J., Li, B., Khademhosseini, A., & Toyserkani, E. (2021). Additively manufactured metallic biomaterials. *Bioactive Materials, 15*, 214–249. https://doi.org/10.1016/j.bioactmat.2021.12.027

[3] Vyas, J., Shah, I., Singh, S., & Prajapati, B. G. (2023). Biomaterials-based additive manufacturing for customized bioengineering in management of otolaryngology: A comprehensive review. *Frontiers in Bioengineering and Biotechnology, 11*, 1234340. https://doi.org/10.3389/fbioe.2023.1234340

[4] Amukarimi, S., Rezvani, Z., Eghtesadi, N., & Mozafari, M. (2022). Smart biomaterials: From 3D printing to 4D bioprinting. *Methods (San Diego, Calif.), 205*, 191–199. https://doi.org/10.1016/j.ymeth.2022.07.006

[5] Zhou, J., See, C. W., Sreenivasamurthy, S., & Zhu, D. (2023). Customized additive manufacturing in bone scaffolds-the gateway to precise bone defect treatment. *Research (Washington, D.C.), 6*, 0239. https://doi.org/10.34133/research.0239

[6] Datta, S. & Barua, R. (2024). 3D Printing in Modern Healthcare: An Overview of Materials, Methods, Applications, and Challenges. In M. Garcia & R. de Almeida (Eds.), *Emerging Technologies for Health Literacy and Medical Practice* (pp. 132–152). IGI Global Scientific Publishing. https://doi.org/10.4018/979-8-3693-1214-8.ch007

[7] Boretti, A. (2024). A perspective on 3D printing in the medical field. *Annals of 3D Printed Medicine, 13*, 100138.

[8] Meyer-Szary, J., Luis, M. S., Mikulski, S., Patel, A., Schulz, F., Tretiakow, D., . . . Kwiatkowska, J. (2022). The role of 3D printing in planning complex medical procedures and training of medical professionals – Cross-sectional multispecialty review. *International Journal of Environmental Research and Public Health, 19*(6), 3331.

[9] Das, S., Datta, S., Barman, A., & Barua, R. (2023). Smart Biodegradable and Bio-Based Polymeric Biomaterials for Biomedical Applications. In A. Kumar, P. Kumar, A. Srivastava, & V. Goyat (Eds.), *Modeling, Characterization, and Processing of Smart Materials* (pp. 56–82). IGI Global Scientific Publishing. https://doi.org/10.4018/978-1-6684-9224-6.ch003.

[10] Bernatoniene, J., Stabrauskiene, J., Kazlauskaite, J. A., Bernatonyte, U., & Kopustinskiene, D. M. (2025). The Future of Medicine: How 3D Printing Is Transforming Pharmaceuticals. *Pharmaceutics, 17* (3), 390. https://doi.org/10.3390/pharmaceutics17030390

[11] Barua, R., Giria, H., Datta, S., Roy Chowdhury, A., & Datta, P. (2020). Force modeling to develop a novel method for fabrication of hollow channels inside a gel structure. *Proceedings of the Institution of Mechanical Engineers, Part H: Journal of Engineering in Medicine, 234*(2), 223–231.

[12] Barua, R., Datta, S., Datta, P., & Chowdhury, A. R. (2021). Scaffolds and Tissue Engineering Applications by 3D Bio-Printing Process: A New Approach. In I. Management Association (Ed.), *Research Anthology on Emerging Technologies and Ethical Implications in Human Enhancement* (pp. 718–733). IGI Global Scientific Publishing. https://doi.org/10.4018/978-1-7998-8050-9.ch037.

[13] Barua, R., Datta, S., Roychowdhury, A., & Datta, P. (2019). Importance of 3D Printing Technology in Medical Fields. In K. Kumar, D. Zindani, & J. Davim (Eds.), *Additive Manufacturing Technologies From an Optimization Perspective* (pp. 21–40). IGI Global Scientific Publishing. https://doi.org/10.4018/978-1-5225-9167-2.ch002

[14] Barua, R., Datta, S., Datta, P., & Roychowdhury, A. (2022). Study and Application of Machine Learning Methods in Modern Additive Manufacturing Processes. In S. Salunkhe, H. Hussein, & J. Davim (Eds.), *Applications of Artificial Intelligence in Additive Manufacturing* (pp. 75–95). IGI Global Scientific Publishing. https://doi.org/10.4018/978-1-7998-8516-0.ch004

[15] Salmi, M. (2020). Additive manufacturing processes in medical applications. *Materials, 14*(1), 191. https://doi.org/10.3390/ma14010191

[16] Ambrogio, G., De Napoli, L., Filice, L., Gagliardi, F., & Muzzupappa, M. (2005). Application of incremental forming process for high customised medical product manufacturing. *Journal of Materials Processing Technology, 162*, 156–162.

[17] Kretzschmar, N., Chekurov, S., Salmi, M., & Tuomi, J. (2018). Evaluating the readiness level of additively manufactured digital spare parts: An industrial perspective. *Applied Sciences, 8*(10), 1837.

[18] Calle, M. A., Salmi, M., Mazzariol, L. M., Alves, M., & Kujala, P. (2020). Additive manufacturing of miniature marine structures for crashworthiness verification: Scaling technique and experimental tests. *Marine Structures, 72*, 102764.

[19] Conner, B. P., Manogharan, G. P., Martof, A. N., Rodomsky, L. M., Rodomsky, C. M., Jordan, D. C., & Limperos, J. W. (2014). Making sense of 3-D printing: Creating a map of additive manufacturing products and services. *Additive Manufacturing, 1*, 64–76.

[20] Baumers, M., Dickens, P., Tuck, C., & Hague, R. (2016). The cost of additive manufacturing: Machine productivity, economies of scale and technology-push. *Technological Forecasting and Social Change, 102*, 193–201.

[21] Thomas D. (2016). Costs, Benefits, and Adoption of Additive Manufacturing: A Supply Chain Perspective. *The International journal, advanced manufacturing technology, 85*(5–8), 1857–1876. https://doi.org/10.1007/s00170-015-7973-6

[22] Ghomi, E. R., Khosravi, F., Neisiany, R. E., Singh, S., & Ramakrishna, S. (2021). Future of additive manufacturing in healthcare. *Current Opinion in Biomedical Engineering*, *17*, 100255.

[23] Kumar, R., Kumar, M., & Chohan, J. S. (2021). The role of additive manufacturing for biomedical applications: A critical review. *Journal of Manufacturing Processes*, *64*, 828–850.

[24] Charbonnier, B., Hadida, M., & Marchat, D. (2021). Additive manufacturing pertaining to bone: Hopes, reality and future challenges for clinical applications. *Acta Biomaterialia*, *121*, 1–28. https://doi.org/10.1016/j.actbio.2020.11.039

[25] Rosenthal, M., Henneberger, C., Gutkes, A., & Bues, C. T. (2018). Liquid Deposition Modeling: A promising approach for 3D printing of wood. *European Journal of Wood and Wood Products*, *76*, 797–799.

[26] Pettersson, A. B. V., Salmi, M., Vallittu, P., Serlo, W., Tuomi, J., & Mäkitie, A. A. (2020). Main clinical use of additive manufacturing (three-dimensional printing) in Finland restricted to the head and neck area in 2016–2017. *Scandinavian Journal of Surgery: SJS: Official Organ for the Finnish Surgical Society and the Scandinavian Surgical Society*, *109*(2), 166–173. https://doi.org/10.1177/1457496919840958

[27] Zadpoor, A. A., & Malda, J. (2017). Additive manufacturing of biomaterials, tissues, and organs. *Annals of Biomedical Engineering*, *45*, 1–11.

[28] Mancilla-De-la-Cruz, J., Rodriguez-Salvador, M., An, J., & Chua, C. K. (2022). Three-Dimensional Printing Technologies for Drug Delivery Applications: Processes, Materials, and Effects. *International journal of bioprinting*, *8*(4), 622. https://doi.org/10.18063/ijb.v8i4.622

[29] Kumbhar, N. N., & Mulay, A. V. (2018). Post processing methods used to improve surface finish of products which are manufactured by additive manufacturing technologies: A review. *Journal of the Institution of Engineers (India): Series C*, *99*, 481–487.

[30] Datta, S., Barua, R., Sarkar, R., Barui, A., Chowdhury, A. R., & Datta, P. (2018, September). Design and development of alginate: Poly-l-lysine scaffolds by 3D bio printing and studying their mechanical, structural and cell viability properties. In *IOP Conference Series: Materials Science and Engineering* (Vol. 402, p. 012113). IOP Publishing.

[31] Jain, N., Waidi, Y. O., Chowdhury, S., Barua, R., Das, S., Prasad, A., & Datta, S. (2024). Medical Devices Tribology. In: Kumar, A., Kumar, A., Kumar, A. (eds) *Applications of Biotribology in Biomedical Systems* (pp. 235–250). Cham: Springer Nature Switzerland.

[32] Chowdhury, S., Jain, N., Waidi, Y. O., Barua, R., Das, S., Prasad, A., & Datta, S. (2024). Biological Smart Biomaterials: Materials for Biomedical Applications. In: Kumar, A. Kumar, A., Kumar, A. (eds) *Applications of Biotribology in Biomedical Systems* (pp. 313–325). Cham: Springer Nature Switzerland.

[33] Herzog, D., Seyda, V., Wycisk, E., & Emmelmann, C. (2016). Additive manufacturing of metals. *Acta Materialia*, *117*, 371–392.

[34] Javaid, M., & Haleem, A. (2018). Additive manufacturing applications in medical cases: A literature based review. *Alexandria Journal of Medicine*, *54*(4), 411–422.

[35] Tuomi, J., Paloheimo, K. S., Vehviläinen, J., Björkstrand, R., Salmi, M., Huotilainen, E., . . . Mäkitie, A. A. (2014). A novel classification and online platform for planning and documentation of medical applications of additive manufacturing. *Surgical Innovation*, *21*(6), 553–559.

[36] Wang, Y. C., Chen, T., & Lin, Y. C. (2019). A Collaborative and Ubiquitous System for Fabricating Dental Parts Using 3D Printing Technologies. *Healthcare (Basel, Switzerland)*, *7*(3), 103. https://doi.org/10.3390/healthcare7030103

[37] Tack, P., Victor, J., Gemmel, P., & Annemans, L. (2016). 3D-printing techniques in a medical setting: A systematic literature review. *Biomedical Engineering Online*, *15*(1), 115. https://doi.org/10.1186/s12938-016-0236-4

[38] Salmi, M. (2016). Possibilities of preoperative medical models made by 3D printing or additive manufacturing. *Journal of Medical Engineering*, *2016*(1), 6191526.

[39] Bücking, T. M., Hill, E. R., Robertson, J. L., Maneas, E., Plumb, A. A., & Nikitichev, D. I. (2017). From medical imaging data to 3D printed anatomical models. *PloS One*, *12*(5), e0178540.

[40] Marro, A., Bandukwala, T., & Mak, W. (2016). Three-dimensional printing and medical imaging: A review of the methods and applications. *Current Problems in Diagnostic Radiology*, *45*(1), 2–9. https://doi.org/10.1067/j.cpradiol.2015.07.009

[41] Ahangar, P., Cooke, M. E., Weber, M. H., & Rosenzweig, D. H. (2018). Current biomedical applications of 3D printing and additive manufacturing. *Applied Sciences*, *9*(8), 1713. https://doi.org/10.3390/app9081713

[42] Fischer, D., Cheng, K. Y., Neto, M. Q., Hall, D., Bijukumar, D., Orías, A. A. E., Pourzal, R., Van Arkel, R. J., & Mathew, M. T. (2022). Corrosion behavior of Selective Laser Melting (SLM) manufactured Ti6Al4V alloy in saline and BCS solution. *Journal of Bio- and Tribo-corrosion*, *8*(2), 63. https://doi.org/10.1007/s40735-022-00657-1

[43] Rana, M., Moellmann, H. L., Schorn, L., Lommen, J., Rana, M., Wilkat, M., & Hufendiek, K. (2022). Primary orbital reconstruction with selective laser melting (SLM) of patient-specific implants (PSIs): An overview of 96 surgically treated patients. *Journal of Clinical Medicine*, *11*(12), 3361. https://doi.org/10.3390/jcm11123361

[44] Moraru, E., Stoica, A. M., Donțu, O., Cănănău, S., Stoica, N. A., Constantin, V., Cioboată, D. D., & Bădiță-Voicu, L. L. (2023). Mechanical and surface characteristics of selective laser melting-manufactured dental prostheses in different processing stages. *Materials (Basel, Switzerland)*, *16*(18), 6141. https://doi.org/10.3390/ma16186141

[45] Ginestra, P., Ferraro, R. M., Zohar-Hauber, K., Abeni, A., Giliani, S., & Ceretti, E. (2020). Selective laser melting and electron beam melting of Ti6Al4V for orthopedic applications: A comparative study on the applied building direction. *Materials (Basel, Switzerland)*, *13*(23), 5584. https://doi.org/10.3390/ma13235584

[46] Hoshi, I., Kawai, T., Kurosu, S., Minamino, T., Onodera, K., Miyamoto, I., & Yamada, H. (2021). Custom-made titanium mesh tray for mandibular reconstruction using an electron beam melting system. *Materials (Basel, Switzerland)*, *14*(21), 6556. https://doi.org/10.3390/ma14216556

[47] Lee, Y. C., Zheng, J., Kuo, J., Acosta-Vélez, G. F., Linsley, C. S., & Wu, B. M. (2022). Binder jetting of custom silicone powder for direct three-dimensional printing of maxillofacial prostheses. *3D Printing and Additive Manufacturing*, *9*(6), 520–534. https://doi.org/10.1089/3dp.2021.0019

[48] Wang, Y., Genina, N., Müllertz, A., & Rantanen, J. (2024). Binder jetting 3D printing in fabricating pharmaceutical solid products for precision medicine. *Basic & Clinical Pharmacology & Toxicology*, *134*(3), 325–332. https://doi.org/10.1111/bcpt.13974

[49] Cailleaux, S., Sanchez-Ballester, N. M., Gueche, Y. A., Bataille, B., & Soulairol, I. (2021). Fused Deposition Modeling (FDM), the new asset for the production of tailored medicines. *Journal of Controlled Release: Official Journal of the Controlled Release Society*, *330*, 821–841. https://doi.org/10.1016/j.jconrel.2020.10.056

[50] Xu, X., Wang, H., Shen, L., Yang, Q., & Yang, Y. (2023). Application and evaluation of fused deposition modeling technique in customized medical products. *International Journal of Pharmaceutics*, *640*, 122999. https://doi.org/10.1016/j.ijpharm.2023.122999

[51] Blake, C., Birch, S., & Brandão, J. (2019). Medical three-dimensional printing in zoological medicine. *The Veterinary Clinics of North America. Exotic Animal Practice*, *22*(3), 331–348. https://doi.org/10.1016/j.cvex.2019.05.004

[52] Xu, X., Goyanes, A., Trenfield, S. J., Diaz-Gomez, L., Alvarez-Lorenzo, C., Gaisford, S., & Basit, A. W. (2021). Stereolithography (SLA) 3D printing of a bladder device for intravesical drug delivery. *Materials Science & Engineering. C, Materials for Biological Applications*, *120*, 111773. https://doi.org/10.1016/j.msec.2020.111773

[53] Lakkala, P., Munnangi, S. R., Bandari, S., & Repka, M. (2023). Additive manufacturing technologies with emphasis on stereolithography 3D printing in pharmaceutical and medical applications: A review. *International Journal of Pharmaceutics: X*, *5*, 100159. https://doi.org/10.1016/j.ijpx.2023.100159

[54] Mohammed, A., Elshaer, A., Sareh, P., Elsayed, M., & Hassanin, H. (2020). Additive manufacturing technologies for drug delivery applications. *International Journal of Pharmaceutics*, *580*, 119245. https://doi.org/10.1016/j.ijpharm.2020.119245

[55] Durga Prasad Reddy, R., & Sharma, V. (2020). Additive manufacturing in drug delivery applications: A review. *International Journal of Pharmaceutics*, *589*, 119820. https://doi.org/10.1016/j.ijpharm.2020.119820

[56] Lind, J., Kälvemark Sporrong, S., Kaae, S., Rantanen, J., & Genina, N. (2017). Social aspects in additive manufacturing of pharmaceutical products. *Expert Opinion on Drug Delivery*, *14*(8), 927–936. https://doi.org/10.1080/17425247.2017.1266336

[57] Rupp, F., Liang, L., Geis-Gerstorfer, J., Scheideler, L., & Hüttig, F. (2018). Surface characteristics of dental implants: A review. *Dental Materials: Official Publication of the Academy of Dental Materials*, *34*(1), 40–57. https://doi.org/10.1016/j.dental.2017.09.007

[58] Saha, S., & Roy, S. (2022). Metallic dental implants wear mechanisms, materials, and manufacturing processes: A literature review. *Materials (Basel, Switzerland)*, *16*(1), 161. doi: https://doi.org/10.3390/ma16010161

[59] De Oliveira, T. T., Valente, M. L. D. C., Simões, I. G., Batalha, R. L., Macedo, A. P., Bolfarini, C., Pauly, S., & Dos Reis, A. C. (2023). Analysis of stress/strain distribution in dental mini-implants manufactured by additive manufacturing and machining. *Journal of Biomedical Materials Research. Part B, Applied Biomaterials*, *111*(10), 1751–1762. https://doi.org/10.1002/jbm.b.35282

[60] Prima, M. D., Coburn, J., Hwang, D., Kelly, J., Khairuzzaman, A., & Ricles, L. (2016). Additively manufactured medical products – The FDA perspective. *3D Printing in Medicine*, *2*, 1. https://doi.org/10.1186/s41205-016-0005-9

[61] Anitua, E., Eguia, A., Staudigl, C., & Alkhraisat, M. H. (2024). Clinical performance of additively manufactured subperiosteal implants: A systematic review. *International Journal of Implant Dentistry*, *10*(1), 4. https://doi.org/10.1186/s40729-024-00521-6

[62] Modaresifar, K., Hamzah Aziz Nurmohamed, F. R., Leeflang, M. A., Fluit, A. C., Fratila-Apachitei, L. E., Apachitei, I., Weinans, H., & Zadpoor, A. A. (2023). In vivo prevention of implant-associated infections caused by antibiotic-resistant bacteria through biofunctionalization of additively manufactured porous titanium. *Journal of Functional Biomaterials*, *14*(10), 520. https://doi.org/10.3390/jfb14100520

[63] Morrison, R. J., Hollister, S. J., Niedner, M. F., Mahani, M. G., Park, A. H., Mehta, D. K., Ohye, R. G., & Green, G. E. (2015). Mitigation of tracheobronchomalacia with 3D-printed personalized medical devices in pediatric patients. *Science Translational Medicine*, *7*(285), 285ra64. https://doi.org/10.1126/scitranslmed.3010825

[64] Les, A. S., Ohye, R. G., Filbrun, A. G., Ghadimi Mahani, M., Flanagan, C. L., Daniels, R. C., Kidwell, K. M., Zopf, D. A., Hollister, S. J., & Green, G. E. (2019). 3D-printed, externally-implanted, bioresorbable airway splints for severe tracheobronchomalacia. *The Laryngoscope*, *129*(8), 1763–1771. https://doi.org/10.1002/lary.27863

[65] Prasad, A., Bhasney, S. M., Prasannavenkadesan, V., Sankar, M. R., & Katiyar, V. (2023). Polylactic acid reinforced with nano-hydroxyapatite bioabsorbable cortical screws for bone fracture treatment. *Journal of Polymer Research*, *30*(5), 177.

[66] Prasad, A., Bhasney, S. M., Prasannavenkadesan, V., Sankar, M. R., & Katiyar, V. (2023). Nano-hydroxyapatite reinforced polylactic acid bioabsorbable cancellous screws for bone fracture fixations. *Journal of Applied Polymer Science*, *140*(43), e54577.

[67] Waidi, Y. O., Jain, N., Chowdhury, S., Barua, R., Das, S., Prasad, A., & Datta, S. (2024). Challenges and Perspective of Manufacturing Techniques in Biomedical Applications. In *Applications of Biotribology in Biomedical Systems* (pp. 433–445). Cham: Springer Nature Switzerland.

[68] Prasad, A. (2022). Biomaterial-based Nanofibers Scaffolds in Tissue Engineering Application. In: Jana, S., Jana, S. (eds) *Functional Biomaterials: Drug Delivery and Biomedical Applications* (pp. 245–264). Singapore: Springer Singapore.

[69] Krishani, M., Shin, W. Y., Suhaimi, H., & Sambudi, N. S. (2023). Development of Scaffolds from Bio-Based Natural Materials for Tissue Regeneration Applications: A Review. *Gels (Basel, Switzerland), 9* (2), 100. https://doi.org/10.3390/gels9020100

[70] Barua, R., Datta, S., Chowdhury, A. R., & Datta, P. (2018). Advances in MEMS and Micro-Scale Technologies for Application in Controlled Drug-Dosing Systems: MEMS-Based Drug Delivery Systems. In A. Bit (Ed.), *Design and Development of Affordable Healthcare Technologies* (pp. 165–179). IGI Global Scientific Publishing. https://doi.org/10.4018/978-1-5225-4969-7.ch007

[71] Datta, S., Barua, R., & Das, S. (2023). Role and Challenges of Bioprinting in Bone Tissue Engineering. In R. Ranjith & J. Davim (Eds.), *Handbook of Research on Advanced Functional Materials for Orthopedic Applications* (pp. 205–218). IGI Global Scientific Publishing. https://doi.org/10.4018/978-1-6684-7412-9.ch012

[72] Barua, R. (2024). Innovations in Minimally Invasive Surgery: The Rise of Smart Flexible Surgical Robots. In M. Garcia & R. de Almeida (Eds.), *Emerging Technologies for Health Literacy and Medical Practice* (pp. 110–131). IGI Global Scientific Publishing. https://doi.org/10.4018/979-8-3693-1214-8.ch006

[73] Anwar, S., Singh, G. K., Miller, J., Sharma, M., Manning, P., Billadello, J. J., Eghtesady, P., & Woodard, P. K. (2018). 3D printing is a transformative technology in congenital heart disease. *JACC. Basic to Translational Science, 3*(2), 294–312. https://doi.org/10.1016/j.jacbts.2017.10.003

[74] Bellia-Munzon, G., Cieri, P., Toselli, L., Cuestas, G., Doormann, F., Gabaldón-Massé, P., Rodriguez, V., & Bellia-Munzon, P. (2021). Resorbable airway splint, stents, and 3D reconstruction and printing of the airway in tracheobronchomalacia. *Seminars in Pediatric Surgery, 30*(3), 151063. https://doi.org/10.1016/j.sempedsurg.2021.151063

[75] Sood, V., Green, G. E., Les, A., & Ohye, R. G. (2021). Advanced therapies for severe tracheobronchomalacia: A review of the use of 3D-printed, patient-specific, externally implanted, bioresorbable airway splints. *Seminars in Thoracic and Cardiovascular Surgery. Pediatric Cardiac Surgery Annual, 24*, 37–43. https://doi.org/10.1053/j.pcsu.2021.02.001

[76] Barua, R. & Das, S. (2022). Improvements of Virtual and Augmented Reality for Advanced Treatments in Urology. In L. Coelho, R. Queirós, & S. Reis (Eds.), *Emerging Advancements for Virtual and Augmented Reality in Healthcare* (pp. 117–131). IGI Global Scientific Publishing. https://doi.org/10.4018/978-1-7998-8371-5.ch008

Tuhin Kar and Arjyajyoti Goswami*

Chapter 11
Laser surface modification of metallic implant materials

Abstract: The demand for transforming biomaterials into excellent medical implants for human body parts has experienced a significant surge. Implants such as screws, drug transport device, cranial devices, dental devices, and bone plates are widely used and considered promising alternatives for enhancing human life survival. Stainless, cobalt-based and titanium-based alloys are the extensively utilized materials for biological implant purposes. Selection of these materials depends on their mechanical characteristics including hardness, yield strength, and wear resistance in biological settings. Nevertheless, these materials experience significant surface deterioration and corrosion when exposed to prolonged durations. Poor osseointegration could result from this. These issues have prompted studies for surface modification of existing materials with potential advantages for biological applications. The objective of surface modification is to conserve the essential characteristics of the material while raising the surface biocompatibility. Laser surface modification is a method of altering the surface profile of a substrate via a high-power laser source. The procedure involves heating the material surface with the laser beam without physical contact between the laser source and workpiece and allowing the material to cool naturally through conduction. Laser surface alteration encompasses several techniques such as cladding, melting, heat treatment, shock peening, and alloying. Laser surface texturing can alter the surface wettability and convert it into superhydrophobic or superhydrophilic. Superhydrophobic titanium-based implants can reduce microbial adherence by minimizing the surface energy. Superhydrophilic surfaces exhibited superior osseointegration characteristics in comparison to conventional surfaces. Similarly, laser surface modification using silver imparts antibacterial properties to titanium orthopedic tools.

Keywords: Laser, surface modification, metallic implant, biocompatibility

---

***Corresponding author: Arjyajyoti Goswami,** Department of Mechanical Engineering, National Institute of Technology, Durgapur 713209, West Bengal, India, e-mail: agoswami.me@nitdgp.ac.in
Tuhin Kar, Department of Mechanical Engineering, National Institute of Technology, Durgapur 713209, India

https://doi.org/10.1515/9783111571423-011

11.1 Introduction

Hip and knee replacements are the most common arthroplasty operations performed. In 2022, there were around 765,558 total hip and 1,299,084 knee replacements performed in the United States [1]. With these procedures, the joint is replaced with an implant material with the goal of reducing pain and improving joint function [2]. Osseointegration depends on the functionality of biomaterials with body fluid inside the human body. The human body can employ a variety of metallic biomaterials, but titanium and its alloys are preferred because of their exceptional mechanical qualities, biocompatibility, and superior fatigue and corrosion resistance, making them the most bone-compatible metals [3]. Osteoarthritis affects 18% of men and 10% of women. It is a common joint condition in which the knee and hip (the articular cartilage of synovial joints) deteriorates [4]. Patients now have more comfort because to cutting-edge implant surgery methods, new biomaterial discoveries. It is anticipated that the orthopedic market will surge to USD 206.64 billion by 2024 from USD 105.18 billion due to global increase in demand [5]. Fewer than 5% of orthopedic implants last longer than 20 years despite significant advancements in implant procedures. Roughly 90% of hip and knee implants have a 10–15 years lifespan [6]. Failure appears shortly after surgery in fewer cases. Five to twenty percent of hip replacements and dental implants fail. With a poor success rate, the revision surgery is costly and traumatic [7]. The main causes for the revision surgeries are osteolysis, infections aseptic loosening, implant wear instability, implant fracture, and so on [8, 9]. Table 11.1 shows main causes of failure including infections and a lack of osseointegration.

Table 11.1: Common reasons of implant failures [10].

Causes of failure	Types of failure	
	Total hip replacement (%)	Total knee replacement (%)
Aseptic loosening	28	29
Infection	15	23
Instability	16	16
Others	41	32

There are drawbacks to the current surface modification techniques such as coating and chemical modification. Surface coating has the potential to be uneven and is brittle. This reduces its stability over a longer period of time. However, changes in surface chemistry could result in chemical reactions [11]. Researchers are putting a lot of effort into creating new biomaterials and altering the surface properties in order to improve the situation. Research on biomaterials requires simultaneous multidisciplinary efforts from a variety of scientific fields including material, chemical, biological, medical and mechanical [12]. One popular technique for modifying the surface properties of implant material is laser surface texturing. It is possible to create micro and nano patterns on implant material using laser processing [13, 14].

11.2 Classification of metallic biomaterials

Implant materials that successfully replace a material typically exhibit similar characteristics to the replaced material. Implant materials for load-bearing and orthopedic applications need to have a high Young's modulus that is compatible with bone in order to prevent load shielding, porosity, low density, and resilience to wear and corrosion [15]. Hence, metals are a good fit for this application, especially titanium, cobalt alloys and stainless steel are good choices, apart from those materials' magnesium alloys, zirconium alloys, zinc alloys, iron, and high-entropy alloys (HEAs) are sometimes used as implant materials. Table 11.2 provides a list of frequently utilized metallic implant materials and their respective benefits and drawbacks in terms of their compatibility with the human body as an orthopedic implant. Figure 11.1 shows different orthopedic implants composed of Ti-6Al-4V alloy and stainless steel.

Table 11.2: List of assessment of metallic implant materials for orthopedic applications [16–18].

Metals	Applications	Mechanical properties	Benefits	Drawbacks
Titanium and its alloys (Cp-Ti and Ti-6Al-4V)	Femoral heads, pacemaker, and nails	Elastic modulus (GPa): 105–110 Density (kg/m^3): 4,500 Tensile stress (MPa): 960–970	Strong osseointegration, excellent resistance to corrosion, and low density	High coefficient friction, high wear rate
Stainless steel (SS316L)	Screws, fracture plates, and hip replacements	Elastic modulus (GPa): 200 Density (kg/m^3): 7,800 Tensile stress (MPa): 465–950	Resistance to fatigue and corrosion	Wear particles produced which triggers an allergic reaction
Cobalt alloys (Co-Cr-Mo, Co-Ni-Cr-Mo)	Dentistry castings and joint replacements	Elastic modulus (GPa): 275–1,585 Density (kg/m^3): 8,500 Tensile stress (MPa): 600–1,795	Durability against wear and corrosion	Releasing toxic compounds

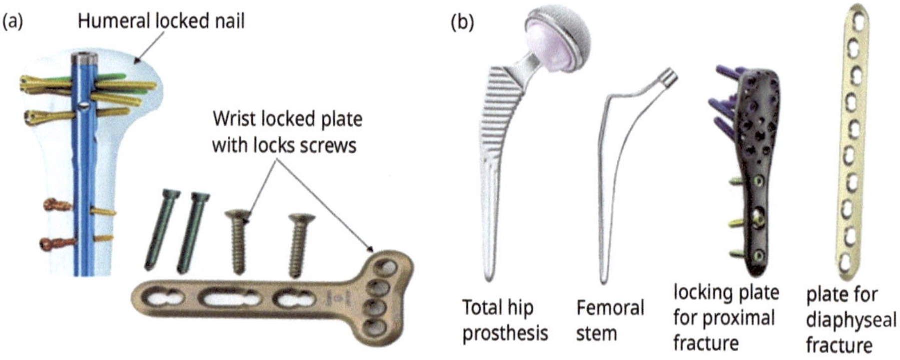

Figure 11.1: Different orthopedic implants composed of (a) titanium alloy and (b) SS316L [19].

11.2.1 Titanium and its alloys

Since the 1970s, titanium and its alloys were extensively utilized in implantation procedures to repair injured hard tissue because of their durability against corrosion and fatigue. A few examples of uses are cardiac valve prostheses, artificial hips and knees, bone plates, corneas, pacemakers, artificial hearts, back plates, screws for fracture repair, and orthopedic and dental implants [20]. Titanium has a relatively low density and exhibits inadequate thermal and electrical conductivity when compared to other noble metals but their remarkable corrosion resistance is attributed to the development of a thin oxide layer [21]. Titanium is a suitable choice for orthopedic implants due to its osteointegration characteristics, which enhance the implants performance [22].

11.2.2 Stainless steel

Stainless steel 316L (austenite) functions as a transient biomaterial. As SS316L contains chromium, which provides excellent resistance to corrosion. The existence of chromium responsible for the formation of an oxide deposition on the metal's surface when exposed to oxygen acts as a protective barrier, preventing further deterioration [23]. Amel-Farzad et al. [24] observed that the implant experienced pitting and corrosion, which develop fatigue in the implant. Stainless steel exhibits poor durability under wear, resulting in the generation of plenty of debris under wear testing. Apart from exhibiting inadequate resistance to wear and corrosion, the utilization of stainless steel as an implant material has been observed to trigger allergic responses in a substantial proportion of patients, causing the removal of the implants [25]. Laser micromachining is employed to alter the surface topography, wettability, morphology, and surface energy which affect cell adhesion and proliferation. These characteristics also aid in making the surface antimicrobial. The combination strategy for producing antibacterial and biocompatible qualities was rarely studied by researchers [19, 26].

11.2.3 Cobalt alloys

Co-Cr-Mo and Co-Ni-Cr-Mo are sometimes used as implant materials. These alloys have good resistant to corrosion, even in the presence of chloride. The formation of oxide coating makes the implant corrosion resistive in biological conditions, which makes them ideal for use as biomaterials [23]. The leaching of Ni, Cr, and Co from the implants is extremely harmful to the human body, particularly chromium, which has been associated with the development of skin cancer. Malignant tumors may develop when cobalt is inhaled by humans [26]. Cobalt alloys exhibit a reduced frequency of osteointegration compared to Ti alloys. Due to their poor compatibility with osseous tissue, cobalt alloy screws can be readily extracted after the healing of the injury. These implants exhibit more rigidity compared to Ti and SS implants, making them suitable for the treatment of spinal deformity [27].

11.2.4 Magnesium alloys

Magnesium alloys are biodegradable in nature having exceptional osseointegration characteristics. The density of magnesium alloys (1.74–2 g/cm^3) are almost equivalent to cortical bone (1.8–2.1 g/cm^3) [28]. Consequently, the implant will display less stress shielding effect. An inherent drawback of magnesium alloys is their limited resistance to corrosion, as they tend to deteriorate within a few months [29].

11.2.5 Zirconium alloys

Zirconium is commonly utilized in dental implants due to its compatibility with the human body and its visually appealing appearance in tooth restorations. Its superiority over Ti implants was attributed to its reduced risk of plaque buildup, effortless connection with the surrounding tissue, and aesthetic attractiveness [30]. Zr implants exhibited reduced bone resorption as a result of their lower Young's modulus, thereby minimizing the stress exerted on the bone. Recently, Zr alloys have attracted significant interest as a replacement implant material. Li et al. [31] demonstrated the biological suitability of Sn, Ta, and Zr in bulk and powder states. Bulk Ti and Nb were suitable for surgical implants, but their powder forms exhibited some toxicity.

11.2.6 Zinc alloys

Zinc alloys show promise as feasible materials for using in biodegradable implants. Zinc alloys can be easily manufactured by melting, forming, and casting processes due to their high reactivity and low melting point. Research indicates that zinc alloys do not exhibit any systemic or local toxicities or other adverse effects on biological compatibility. The significant drawback of zinc alloys as a sustainable implant material is its low mechanical strength [32]. The cast and wrought states of Zn exhibit ultimate tensile strength values of 20 and 120 MPa, respectively.

11.2.7 High-entropy alloys

HEAs are attracting considerable interest as metal alloys in recent time. HEA is a stable solid solution alloy, which mixing entropy is enhanced through the development of a single-phase solid. Hence, they are sometimes referred to as multi-principal element alloys or compound alloys due to their high concentration of distinctive elements. HEAs can have either single or multiphase structures, irrespective of the complexity of their structure [33]. HEAs consist of a minimum of five elements where all lattice sites have an equal likelihood of appearing. The difference in atomic radius induces a conformational deformation in the crystal lattice on the process of solid solution formation. Elevated lattice distortion energy leads to the loss of crystallographic stability and the formation of amorphous phases or intermetallic complexes. This distortion will compromise the functional characteristics of the material, regardless of whether it is amorphous or crystalline [34].

11.3 Surface modification techniques for implant material

Surface modification refers to the alteration of the surface characteristics of a material to enhance its functionality, life expectancy, and utility [35]. Metallic implants are mostly affected by bioenvironmental factors on their outer layers. Improving surface qualities is essential to lowering implant failure rates. Enhancing mechanical, physical, biocompatible, and chemical properties of implant materials is done by surface modification. The bone retrieval study suggested that bone to implant contact in orthopedic implants is around 70–80% [36]. It is possible to minimize infections by 70–90% by enhancing cell proliferation by 30% with improved osseointegration [37]. As a result, there will be plenty of opportunity to improve functionality of biomaterials by modifying the surface. In general, surface modification is classified into two parts: surface treatment and surface coating. As demonstrated in Figure 11.2, the sur-

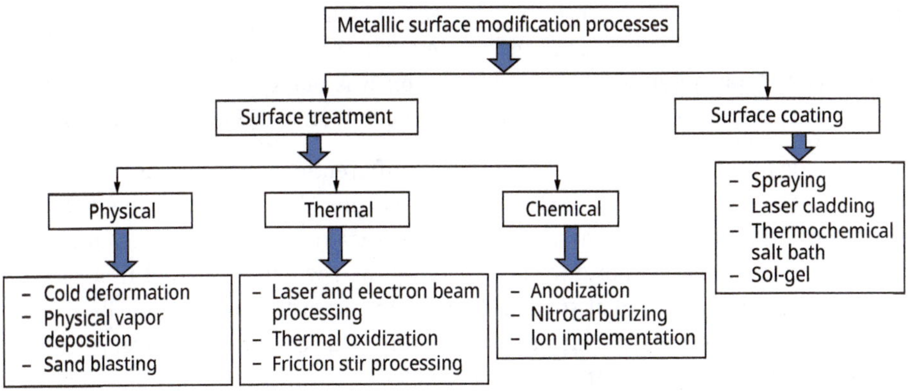

Figure 11.2: Different metallic surface modification processes.

face treatment can be classified into four types: chemical, thermal, physical, and coatings. Thermal methods utilize heat to remove the material from surface. Physical or chemical methods utilize material deposition on the surface. These techniques aim to alter the density and porosity of implants. The contact area is increased by these porous surfaces, which enhances cell proliferation [38].

Cold deformation: The cold deformation approach enhanced both surface diffusion and corrosion resistance, leading to the production of passive layers on the exterior surface [39]. Yang et al. [40] devised a straightforward and eco-friendly method for water treatment aimed at addressing the excessive nitrogen content in nickel-free stainless steels. This technique enhanced the formation of passive coatings and yielded a considerable reduction in the corrosion rate of stainless steels to 95%.

Physical vapor deposition: It is a process of physically depositing molecules or atoms from a material source to a substrate surface. Typically, films with certain properties can be applied to the surface of a substrate in a vacuum environment [41]. The vapor phase gets moving to the material, thereby resulting the formation of compact coatings upon condensation. This approach allows for the deposition of silica coatings onto the exterior of dental implants made from zirconia to regulate the degradation of the implants at low temperatures [42].

Sandblasting: Sandblasting is a technique that uses compressed air to propel solid particles (silica, alumina) at high velocities across a hard surface in order to create surface texture. Balza et al. [43] employed sandblasting as a method of altering the surface properties of titanium grade 5 for orthopedic implants.

Laser beam processing: Laser surface modification refers to the method of transforming the surface property of a material by the application of laser beam. The procedure involves heating the substate with the laser pulses and allowing it to cool naturally through conduction [44].

Thermal oxidization: It is a process of heating the metal at high temperature (around 600 °C) to create oxygen diffusion zone on metal. After subjecting the Ti alloy to oxidation at 600 °C for a period of 60 h, observation was made that the alloy displayed superior resistance to wear due to the development of an exterior film of TiO_2 [45].

Friction stir processing: Friction stir welding is differentiated from this process by the use of a stir tool that comprises a shoulder and pin. The interaction between the metal surface and stir tool generates significant friction, resulting in substantial plastic deformation. The temperature produced by the revolving stir tool in a specific area which causes the metal surface to melt and improves hardness. The molten metal surface undergoes solidification once more. This enhances the strength of the titanium due to its ultrafine-grained structure [46].

Thermal spraying: Thermal spraying refers to a group of coating deposition processes where molten material is sprayed on the metal surface for coating. The coating substrate is heated to a temperature over its melting point using a torch. The torch can be heated using flames or hot gasses. The particles that come from the process are propelled at high speed via a stream of gas, creating a thin layered of coating onto the substrate [47].

Laser cladding: This is one kind of additive manufacturing technique that uses heat to apply a protective coating to metallic biomaterials, improving their resistance to corrosion and wear. The approach employs a laser that is intentionally out of focus, combined with either a prepositioned or synchronized supply of powder. Typically, ceramic powders like TiC are utilized [48].

Thermochemical salt bath: This is a thermochemical surface modification process that involves application of high temperatures for thermal diffusion of implant materials. This method can diffuse both metal and nonmetal implant materials, thereby changing their chemistry and microstructure. This therapy is conducted in a media that can be solid, liquid, or gaseous and contains many active chemical components [49].

Apart from these approaches, some machining techniques like milling [50, 51], hot embossing [52], and electrodischarge machining [53] can also be employed for surface modification.

11.4 Surface modification of implant materials by laser

The goal of surface modification is to change a metal's surface behavior without changing the material's overall characteristics. These alterations have the potential to enhance bulk characteristics like corrosion resistance and customize the host response like osseointegration [54]. Surface coating and surface treatment with heat are two beneficial, effective surface treatment processes for metallic implants. Thermal surface treatment provides better stability against wear and corrosion. This technique is usually the fastest to execute with better control [55]. Thermal surface modification methods typically exhibit stronger interfacial stress between the coating and material compared to physical approaches. Additionally, they offer a more cost-effective alternative to chemical procedures [56].

Over time, a number of thermal surface modification methods, such as laser, electron beam, and friction stir processing, have been developed. Among these techniques, laser surface alteration has generated momentous importance due to its special advantages, which include being fast, efficient, nontoxic, and able to create surface textures with a high degree of intricacy and accuracy [57]. Laser beam processing is a noncontact-type thermal energy-based advance methodology for surface alteration that can be used with nearly any kind of materials. The substance from the parent material is melted and evaporated by a focused laser beam to form desired surface [58]. As shown in Figure 11.3, implant material surface texturing study goes through the following steps (according to ISO 10993). Orthopedic replacement mainly performed using metal and alloys to increase load-bearing capacity. Major biomaterials that are frequently utilized, including stainless steel, Co-Cr alloys, Ti alloys, Al_2O_3, and ZrO_2 are explored in regard to improving surface qualities. The following section discusses the laser surface modification of two most important implant materials namely stainless steel and titanium and its alloys.

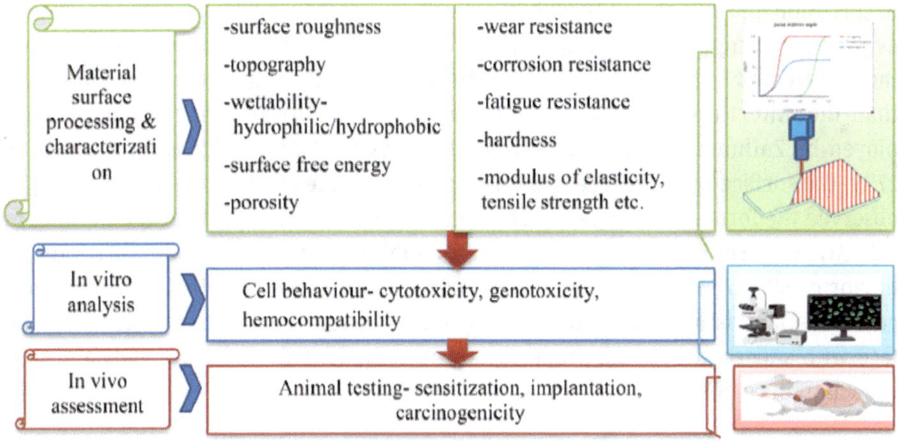

Figure 11.3: The process of characterizing laser-textured implant materials [59, 60].

11.4.1 Titanium and its alloys

Sun et al. [61] employed an Nd:YAG laser to machine CP-Ti, improving its corrosion resistance by varying laser power between 189 and 310 W. The rapid transformation of the homogeneous to a higher hardness martensitic structure was shown which was quenched quickly. Balla et al. [62] suggested that laser processing of the Ti-6Al-4V alloy surface by an Nd:YAG laser (power: 400 W), which will enhance wear property and hardness of alloy. The hardness was increased by 15–22% after laser processing. Yue et al. [63] decreased the pitting tendency of the Ti alloy after using laser surface remelting. This procedure involved melting and quickly solidifying the Ti-6Al-4V alloy's surface with an ns pulsing laser. This caused the alloy to become stronger and more corrosion-resistant through surface remelting. Braga et al. [64] enhanced the osseointegration of the dental implant material by pulse laser surface texturing. Pulse energy and scanning speed were taken as input variables and formation of TiO and Ti_3O phases were also examined. Patil et al. [65] created superhydrophobic surfaces utilizing nanosecond pulsed laser technology to prevent the growth of *E. coli* on orthopedic and dental implants. Pit shape on titanium alloy (Grade 5) by laser, followed by annealing at 300 °C, was performed to shorten transition time for the formation of superhydrophobic surface from hydrophilic.

11.4.2 Stainless steel

In the biomedical implant and surgical sectors, austenitic stainless steels (SS316L) are considered to be indispensable materials [66]. By virtue of their exceptional resistance to corrosion and compatibility with living organisms, they have been extensively used as casings for implants and scalpels [67]. Wang et al. [68] demonstrated that introducing texture and enhancing surface roughness will result in a higher absorption rate than the untextured substrate. The laser surface modification technique was employed by Zaifuddin et al. [69] to create textures on SS316L. With a laser power of 27 W, the surface roughness reached its maximum value of 28.98 µm. Trdan et al. [70] observed the formation a superhydrophobic surface from a superhydrophilic state by Nd:YAG laser. The formation of superhydrophobic surface improves the corrosion behavior of SS316L. Such functionalized surfaces exhibit various distinctive advantageous characteristics such as self-healing and enhanced resistance to corrosion. Surface with superhydrophobic properties (contact angle: $168 \pm 3.0°$) promotes a greater reduction in corrosion.

11.5 Failure of implant materials

The average lifespan of titanium orthopedic implantations ranges from 15 to 20 years [71]. This implies that the implants are prone to failure during the recipient's lifespan, necessitating the need for a second surgery and replacement implants. The primary causes for revision surgery include inadequate osseointegration, formation of biofilms, corrosion, and wear.

11.5.1 Formation of biofilms

Bacterial adhesion to implant materials can lead toward the development of biofilms. Biofilms inhibit the attachment of host cells to the metal surface, thereby impeding the implantation into the body. As a result, there is space for mobility among the bone and implant [72].

11.5.2 Weak osseointegration

Osseointegration denotes the straight and complete integration of bone and metallic implant, without the participation of any other soft connective tissue [73]. The procedure entails osteoblasts migrating to the biomaterial. The cells release a collagen matrix and then deposit bone apatite onto the surface of the metals. This promotes the connection of the bone with the implant, hence decreasing the chances of loosening [74].

11.5.3 Wear

The main form of wear those orthopedic biomaterials experience is oxidative wear. This process entails the elimination of the original oxide layer by the interaction with a rough surface [75]. The inherent wear resistance of titanium is attributed to its strong affinity for oxygen. A layer of oxide grows on the metal surface, acting as an obstacle that inhibits the movement of ions between the metal and its surroundings, thereby preventing corrosion [76].

11.5.4 Corrosion

Corrosion is responsible for approximately 42% of failures in orthopedic implants, rendering it the primary cause for recipients requiring modification of the procedures [77]. Corrosion is the process in which a metal reacts chemically with its surroundings to create a more stable substance [78]. When the oxide layer is compromised, the underlying metal experiences exposure to an extremely corrosive atmosphere. Body fluid comprises corrosive elements including water, chlorine, salt, and proteins [79].

11.6 Biocompatibility testing

Laser engraving of surfaces is utilized to manipulate cell responses in biomaterials [80]. Cell responses can be analyzed by studying fundamental cellular processes like cell proliferation adhesion, maturation, and differentiation [81] (Table 11.3). Different levels of implant surface roughness at the micro-nano scales have varying effects on the process of osseointegration. Macroscale surface roughness, ranging from millimeters to a few microns, enhances stability through the formation of biomechanical interlocking. This aids in the enduring stability of the implant. Implant osseointegration is enhanced by microlevel surface roughness ranging from 1 to 10 mm, which promotes improved cell proliferation and interlocking [82]. The presence of nanoscale surface roughness (measuring between 1 and 100 nm) enhances the attachment of osteoblast cells and the adsorption of proteins on the surface layer [83] (see Figure 11.4). The laser-patterned specimens exhibited superior cell adhesion and proliferation in comparison with surfaces created using conventional machining techniques [84, 85]. Cell growth and differentiation are enhanced by thin and acute edges [86]. The findings indicate that low surface temperature (LST) specimens facilitate the first development of osteoblast cells, hence promoting early healing [87]. Qin et al. [88] demonstrated that the behavior of osteoblast cells in response to Co-Cr-Mo is influenced by changes in surface topographies such as circle, triangle, and square shapes. Ma et al. [89] discovered that the Mg-Gd-Ca surface, which exhibited cytotoxicity due to its significant deterioration, underwent a transformation into a biocompatible material following laser surface treatment. Calderon et al. [90] successfully attained accurate manipulation of stem cell adhesion and differentiation by creating micro-nanopatterns. Nanostructured surfaces can effectively decrease bacterial attachment [91]. The surface roughness, chemistry, geometry, and wettability have significant roles in rendering a surface antibacterial [92]. The fabrication of micro- and nanoscale surface textures using laser surface texturing can enhance the antibacterial properties [93]. Schroder et al. [94] fabricated a spike pattern on Ti-6Al-4V in order to improve biocompatibility as well as antibacterial properties. The rod-structured surface was inserted into the tibia of the rat, along with the introduction of *Staphylococcus aureus* bacteria. The findings were evaluated using radiological and histological techniques after a period of 21 days. The adhesion formed with adjacent tissues was highly satisfactory. However, the antibacterial impact was insufficient to validate the premise. Yasuno et al. [95] utilized fiber and YVO_4 laser to create laser-textured surfaces. A patterned zirconia implant was inserted into the tibia of rats. Samples were obtained to calculate the bone implant contact ratio and implant removal torque after 28 days of plantation. The findings indicate that each of the laser-textured implants was enhancing the osseointegration. The implant treated with fiber laser exhibited improved performance compared to the material treated with YVO_4.

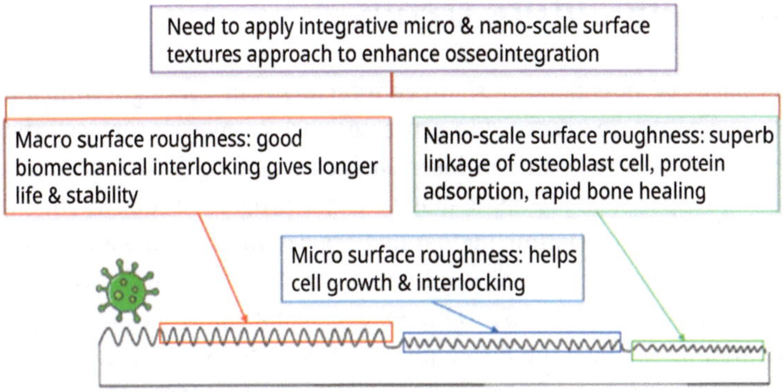

Figure 11.4: The impact of surface texture on the osseointegration.

Table 11.3: Biocompatibility investigation of laser-textured surfaces.

Reference	Material	Surface structure	Compatibility	Findings
[96]	SS316L	Nanopillars	Antibacterial	Impact on wettability and surface morphology
[97]	Gold	Spherical rod-like shape	Antibacterial	Excellent antibacterial surface that inhibits the adherence of bacteria
[98]	Pure titanium	Nanoripples	Biocompatible	Biocompatibility is enhanced with decreased biofilm
[88]	Co-Cr-Mo	Microstructure	Biocompatible	The shape and size of cells are controlled by surface morphology
[99]	Titanium grade 5	Spike shape	Biocompatible	Biocompatible but do not exhibit in vivo antibacterial resistance
[100]	Ti-6Al-4V	Microdimples	Biocompatible	Corners and ridges encourage cell adhesion, while laser surface texturing lessens wear
[101]	Titanium grade 2	Nanopillars and LIPSS	Antibacterial	Minimize the creation of biofilm (roughness <0.3 μm)
[102]	AISI 316L	Nanostructured	Toxicity	Production of toxic hexavalent chromium
[103]	AISI 316L	Nanoripples	Antibacterial	Modeling the relationship between textured surface and bacterial cell
[89]	Mg-Gd-Ca	Microstructure	Biocompatible	Improved biocompatibility and resistance to corrosion
[104]	AISI304	Nanotextured shapes	Biocompatible	Exact regulation of cell differentiation

11.7 Challenges and future trends

Aseptic release and bacterial infection are significant factors contributing to implant failure. Long-term constraints are observed in processes such as coating surfaces and the utilization of antibiotics. The results of laser surface texturing in this scenario show promise due to its capacity to modulate many surface properties such as roughness, wettability, and topography. Optimizing parameters can improve the antibacterial properties and biocompatibility. There is a limited amount of research available on the increase of combined effects by LST. Nanotexture, nanoripples, and LIPSS patterns have the potential to enhance the development of bone material surfaces. We are inspired by the natural patterns found in Taro leaves, lotus leaves, Gecko skin, Sharkskin, and Cicada wings, which push us to create micro- and nanotextured surface designs for implants [105] (see Figure 11.5).

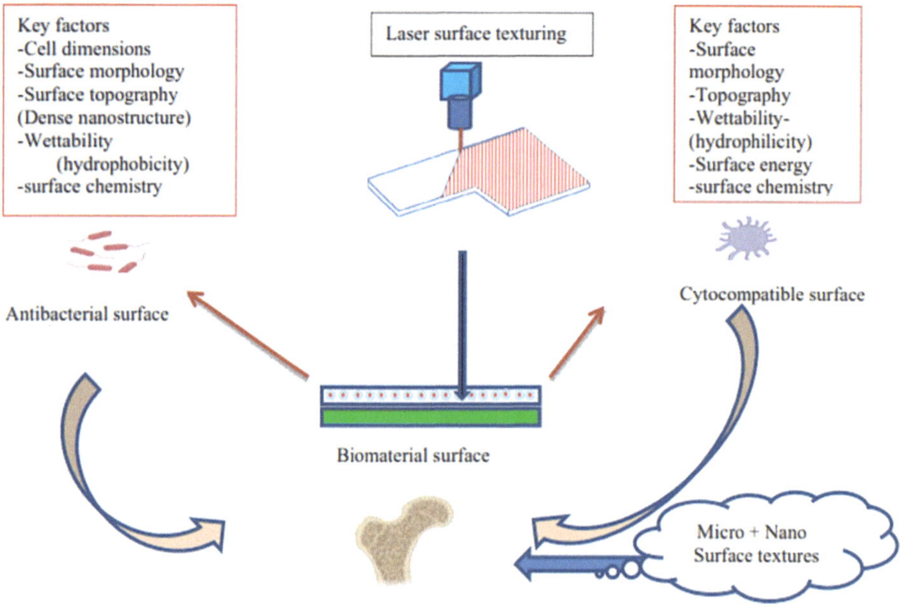

Figure 11.5: Future perspectives of research in laser-textured implant material [106].

The following research problems still need to be resolved despite the improvements in laser surface texturing:

- Different cell responses are generated by micro- and nanotextures. When combined with nanoscale topography, long-term bone fixing, and rapid bone healing advantages might be achieved.
- Two major factors that contribute to implant failure are infections and poor osseointegration. Nonetheless, the bulk of scientists are working to improve either antibacterial surface qualities or cell adhesion and mechanism. There are not many study works on the integrative strategy of biocompatibility, antimicrobial action, and early bone healing.
- Metallic biomaterials can have their improved surface qualities using a variety of materials and techniques. The focus is on improving the substrate's wear resistance and hardness. Further investigation into the human cell interaction of these materials is necessary before they can be fully considered as a viable material for metal implant purposes like knee- or hip-replacing devices.
- Require optimization of surface pattern and size to provide the desired impact.
- The high temperature experienced during processing might results in the creation of undesirable surface microstructure, resulting in the development of surface cracks. The presence of cracks decreasing the fatigue strength of materials, which is the primary issue in laser surface modification. So, optimization of laser input factors is essential to achieve specific geometric features.
- The future research should prioritize the development of fundamental novel techniques or the integration of several surface modification methods to achieve a collaborative impact and leverage their respective strengths to surmount the limitations. However, it is possible to customize the structure and composition of the composite covering to optimize its biomedical performance.

11.8 Summary

The substantial growth in orthopedic surgery and the resulting increase in the proportion of revision procedures highlight the need of examining the reasons behind failure. Surface texturing using laser technology is becoming recognized as the most advanced method to address these problems. Laser surface texturing has demonstrated superior efficacy compared to traditional biomaterial structuring techniques such as sandblasting, PVD, and chemical etching. Laser processing allows for optimized control to produce accurate and consistent geometry. This is the most effective method for managing surface characteristics such as wettability, surface roughness, and topography. In general, laser surface modification has been effective in enhancing surface wear resistance, hardness, and mechanical characteristics such as tensile strength and modulus of elasticity to a certain degree.

This chapter provides an overview of the laser surface modification of several metallic implant materials. This work also provides an overview of laser surface texturing of biomaterials at the laboratory scale using experiments. It offers guidance to researchers on enhancing biocompatibility and inhibiting bacterial activity. Ensuring strong mechanical attachment between bone and implant, as well as achieving fast curing without infections, are essential factors to prevent dysfunction of artificial implants. Specifically, these findings are aimed at advancing the field of micro- and nanoscale laser surface texturing.

References

[1] Curvo Labs, I. (2023). 2023 hip and knee implant review. *Orthopedic Network News*, 1–26.

[2] Uwais, Z. A., Hussein, M. A., Samad, M. A., & Al-Aqeeli, N. (2017). Surface modification of metallic biomaterials for better tribological properties: A review. *Arabian Journal for Science and Engineering*, *42*, 4493–4512. https://doi.org/10.1007/s13369-017-2624-x

[3] Thu, M. K., Kang, Y. S., Kwak, J. M., et al. (2023). Comparison between bone–implant interfaces of microtopographically modified zirconia and titanium implants. *Scientific Reports*, *13*, 11142. https://doi.org/10.1038/s41598-023-38432-y

[4] Glyn-Jones, S., Palmer, A. J. R., Agricola, R., et al. (2015). Osteoarthritis. *The Lancet*, *386*, 376–387. https://doi.org/10.1016/S0140-6736(14)60802-3

[5] Biomaterials Market, https://www.researchandmarkets.com/reports/5238627/biomaterials-market-by-type-metallic-gold (March 2024, accessed 11 May 2025).

[6] Zhang, L., Sirivisoot, S., Balasundaram, G., & Webster, T. J. (2009). Nanomaterials for Improved Orthopedic and Bone Tissue Engineering Applications. In *Advanced Biomaterials* (pp. 205–241). Wiley.

[7] Kaur, M., & Singh, K. (2019). Review on titanium and titanium based alloys as biomaterials for orthopaedic applications. *Materials Science and Engineering: C*, *102*, 844–862. https://doi.org/10.1016/j.msec.2019.04.064

[8] Luo, X., Yao, S., Zhang, H., et al. (2020). Biocompatible nano-ripples structured surfaces induced by femtosecond laser to rebel bacterial colonization and biofilm formation. *Optics & Laser Technology*, *124*, 105973. https://doi.org/10.1016/j.optlastec.2019.105973

[9] Florea, D. A., Albuleț, D., Grumezescu, A. M., & Andronescu, E. (2020). Surface modification – A step forward to overcome the current challenges in orthopedic industry and to obtain an improved osseointegration and antimicrobial properties. *Materials Chemistry and Physics*, *243*, 122579. https://doi.org/10.1016/j.matchemphys.2019.122579

[10] Sirdeshmukh, N., & Dongre, G. (2021). Laser micro & nano surface texturing for enhancing osseointegration and antimicrobial effect of biomaterials: A review. *Materials Today: Proceedings*, *44*, 2348–2355. https://doi.org/10.1016/j.matpr.2020.12.433

[11] Hasan, J., Crawford, R. J., & Ivanova, E. P. (2013). Antibacterial surfaces: The quest for a new generation of biomaterials. *Trends in Biotechnology*, *31*, 295–304. https://doi.org/10.1016/j.tibtech.2013.01.017

[12] Basu, B., Katti, D. S., & Kumar, A. (2009). Advanced biomaterials: Fundamentals, processing, and applications.

[13] Kar, T., & Goswami, A. (2022). Mathematical modeling approaches and new development in laser micro machining process: A review. *Lasers in Manufacturing and Materials Processing*, *9*, 532–568. https://doi.org/10.1007/s40516-022-00189-z

[14] Kar, T., Deshmukh, S. S., & Goswami, A. (2024). Fabrication of micro holes using low power fiber laser: Surface morphology, modeling and soft-computing based optimization. *Advances in Manufacturing*. https://doi.org/10.1007/s40436-024-00484-2
[15] Pilliar, R. M. (2009). Metallic Biomaterials. In *Biomedical Materials* (pp. 41–81). Boston, MA: Springer US.
[16] Tudor-Locke, C., Craig, C. L., Beets, M. W., et al. (2011). How many steps/day are enough? For children and adolescents. *International Journal of Behavioral Nutrition and Physical Activity*, *8*, 78. https://doi.org/10.1186/1479-5868-8-78
[17] Aljenaei, F., Catelas, I., Louati, H., et al. (2017). Effects of hip implant modular neck material and assembly method on fatigue life and distraction force. *Journal of Orthopaedic Research*, *35*, 2023–2030. https://doi.org/10.1002/jor.23481
[18] Chen, Q., & Thouas, G. A. (2015). Metallic implant biomaterials. *Materials Science and Engineering: R: Reports*, *87*, 1–57. https://doi.org/10.1016/j.mser.2014.10.001
[19] Rony, L., Lancigu, R., & Hubert, L. (2018). Intraosseous metal implants in orthopedics: A review. *Morphologie*, *102*, 231–242. https://doi.org/10.1016/j.morpho.2018.09.003
[20] Sidambe, A. (2014). Biocompatibility of advanced manufactured titanium implants – A review. *Materials*, *7*, 8168–8188. https://doi.org/10.3390/ma7128168
[21] Eliaz, N. (2019). Corrosion of metallic biomaterials: A review. *Materials*, *12*, 407. https://doi.org/10.3390/ma12030407
[22] Lee, B.-H., Lee, C., Kim, D.-G., et al. (2008). Effect of surface structure on biomechanical properties and osseointegration. *Materials Science and Engineering: C*, *28*, 1448–1461. https://doi.org/10.1016/j.msec.2008.03.015
[23] Narendrakumar, K., Kulkarni, M., Addison, O., et al. (2015). Adherence of oral streptococci to nanostructured titanium surfaces. *Dental Materials*, *31*, 1460–1468. https://doi.org/10.1016/j.dental.2015.09.011
[24] Amel-Farzad, H., Peivandi, M. T., & Yusof-Sani, S. M. R. (2007). In-body corrosion fatigue failure of a stainless steel orthopaedic implant with a rare collection of different damage mechanisms. *Engineering Failure Analysis*, *14*, 1205–1217. https://doi.org/10.1016/j.engfailanal.2006.11.037
[25] Navarro, M., Michiardi, A., Castaño, O., & Planell, J. A. (2008). Biomaterials in orthopaedics. *Journal of the Royal Society Interface*, *5*, 1137–1158. https://doi.org/10.1098/rsif.2008.0151
[26] Czarnek, K., Terpiłowska, S., & Siwicki, A. K. (2015). Review paper selected aspects of the action of cobalt ions in the human body. *Central European Journal of Immunology*, *2*, 236–242. https://doi.org/10.5114/ceji.2015.52837
[27] Wang, X., Yeung, K., Cheung, J. P. Y., et al. (2020). A novel scoliosis instrumentation using special superelastic nickel–titanium shape memory rods: A biomechanical analysis using a calibrated computer model and data from a clinical trial. *Spine Deformity*, *8*, 369–379. https://doi.org/10.1007/s43390-020-00075-8
[28] Meng, S.-J., Yu, H., Fan, S.-D., et al. (2019). Recent progress and development in extrusion of rare Earth free Mg alloys: A review. *Acta Metallurgica Sinica (English Letters)*, *32*, 145–168. https://doi.org/10.1007/s40195-018-00871-2
[29] Hanas, T., Sampath Kumar, T. S., Perumal, G., et al. (2018). Electrospun PCL/HA coated friction stir processed AZ31/HA composites for degradable implant applications. *Journal of Materials Processing Technology*, *252*, 398–406. https://doi.org/10.1016/j.jmatprotec.2017.10.009
[30] Wang, S.-P., & Xu, J. (2017). TiZrNbTaMo high-entropy alloy designed for orthopedic implants: As-cast microstructure and mechanical properties. *Materials Science and Engineering: C*, *73*, 80–89. https://doi.org/10.1016/j.msec.2016.12.057
[31] Li, Y., Wong, C., Xiong, J., et al. (2010). Cytotoxicity of titanium and titanium alloying elements. *Journal of Dental Research*, *89*, 493–497. https://doi.org/10.1177/0022034510363675

[32] Mostaed, E., Sikora-Jasinska, M., Drelich, J. W., & Vedani, M. (2018). Zinc-based alloys for degradable vascular stent applications. *Acta Biomaterialia*, *71*, 1–23. https://doi.org/10.1016/j.actbio.2018.03.005

[33] Kumar, J. S., Raju, C. V. S., Naik, A. B., et al. (2022). Effect of reinforcement and sintering on dry sliding wear and hardness of titanium – (AlSi)0.5CoFeNi based composite. *Surface Topography*, *10*, 035027. https://doi.org/10.1088/2051-672X/ac8ab7

[34] Ma, Y., Ma, Y., Wang, Q., et al. (2021). High-entropy energy materials: Challenges and new opportunities. *Energy and Environmental Science*, *14*, 2883–2905. https://doi.org/10.1039/D1EE00505G

[35] Deshmukh, S. S., Kar, T., & Goswami, A. (2023). Investigation and multi-objective optimization of replication accuracy of microfluidic chip fabricated through induction-aided hot embossing. *Journal of the Brazilian Society of Mechanical Sciences and Engineering*, *45*, 628. https://doi.org/10.1007/s40430-023-04538-7

[36] Davis, R., Singh, A., Jackson, M. J., et al. (2022). A comprehensive review on metallic implant biomaterials and their subtractive manufacturing. *The International Journal of Advanced Manufacturing Technology*, *120*, 1473–1530. https://doi.org/10.1007/s00170-022-08770-8

[37] Tiainen, L., Abreu, P., Buciumeanu, M., et al. (2019). Novel laser surface texturing for improved primary stability of titanium implants. *Journal of the Mechanical Behavior of Biomedical Materials*, *98*, 26–39. https://doi.org/10.1016/j.jmbbm.2019.04.052

[38] da Silva, L. R. R., Sales, W. F., Campos, F. A. R., et al. (2021). A comprehensive review on additive manufacturing of medical devices. *Progress in Additive Manufacturing*, *6*, 517–553. https://doi.org/10.1007/s40964-021-00188-0

[39] Talha, M., Ma, Y., Lin, Y., et al. (2019). Corrosion performance of cold deformed austenitic stainless steels for biomedical applications. *Corrosion Reviews*, *37*, 283–306. https://doi.org/10.1515/corrrev-2019-0004

[40] Yang, Y., Wang, Q., Li, J., et al. (2019). Enhancing general corrosion resistance of biomedical high nitrogen nickel-free stainless steel by water treatment. *Material Letters*, *251*, 196–200. https://doi.org/10.1016/j.matlet.2019.05.081

[41] Jiang, L., Bao, M., Dong, Y., et al. (2023). Processing, production and anticorrosion behavior of metallic glasses: A critical review. *Journal of Non-Crystalline Solids*, *612*, 122355. https://doi.org/10.1016/j.jnoncrysol.2023.122355

[42] Cakir-Omur, T., Gozneli, R., & Ozkan, Y. (2019). Effects of silica coating by physical vapor deposition and repeated firing on the low-temperature degradation and flexural strength of a zirconia ceramic. *Journal of Prosthodontics*, *28*. https://doi.org/10.1111/jopr.12618

[43] Balza, J. C., Zujur, D., Gil, L., et al. (2013). Sandblasting as a surface modification technique on titanium alloys for biomedical applications: Abrasive particle behavior. *IOP Conference Series: Materials Science and Engineering*, *45*, 012004. https://doi.org/10.1088/1757-899X/45/1/012004

[44] Pereira, H., Moura, C. G., Miranda, G., & Silva, F. S. (2021). Influence of liquid media and laser energy on the production of MgO nanoparticles by laser ablation. *Optics & Laser Technology*, *142*, 107181. https://doi.org/10.1016/j.optlastec.2021.107181

[45] Guleryuz, H., & Cimenoglu, H. (2005). Surface modification of a Ti–6Al–4V alloy by thermal oxidation. *Surface and Coatings Technology*, *192*, 164–170. https://doi.org/10.1016/j.surfcoat.2004.05.018

[46] Zhang, L.-C., Chen, L.-Y., & Wang, L. (2020). Surface modification of titanium and titanium alloys: Technologies, developments, and future interests. *Advanced Engineering Materials*, *22*. https://doi.org/10.1002/adem.201901258

[47] Tejero-Martin, D., Rezvani Rad, M., McDonald, A., & Hussain, T. (2019). Beyond traditional coatings: A review on thermal-sprayed functional and smart coatings. *Journal of Thermal Spray Technology*, *28*, 598–644. https://doi.org/10.1007/s11666-019-00857-1

[48] Boshoman, S. B., Fatoba, O. S., & Jen, T.-C. (2023). Laser Cladding of Titanium Alloys: A Review. In *2023 14th International Conference on Mechanical and Intelligent Manufacturing Technologies (ICMIMT)* (pp. 158–163). IEEE.

[49] Galey, B., Gautier, M., Kim, B., et al. (2022). Trace metal elements vaporization and phosphorus recovery during sewage sludge thermochemical treatment – A review. *Journal of Hazardous Materials, 424*, 127360. https://doi.org/10.1016/j.jhazmat.2021.127360

[50] Kar, T., Mandal, N. K., & Singh, N. K. (2020). Multi-response optimization and surface texture characterization for CNC milling of inconel 718 alloy. *Arabian Journal for Science and Engineering, 45*, 1265–1277. https://doi.org/10.1007/s13369-019-04324-5

[51] Kar, T., Goswami, A., & Mandal, N. K. (2023). Multi-response optimization of CNC milling parameters for improving performance characteristics using RSM and MOGA. *Advances in Materials and Processing Technologies*, 1–15. https://doi.org/10.1080/2374068X.2023.2208022

[52] Datta, S., Deshmukh, S. S., Kar, T., & Goswami, A. (2023). A review on modelling and numerical simulation of micro hot embossing process: Fabrication, mold filling behavior, and demolding analysis. *Engineering Research Express, 5*, 012006. https://doi.org/10.1088/2631-8695/acc1c1

[53] Kumar, A., Sijina, K. P., Rajanikant, G. K., & Kuriachen, B. (2022). Experimental investigation into the effects of electric discharge surface modification process parameters on the biocompatibility of Ti6Al4V. *Biomedical Engineering Advances, 4*, 100063. https://doi.org/10.1016/j.bea.2022.100063

[54] Qin, P., Chen, Y., Liu, Y.-J., et al. (2019). Resemblance in corrosion behavior of selective laser melted and traditional monolithic β Ti-24Nb-4Zr-8Sn alloy. *ACS Biomaterials Science & Engineering, 5*, 1141–1149. https://doi.org/10.1021/acsbiomaterials.8b01341

[55] Zhang, L.-C., Chen, L.-Y., & Wang, L. (2020). Surface modification of titanium and titanium alloys: Technologies, developments, and future interests. *Advanced Engineering Materials, 22*. https://doi.org/10.1002/adem.201901258

[56] Xue, T., Attarilar, S., Liu, S., et al. (2020). Surface modification techniques of titanium and its alloys to functionally optimize their biomedical properties: Thematic review. *Frontiers in Bioengineering and Biotechnology, 8*. https://doi.org/10.3389/fbioe.2020.603072

[57] Kar, T., & Goswami, A. (2024). Experimental investigation on fiber laser micro drilling of Titanium grade 5: Fabrication, nature-inspired optimization and analysis through image processing. *Engineering Research Express, 6*, 025411. https://doi.org/10.1088/2631-8695/ad4436

[58] Kar, T., Deshmukh, S. S., & Goswami, A. (2022). Investigation of fiber laser micro-channel depth on silicon wafer. *Materials Today: Proceedings, 60*, 2105–2110. https://doi.org/10.1016/j.matpr.2022.02.024

[59] Watanabe, I., McBride, M., Newton, P., & Kurtz, K. S. (2009). Laser surface treatment to improve mechanical properties of cast titanium. *Dental Materials, 25*, 629–633. https://doi.org/10.1016/j.dental.2008.11.006

[60] Sirdeshmukh, N., & Dongre, G. (2021). Laser micro & nano surface texturing for enhancing osseointegration and antimicrobial effect of biomaterials: A review. *Materials Today: Proceedings, 44*, 2348–2355. https://doi.org/10.1016/j.matpr.2020.12.433

[61] Sun, Z., Annergren, I., Pan, D., & Mai, T. A. (2003). Effect of laser surface remelting on the corrosion behavior of commercially pure titanium sheet. *Materials Science and Engineering: A, 345*, 293–300. https://doi.org/10.1016/S0921-5093(02)00477-X

[62] Balla, V. K., Soderlind, J., Bose, S., & Bandyopadhyay, A. (2014). Microstructure, mechanical and wear properties of laser surface melted Ti6Al4V alloy. *Journal of the Mechanical Behavior of Biomedical Materials, 32*, 335–344. https://doi.org/10.1016/j.jmbbm.2013.12.001

[63] Yue, T. M., Yu, J. K., Mei, Z., & Man, H. C. (2002). Excimer laser surface treatment of Ti–6Al–4V alloy for corrosion resistance enhancement. *Material Letters, 52*, 206–212. https://doi.org/10.1016/S0167-577X(01)00395-0

[64] Braga, F. J. C., Marques, R. F. C., de A Filho, E., & Guastaldi, A. C. (2007). Surface modification of Ti dental implants by Nd:YVO4 laser irradiation. *Applied Surface Science*, *253*, 9203–9208. https://doi.org/10.1016/j.apsusc.2007.05.048

[65] Patil, D., Aravindan, S., Kaushal Wasson, M., et al. (2018). Fast fabrication of superhydrophobic titanium alloy as antibacterial surface using nanosecond laser texturing. *Journal of Micro and Nano-Manufacturing*, 6. https://doi.org/10.1115/1.4038093

[66] Lu, Y., Guan, Y., Li, Y., et al. (2020). Nanosecond laser fabrication of superhydrophobic surface on 316L stainless steel and corrosion protection application. *Colloids and Surfaces A: Physicochemical and Engineering Aspects*, *604*, 125259. https://doi.org/10.1016/j.colsurfa.2020.125259

[67] Dashtbozorg, B., Li, X., Romano, J., et al. (2020). A study on the effect of ultrashort pulsed laser texturing on the microstructure and properties of metastable S phase layer formed on AISI 316L surfaces. *Applied Surface Science*, *511*, 145557. https://doi.org/10.1016/j.apsusc.2020.145557

[68] Wang, Y., Ke, C., Wu, T., et al. (2019). Nanosecond laser texturing with hexagonal honeycomb micro-structure on Titanium for improved wettability and optical properties. *Optik (Stuttgart)*, *192*, 162953. https://doi.org/10.1016/j.ijleo.2019.162953

[69] Zaifuddin, A. Q., Afiq, M. D., Aiman, M. H., et al. (2023). *Effect of Laser Surface Modification on SS316L Surface Roughness and Laser Heating Temperature* (pp. 959–969).

[70] Trdan, U., Hočevar, M., & Gregorčič, P. (2017). Transition from superhydrophilic to superhydrophobic state of laser textured stainless steel surface and its effect on corrosion resistance. *Corrosion Science*, *123*, 21–26. https://doi.org/10.1016/j.corsci.2017.04.005

[71] Geetha, M., Singh, A. K., Asokamani, R., & Gogia, A. K. (2009). Ti based biomaterials, the ultimate choice for orthopaedic implants – A review. *Progress in Materials Science*, *54*, 397–425. https://doi.org/10.1016/j.pmatsci.2008.06.004

[72] Lopez, D., Vlamakis, H., & Kolter, R. (2010). Biofilms. *Cold Spring Harbor Perspectives in Biology*, *2*, a000398–a000398. https://doi.org/10.1101/cshperspect.a000398

[73] Guo, C. Y., Matinlinna, J. P., & Tang, A. T. H. (2012). Effects of surface charges on dental implants: Past, present, and future. *International Journal of Biomaterials*, *2012*, 1–5. https://doi.org/10.1155/2012/381535

[74] Rahyussalim, A. J., Marsetio, A. F., Saleh, I., et al. (2016). The needs of current implant technology in orthopaedic prosthesis biomaterials application to reduce prosthesis failure rate. *Journal of Nanomaterials*, *2016*, 1–9. https://doi.org/10.1155/2016/5386924

[75] Tanzi, M. C., Farè, S., & Candiani, G. (2019). Sterilization and Degradation. In *Foundations of Biomaterials Engineering* (pp. 289–328). Elsevier.

[76] Kaur, M., & Singh, K. (2019). Review on titanium and titanium based alloys as biomaterials for orthopaedic applications. *Materials Science and Engineering: C*, *102*, 844–862. https://doi.org/10.1016/j.msec.2019.04.064

[77] Aksakal, B., Yildirim, Ö. S., & Gul, H. (2004). Metallurgical failure analysis of various implant materials used in orthopedic applications. *Journal of Failure Analysis and Prevention*, *4*, 17–23. https://doi.org/10.1007/s11668-996-0007-9

[78] Eliaz, N. (2019). Corrosion of metallic biomaterials: A review. *Materials*, *12*, 407. https://doi.org/10.3390/ma12030407

[79] Kaur, M., & Singh, K. (2019). Review on titanium and titanium based alloys as biomaterials for orthopaedic applications. *Materials Science and Engineering: C*, *102*, 844–862. https://doi.org/10.1016/j.msec.2019.04.064

[80] Saptaji, K., Gebremariam, M. A., & Azhari, M. A. B. M. (2018). Machining of biocompatible materials: A review. *The International Journal of Advanced Manufacturing Technology*, *97*, 2255–2292. https://doi.org/10.1007/s00170-018-1973-2

[81] Tiainen, L., Abreu, P., Buciumeanu, M., et al. (2019). Novel laser surface texturing for improved primary stability of titanium implants. *Journal of the Mechanical Behavior of Biomedical Materials*, *98*, 26–39. https://doi.org/10.1016/j.jmbbm.2019.04.052
[82] Barfeie, A., Wilson, J., & Rees, J. (2015). Implant surface characteristics and their effect on osseointegration. *British Dental Journal*, *218*, E9–E9. https://doi.org/10.1038/sj.bdj.2015.171
[83] Kang, C.-W., & Fang, F.-Z. (2018). State of the art of bioimplants manufacturing: Part II. *Advances in Manufacturing*, *6*, 137–154. https://doi.org/10.1007/s40436-018-0218-9
[84] Hsiao, W.-T., Chang, H.-C., Nanci, A., & Durand, R. (2016). Surface microtexturing of Ti–6Al–4V using an ultraviolet laser system. *Materials and Design*, *90*, 891–895. https://doi.org/10.1016/j.matdes.2015.11.039
[85] Villapún, V. M., Man, K., Carter, L., et al. (2023). Laser texturing of additively manufactured implants: A tool to programme biological response. *Biomaterials Advances*, *153*, 213574. https://doi.org/10.1016/j.bioadv.2023.213574
[86] Mukherjee, S., Dhara, S., & Saha, P. (2015). Enhancing the biocompatibility of Ti6Al4V implants by laser surface microtexturing: An in vitro study. *The International Journal of Advanced Manufacturing Technology*, *76*, 5–15. https://doi.org/10.1007/s00170-013-5277-2
[87] Pires, L. C., Guastaldi, F. P. S., Nogueira, A. V. B., et al. (2019). Physicochemical, morphological, and biological analyses of Ti-15Mo alloy surface modified by laser beam irradiation. *Lasers in Medical Science*, *34*, 537–546. https://doi.org/10.1007/s10103-018-2626-2
[88] Qin, L., Zeng, Q., Wang, W., et al. (2014). Response of MC3T3-E1 osteoblast cells to the microenvironment produced on Co–Cr–Mo alloy using laser surface texturing. *Journal of Materials Science*, *49*, 2662–2671. https://doi.org/10.1007/s10853-013-7972-7
[89] Ma, C., Peng, G., Nie, L., et al. (2018). Laser surface modification of Mg-Gd-Ca alloy for corrosion resistance and biocompatibility enhancement. *Applied Surface Science*, *445*, 211–216. https://doi.org/10.1016/j.apsusc.2018.03.174
[90] Martínez-Calderon, M., Manso-Silván, M., Rodríguez, A., et al. (2016). Surface micro- and nano-texturing of stainless steel by femtosecond laser for the control of cell migration. *Scientific Reports*, *6*, 36296. https://doi.org/10.1038/srep36296
[91] Barfeie, A., Wilson, J., & Rees, J. (2015). Implant surface characteristics and their effect on osseointegration. *British Dental Journal*, *218*, E9–E9. https://doi.org/10.1038/sj.bdj.2015.171
[92] Wahab, J. A., Ghazali, M. J., Yusoff, W. M. W., & Sajuri, Z. (2016). Enhancing material performance through laser surface texturing: A review. *Transactions of the IMF*, *94*, 193–198. https://doi.org/10.1080/00202967.2016.1191141
[93] Ionescu, A. C., Brambilla, E., Azzola, F., et al. (2018). Laser microtextured titanium implant surfaces reduce in vitro and in situ oral biofilm formation. *PLoS One*, *13*, e0202262. https://doi.org/10.1371/journal.pone.0202262
[94] Schröder, M.-L., Angrisani, N., Fadeeva, E., et al. (2020). Laser-structured spike surface shows great bone integrative properties despite infection in vivo. *Materials Science and Engineering: C*, *109*, 110573. https://doi.org/10.1016/j.msec.2019.110573
[95] Yasuno, K., Kakura, K., Taniguchi, Y., et al. (2014). Zirconia implants with laser surface treatment: Peri-implant bone response and enhancement of osseointegration. *Journal of Hard Tissue Biology*, *23*, 93–100. https://doi.org/10.2485/jhtb.23.93
[96] Lutey, A. H. A., Gemini, L., Romoli, L., et al. (2018). Towards laser-textured antibacterial surfaces. *Scientific Reports*, *8*, 10112. https://doi.org/10.1038/s41598-018-28454-2
[97] Jalil, S. A., Akram, M., Bhat, J. A., et al. (2020). Creating superhydrophobic and antibacterial surfaces on gold by femtosecond laser pulses. *Applied Surface Science*, *506*, 144952. https://doi.org/10.1016/j.apsusc.2019.144952
[98] Li, D., Chen, X., Guo, C., et al. (2017). Micro surface texturing of alumina ceramic with nanosecond laser. *Procedia Engineering*, *174*, 370–376. https://doi.org/10.1016/j.proeng.2017.01.155

[99] Capitanu, L. (2019). Increasing the wear resistance of hip prosthesis by laser surface microtexturing of the femoral head. *Acta Electrotehnica, 60.*

[100] Kumari, R., Scharnweber, T., Pfleging, W., et al. (2015). Laser surface textured titanium alloy (Ti–6Al–4V) – Part II – Studies on bio-compatibility. *Applied Surface Science, 357*, 750–758. https://doi.org/10.1016/j.apsusc.2015.08.255

[101] Cunha, A., Elie, A.-M., Plawinski, L., et al. (2016). Femtosecond laser surface texturing of titanium as a method to reduce the adhesion of *Staphylococcus aureus* and biofilm formation. *Applied Surface Science, 360*, 485–493. https://doi.org/10.1016/j.apsusc.2015.10.102

[102] Hočevar, M., Šetina Batič, B., Godec, M., et al. (2020). The interaction between the osteosarcoma cell and stainless steel surface, modified by high-fluence, nanosecond laser pulses. *Surface and Coatings Technology, 394*, 125878. https://doi.org/10.1016/j.surfcoat.2020.125878

[103] Lazzini, G., Romoli, L., Lutey, A. H. A., & Fuso, F. (2019). Modelling the interaction between bacterial cells and laser-textured surfaces. *Surface and Coatings Technology, 375*, 8–14. https://doi.org/10.1016/j.surfcoat.2019.06.078

[104] Martínez-Calderon, M., Manso-Silván, M., Rodríguez, A., et al. (2016). Surface micro- and nano-texturing of stainless steel by femtosecond laser for the control of cell migration. *Scientific Reports, 6*, 36296. https://doi.org/10.1038/srep36296

[105] Jaggessar, A., Shahali, H., Mathew, A., & Yarlagadda, P. K. D. V. (2017). Bio-mimicking nano and micro-structured surface fabrication for antibacterial properties in medical implants. *Journal of Nanobiotechnology, 15*, 64. https://doi.org/10.1186/s12951-017-0306-1

[106] Sirdeshmukh, N., & Dongre, G. (2021). Laser micro & nano surface texturing for enhancing osseointegration and antimicrobial effect of biomaterials: A review. *Materials Today: Proceedings, 44*, 2348–2355. https://doi.org/10.1016/j.matpr.2020.12.433

Ricardo Luiz Perez Teixeira

Chapter 12
Electrochemical investigations and corrosion stability of the metallic implants

Abstract: This integrative literature review examines the impact of electrochemical investigations and corrosion stability on enhancing the safety and effectiveness of metallic implants. It demonstrates that advanced surface modifications, such as the formation of protective titanium dioxide films and mechanical surface treatments, significantly improve the corrosion resistance of implants. These enhancements reduce the release of toxic metal ions and particulate debris, thereby improving biocompatibility and overall implant performance. The review evaluates key electrochemical techniques, including chronoamperometry, cyclic voltammetry, potentiodynamic polarization, and impedance spectroscopy, for their efficacy in assessing corrosion rates and mechanisms. By highlighting the importance of understanding these electrochemical processes, the study underscores the need for continuous research and development of novel materials to enhance long-term stability and biological integration, ultimately advancing medical technology and patient outcomes.

Keywords: Corrosion resistance, electrochemical techniques, metallic implants, implant degradation

12.1 Introduction

Metallic implants are essential in modern medicine due to their ability to prevent corrosion through the formation of protective surface films, such as titanium dioxide (TiO_2) on titanium implants. This TiO_2 film serves as a robust barrier between the implant and the biological environment, significantly reducing the risk of corrosion by minimizing the chemical reactivity of the implant with surrounding tissues and fluids [1]. Maintaining the structural integrity and longevity of implants is crucial for their effectiveness over time.

Acknowledgment: Gratitude is extended to the Universidade Federal de Itajubá and Dr. Arbind Prasad for the opportunity to share this theoretical work in this book published by De Gruyter Press, Germany.

Ricardo Luiz Perez Teixeira, Federal University of Itajubá Campus Theodomiro Carneiro Santiago, 200 Irmã Ivone Drumond Street, Industrial District II, Itabira, Minas Gerais 35903-087, Brazil, e-mail: ricardo.luiz@unifei.edu.br

https://doi.org/10.1515/9783111571423-012

Mechanical surface treatments, such as deep rolling and microblasting, are also applied to enhance the performance of metallic implants. These treatments introduce compressive residual stresses, improving resistance to fretting and fatigue, and subsequently reducing susceptibility to corrosion [2, 3]. Given the inherently corrosive physiological environment in the human body, strategies that minimize corrosion impact on implant longevity are essential [4–6].

Corrosion of metallic implants leads to the release of metallic species, which poses potential health risks due to wear debris, including nanoparticles and metal ions, accumulating in surrounding tissues and bodily fluids. This can result in adverse biological responses and raise concerns about long-term implant failure [10–12]. The degradation of metallic biomaterials affects both local tissue environments and systemic health due to metal ion and particulate debris release [13]. Therefore, controlling corrosion processes is critical for ensuring the safety and effectiveness of metallic implants.

The chemical form of metal species, whether particulate or ionic, influences their local nanotoxicity effects, with the nature of corrosion products playing a crucial role in determining biological impact [14]. Electrochemical investigations are vital for evaluating the stability and degradation mechanisms of metallic implants, employing techniques such as chronoamperometry, cyclic voltammetry (CV), potentiodynamic polarization curves, and Nyquist plots to measure corrosion rates and understand degradation mechanisms under various conditions [15–18]. These methods provide detailed insights into these processes, facilitating the development of more durable and reliable implants.

Interactions between metallic implants and the body's electrochemical environment can affect cellular behaviors and implant performance. Adjustments to the Volta potential distribution on metallic surfaces can influence cell attachment, deformation, and osteogenic responses, offering promising avenues for designing bone-repairing implants [19]. Materials like Ti6Al4V have shown superior biocompatibility in corrosive environments like synovial fluids compared to alternatives like CoCrMo, due to reduced electrochemical damage and enhanced mechanical properties [20, 21]. Electrochemical methods also improve osseointegration and infection control, underscoring the potential of metallic prostheses to function as stimulating electrodes [22].

Understanding the interplay of electrochemical, mechanical, and biological processes in metallic implant degradation is crucial for advancing implant technology. Accurate modeling of metal implants is essential for precise dosimetry assessments, as incorrect representations can lead to significant errors in estimating induced electric fields. Metallic implants can act as insulators or conductors, complicating dosimetry evaluations [14]. The presence of metal implants, such as titanium, can notably affect dose calculations in radiotherapy planning [23]. Inaccurate modeling can exacerbate errors in two-dimensional entrance fluences, highlighting the need for precise representation in dosimetric models [24].

Furthermore, metal implants interact with electromagnetic fields in various scenarios. For instance, exposure to magnetic fields generated by electric vehicles can induce higher electric field values in individuals with metal implants compared to the general population. This necessitates realistic human models incorporating metal implants when evaluating electromagnetic field exposure [25, 26]. Accurate integration of metal implants into electrical models ensures precise and reliable dosimetry assessments across diverse exposure conditions, contributing to enhanced safety and effectiveness in clinical applications.

12.2 Methodology

This study employs an integrative literature review model to systematically collect data on electrochemical investigations and the corrosion stability of metallic implants, and their role in promoting safer and more effective medical implants, enhancing quality of life, and driving technological progress [27, 28]. The integrative review approach is designed to synthesize existing research, enabling a comprehensive understanding of how these investigations contribute to advancements in medical technology.

The primary research question guiding this study is: "How do electrochemical investigations and the corrosion stability of metallic implants contribute to enhancing the safety and effectiveness of medical implants?" This question aims to explore the mechanisms through which corrosion resistance can be improved, thereby increasing the longevity and biocompatibility of metallic implants. The focus is on identifying the factors influencing corrosion processes and understanding the role of protective surface films and mechanical surface treatments in mitigating these effects.

The hypothesis posited by this research is that electrochemical investigations will demonstrate that advanced surface modifications, such as the formation of protective titanium dioxide films and the application of mechanical treatments, significantly enhance the corrosion stability of metallic implants. This improved stability is expected to reduce the release of toxic metal ions and particulate debris, thereby minimizing adverse biological responses and improving the overall performance and safety of the implants. The study further hypothesizes that these enhancements will contribute to the development of more durable and biocompatible implants, leading to better patient outcomes and advancing technological progress in the field of biomedical materials.

Utilizing the PICO/PIO/PEO strategy, the research question directs an extensive exploration of esteemed databases such as Web of Science, Springer Nature Librarian, and Wiley Online Library. The inclusion criteria are stringent, specifying scientific articles and chapters published between 2010 and 2024 in English, and peer-reviewed through a blind review process [29, 30]. This meticulous selection process ensures the relevance and quality of the data gathered for analysis.

The integrative review methodology facilitates the synthesis of selected studies by utilizing comparative tables to systematically organize data. This method allows for the identification of patterns and critical insights that are essential for understanding the current state and future directions of research in electrochemical investigations and corrosion stability of metallic implants [31, 32]. The structured organization of data aids in highlighting significant trends and gaps within the existing literature.

Additionally, a systematic literature review approach is employed, guided by PRISMA and PRISMA-P standards. This approach aims to provide a comprehensive overview of the research theme, identify knowledge gaps, and contribute to evidence-based clinical practice and scientific progress [33, 34]. By adhering to these rigorous guidelines, the study ensures that the review process is thorough and transparent.

The systematic approach adopted in this study underscores the importance of continual research efforts in advancing healthcare technologies. It highlights the significance of electrochemical investigations and corrosion stability in addressing evolving healthcare needs and improving patient care outcomes [35, 36]. The continuous development and refinement of these techniques are crucial for enhancing the safety and effectiveness of medical implants.

By meticulously collating relevant evidence and critically evaluating studies, this research aims to provide valuable insights into the development of novel materials. The focus on fostering safer medical implants and contributing to scientific advancements in the field is paramount [37, 38]. This comprehensive synthesis of literature not only informs evidence-based clinical practice but also propels scientific innovation in the realm of biomedical materials and implantology.

12.3 Results

Table 12.1 provides a comprehensive overview of the current understanding and research directions in the field of metallic implants, emphasizing the importance of corrosion control, accurate electrochemical evaluation, and precise modeling for dosimetry assessments.

Table 12.1 outlines that metallic implants are crucial in corrosion prevention through the formation of protective surface films, such as titanium dioxide (TiO_2), which act as barriers to reduce reactivity with the surrounding biological environment [1–3]. These protective layers are fundamental in enhancing the durability of implants and preventing corrosion. Additionally, mechanical surface treatments such as deep rolling and microblasting introduce compressive residual stresses in metallic biomaterials, which not only improve their mechanical performance in terms of fretting and fatigue but also potentially decrease their susceptibility to corrosion [4–6].

Table 12.1: Corrosion behavior, electrochemical investigations, and dosimetry implications of metallic implants.

Aspect	Key points	References
Formation of protective films	Metallic implants, such as titanium implants, form protective surface films like titanium dioxide (TiO_2) which act as barriers to prevent corrosion by reducing reactivity with the biological environment.	[1–3]
Mechanical surface treatments	Mechanical treatments such as deep rolling and microblasting introduce compressive residual stresses that enhance the mechanical properties of implants and potentially reduce susceptibility to corrosion.	[4–6]
Corrosive environment	The human body's physiological environment is inherently corrosive to metallic surfaces, making corrosion control essential for maintaining implant functionality and durability.	[7, 8]
Strategies for improvement	To improve dental implants, strategies include developing corrosion-resistant alloys and coatings, altering physicochemical properties, and incorporating functional compounds to enhance corrosion resistance and biological responses.	[9]
Corrosion and health risks	Corrosion of metallic implants leads to the release of metallic species, including nanoparticles and metal ions, which can raise health risks and lead to implant failure. The chemical form of these species affects local nanotoxicity.	[10–14]
Electrochemical Investigations	Electrochemical techniques such as chronoamperometry, cyclic voltammetry, potentiodynamic polarization curves, and Nyquist plots are crucial for evaluating the corrosion stability and degradation mechanisms of metallic implants.	[15–18]
Impact on cellular behavior	Adjusting the Volta potential distribution on metallic surfaces can influence cell attachment and osteogenic responses, which is relevant for bone-repairing implants.	[19]
Material comparison	Materials such as Ti6Al4V have shown superior biocompatibility compared to CoCrMo in corrosive synovial environments due to reduced electrochemical damage and better mechanical responses.	[20, 21]
Regulation of interfacial microenvironment	Electrochemical methods can enhance osseointegration and infection control by regulating the interfacial microenvironment of titanium implants.	[22]

Table 12.1 (continued)

Aspect	Key points	References
Modeling and dosimetry assessments	Accurate representation of metal implants in electrical models is crucial for precise dosimetry assessments. Incorrect modeling can cause significant errors in estimating induced electric fields and dose calculations in radiotherapy planning.	[23, 24]
Impact of electromagnetic fields	Metal implants can influence electromagnetic field interactions, with potential increased induced electric field values in individuals exposed to magnetic fields from electric vehicles. Realistic human models are needed to assess such exposures.	[25, 26]

The corrosive nature of the human body underscores the necessity of effective corrosion control strategies for ensuring the longevity and functionality of implants [7, 8].

In addressing the interaction of metallic implants with body tissues, Table 12.1 highlights that corrosion processes lead to the release of metallic species into the body, including nanoparticles and metal ions. This release can have significant implications for health, raising concerns about potential risks and implant failure [10, 11, 13, 14]. The chemical form of the released metal species – whether particulate or ionic – plays a crucial role in determining the local nanotoxicity effects within the body, underscoring the need for a thorough understanding of metallic implant degradation mechanisms.

Electrochemical investigations are essential for evaluating the stability and performance of metallic implants. Techniques such as chronoamperometry, CV, and potentiodynamic polarization are employed to assess the corrosion rates and degradation mechanisms under various conditions [15–18]. These methods provide valuable insights into enhancing the long-term stability of implants. The table also discusses the impact of electrochemical environments on cellular behaviors and implant performance. Manipulating the Volta potential distribution on metallic surfaces can affect cell attachment and osteogenic responses, which is particularly relevant for designing implants intended for bone repair [19]. Moreover, the biocompatibility of materials like Ti6Al4V has been shown to be superior to CoCrMo in corrosive environments due to reduced electrochemical damage [20, 21].

The regulation of the interfacial microenvironment using electrochemical methods can improve osseointegration and infection control, showcasing the potential of metallic prostheses as stimulating electrodes for enhanced integration and monitoring [22]. Furthermore, accurate representation of metal implants in electrical models is crucial for precise dosimetry assessments. Improper modeling can lead to significant errors in estimating induced electric fields and dose calculations in radiotherapy planning [23, 24]. The impact of metal implants extends to electromagnetic field interactions, where individuals with implants may experience higher induced electric field values when exposed to magnetic fields such as those from electric vehicles [25, 26]. This highlights the necessity of incorporating realistic human models in electromagnetic field evaluations to ensure accurate dosimetric assessments.

Another important aspect is the understanding of metallic biomaterials through metallography. Table 12.2 provides a comprehensive review of the key aspects and findings from various studies on metallography and electrochemical methods in the context of biomedical metallic materials.

Table 12.2: Metallography and electrochemical methods in biomedical metallic materials.

Aspect	Description	References
Understanding of metallic biomaterials	Metallography is crucial for understanding the microstructural characteristics of metallic biomedical materials. Metallic nanomaterials offer enhanced mechanical properties, corrosion resistance, and biocompatibility.	[39, 40]
Lanthanide-based nitride anti-perovskites	Investigation of structural, electrical, and magneto-elastic properties of lanthanide-based nitride anti-perovskites. Their mechanical stability and load-bearing capabilities make them suitable for implants.	[41]
Mechanical properties of metallic biomaterials	The mechanical properties of stainless steel, titanium alloys, and cobalt-chromium alloys are controlled by their crystal structure, dislocation motion, and phase transformations. This is essential for optimizing implantation performance.	[42, 43]
Electrochemical metallography	Electrochemical metallography advances novel metallic biomedical materials for orthopedic and cardiovascular applications. Electrochemical studies on titanium alloys evaluate corrosion resistance and biocompatibility.	[44, 45]
Surface modifications	Electrochemical methods, such as anodization and electrochemical deposition, enhance the properties of titanium nanotubes, including hardness and corrosion resistance. These methods also address wear, biocompatibility, and corrosion.	[46–50]
Antimicrobial properties	Deposition of silver grains onto implant materials through pulse current techniques enhances antimicrobial properties by releasing silver ions over time. This contributes to the long-term success of orthopedic implants.	[51]

Metallography plays a crucial role in understanding the microstructural characteristics of metallic biomedical materials. Metallic nanomaterials, including nanocrystalline metallic materials, offer enhanced mechanical properties, corrosion resistance, and biocompatibility, making them ideal for biomedical applications [39, 40]. Figure 12.1 shows a metallographic micrograph of the oxalic acid-etched microstructures of AISI

316L austenitic stainless-steel grade F138, electro-etched with 10% oxalic acid, observed using scanning electron microscopy (SEM) with secondary electron imaging. This stainless steel exhibits excellent resistance to localized corrosion, including pitting and crevice corrosion, in physiological environments. Additionally, it is paramagnetic and is used as a metallic biomaterial.

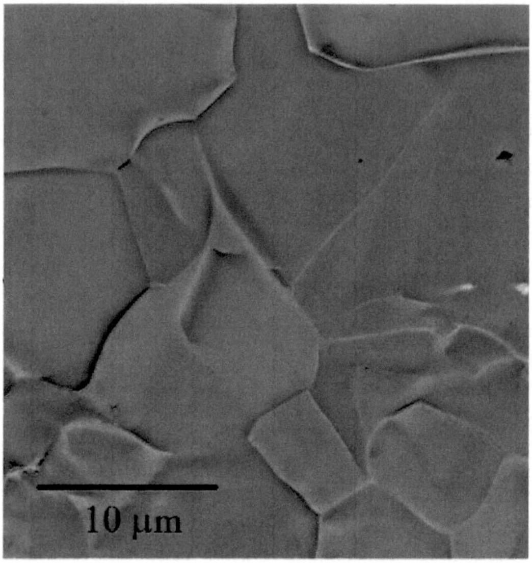

Figure 12.1: Oxalic acid etched microstructures of AISI 316L stainless steel grade F138, electro-etched with 10% oxalic acid.
Source: Author data (2024).

Additionally, the structural, electrical, and magneto-elastic properties of lanthanide-based nitride anti-perovskites have been investigated. These studies highlight their metallic nature and suitability as active biomaterials for implants due to their mechanical stability and load-bearing capabilities [41].

Furthermore, the mechanical properties of metallic biomaterials, such as stainless steel, titanium alloys, and cobalt-chromium alloys, are controlled by their crystal structure, dislocation motion, and phase transformations. This emphasizes the importance of understanding metallography in optimizing their performance for implantation and bone integration [42, 43]. Figure 12.2 shows the metallography of the porous surface of nitinol (NiTi) etched with a metallographic solution of 30 mL glacial acetic acid, 5 mL nitric acid, and 2 mL hydrofluoric acid. Nitinol is a shape memory alloy, meaning it can return to a predetermined shape after deformation due to a reversible phase transformation between martensitic and austenitic phases. When heated above a critical temperature, nitinol returns to its original shape. Additionally, nitinol exhibits superelasticity, allowing for large reversible deformations under mechanical stress. These unique properties make nitinol widely used in medical devices such as vascular stents, orthodontic wires, vena cava filters, and cardiac defect closure devices. Its biocompatibility and corrosion resistance are crucial for its use in medical implants, with a titanium oxide (TiO_2) layer on its surface providing excellent corrosion resistance and minimizing nickel ion release. Surface modifications, such as anodization and electrochemical deposition, further enhance its properties, making nitinol suitable for long-term biomedical applications.

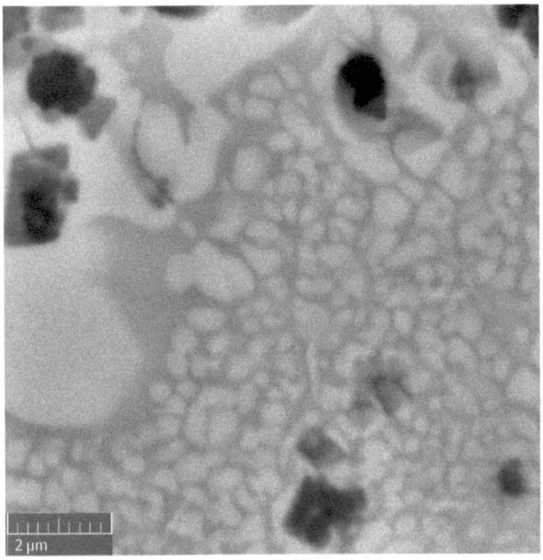

Figure 12.2: Metallography of the porous surface of nitinol (NiTi) etched with a metallographic solution of 30 mL glacial acetic acid, 5 mL nitric acid, and 2 mL hydrofluoric acid.
Source: Author data (2024).

Electrochemical metallography plays a crucial role in advancing novel metallic biomedical materials for orthopedic and cardiovascular applications. Research on titanium alloys, such as Ti-Nb-Zr-Si (TNZS) and Ti-20Zr, underscores the importance of electrochemical studies in evaluating corrosion resistance and biocompatibility [44, 45].

Surface modifications through electrochemical methods, as demonstrated in the case of titanium nanotubes [46], enhance the material's properties by increasing hardness and corrosion resistance. Additionally, the development of new β-titanium alloy compositions, like Ti-35Nb-XSn via electrical resistance sintering [47], highlights the significance of innovative fabrication techniques in tailoring the structure and properties of biomedical alloys.

Electrochemical metallography plays a crucial role in orthopedic implant design by offering methods to enhance biocompatibility, corrosion resistance, and surface properties of metallic implants. Techniques like electrochemical anodization can modify implant surfaces at the nanoscale, improving osseointegration and biocompatibility [48].

Additionally, surface modification methods such as electrochemical deposition and ion implantation are preferred for their ability to address challenges like wear, biocompatibility, and corrosion in orthopedic metallic implants [49]. Studies have shown that anodization of titanium surfaces can create optimal topographic and electrical properties for cellular growth, promoting osteointegration and corrosion resistance [50]. Figure 12.3 shows SEM images of secondary electrons of titanium grade 5 (Ti-6Al-4V), a bioactive metallic material, after immersion in simulated body fluid (SBF) using the Kokubo method. This analysis highlights the surface morphology and potential bioactivity of the titanium alloy, which is crucial for its application in biomedical implants. The immersion in SBF simulates physiological conditions, allowing for the assessment of the material's performance and interaction with biological environments. The SEM images provide detailed insights into the surface characteristics and any changes that occur due to the SBF exposure.

Figure 12.3: Ti-6Al-4V in simulated body fluid (SBF) by Kokubo Method.
Source: Author data (2024).

Furthermore, the deposition of silver grains onto implant materials through pulse current techniques can enhance antimicrobial properties by releasing silver ions over time. This approach contributes to the long-term success of orthopedic implants by preventing infections and ensuring sustained performance [51].

Several electrochemical techniques are applied for the microstructural characterization of metallic biomaterials, offering detailed insights into their properties and performance. Table 12.3 presents a brief systematic review of key electrochemical techniques used for the microstructural characterization of metallic biomaterials, along with their descriptions and some relevant references.

Table 12.3: Electrochemical techniques for microstructural characterization and corrosion stability in metallic biomaterials.

Technique	Description	References
Electrochemical impedance spectroscopy (EIS)	Assesses impedance over a range of frequencies to evaluate corrosion resistance and charge transfer processes.	[52]
Potentiodynamic polarization	Sweeps potential and measures current to evaluate corrosion resistance, passivation, and susceptibility.	[53]
Cyclic voltammetry (CV)	Measures current response to cyclically varying potential to analyze redox processes and stability.	[54]
Electrochemical noise analysis (ENA)	Analyzes spontaneous fluctuations in current and potential to understand corrosion mechanisms and stability.	[55]
Anodic and cathodic polarization	Applies positive or negative potentials to evaluate corrosion resistance and material behavior.	[56]
Chronoamperometry	Measures current as a function of time after a potential step to understand reaction kinetics.	[57]
Chronopotentiometry	Applies constant current and measures potential over time to study electrode kinetics and corrosion.	[58]
Electrochemical quartz crystal microbalance (EQCM)	Measures mass and thickness changes of films on quartz crystals in response to electrochemical reactions.	[59]
Electrochemical scanning tunneling microscopy (ECSTM)	Combines scanning tunneling microscopy with electrochemical techniques to visualize surface structures at the atomic level.	[60]
Electrochemical atomic force microscopy (EC-AFM)	Integrates atomic force microscopy with electrochemical techniques to investigate surface topography and electrochemical reactions at the nanoscale.	[61]

Table 12.3 (continued)

Technique	Description	References
Electrochemical potentiokinetic reactivation (EPR-DL)	EPR-DL can characterize the corrosion resistance and sensitization of metallic biomaterials, demonstrating its efficacy in determining the degree of sensitization in stainless steels with results varying according to tempering temperatures.	[62]
	Optimization of EPR-DL conditions, including acid concentration, scan rate, and temperature, significantly influences the accuracy and resolution in evaluating sensitization levels in austenitic stainless steels.	[63]
Electrochemical potentiokinetic reactivation (EPR-DL) vs. linear sweep voltammetry (LSV-KOH)	LSV-KOH may provide superior accuracy in quantifying deleterious phases in super duplex stainless steels compared to EPR-DL, highlighting the importance of selecting appropriate techniques for specific applications.	[64]
Electrophoretic deposition (EPD)	EPD is a versatile technique for producing a wide range of microstructures and nanostructures, including bioactive coatings such as hydroxyapatite and bioactive glass on orthopedic implants, and biomedical nanostructures like biopolymer-ceramic nanocomposites and carbon nanotube coatings.	[65]
Electrophoretic deposition (EPD)	EPD can be used for the deposition of proteins and other biological entities for sensors and functional coatings, and in combination with 3D printing technology, it enables the creation of patient-specific bioengineering scaffolds with enhanced mechanical properties and antibacterial capabilities for bone tissue engineering.	[66]

Several electrochemical techniques, briefly detailed in Table 12.3, are essential for the microstructural characterization of metallic biomaterials, providing comprehensive insights into their properties and performance. Electrochemical impedance spectroscopy (EIS) is employed to assess impedance across a range of frequencies, delivering crucial information on corrosion resistance and material stability [52]. Potentiodynamic polarization evaluates corrosion resistance and passivation by varying the potential and measuring the resulting current [53]. CV is utilized to analyze redox processes and electrochemical reactivity through cyclic potential variations [54]. Electrochemical noise analysis (ENA) helps identify corrosion mechanisms by examining spontaneous fluctuations in current and potential [55]. Anodic and cathodic polarization studies provide insights into material behavior under different applied potentials [56]. Chronoamperometry measures the current as a function of time following a potential step, thus revealing the kinetics of electrochemical reactions [57]. Chronopotentiometry, which involves applying a constant current, offers insights into potential changes during electrochemical reactions [58]. The electrochemical quartz crystal microbalance (EQCM) measures mass changes on a quartz crystal in response to electrochemical reactions [59]. Electrochemical scanning tunneling microscopy (ECSTM) enables visualization of surface structures while analyzing electrochemical processes at the atomic level [60]. Finally, electrochemical atomic force microscopy (EC-AFM) combines AFM with electrochemical techniques to explore surface topography and electrochemical reactions at the nanoscale [61].

EIS is a sophisticated technique used to measure the impedance of metallic biomaterials across a wide frequency range. This method is instrumental in elucidating the electrochemical processes occurring at the material's surface and interfaces. EIS involves applying an alternating current signal to the material and analyzing the resulting voltage response to determine impedance, which is a complex function of resistance and reactance. The impedance data are often represented in Nyquist or Bode plots, which provide information on corrosion resistance, charge transfer kinetics, and film characteristics. For example, Hou et al. [52] demonstrated the application of EIS in evaluating TiO_2 nanotube arrays, revealing its effectiveness in assessing material stability and resistance to corrosion under physiological conditions.

Potentiodynamic polarization is a critical electrochemical technique used to evaluate the corrosion resistance and passivation properties of metallic biomaterials. This method involves sweeping the electrode potential at a constant rate and measuring the resulting current to construct polarization curves. These curves provide insights into the material's anodic and cathodic reactions, including its susceptibility to pitting and localized corrosion. The technique is especially valuable for determining the onset of passivation and the stability of passive films. Eliaz and Metoki [53] conducted a comprehensive review of various calcium phosphate bioceramics using potentiodynamic polarization, highlighting its role in characterizing the corrosion behavior and effectiveness of protective coatings in biomedical applications.

CV is a dynamic electrochemical technique that analyzes the current response of a metallic material subjected to a cyclically varying potential. During a CV experiment, the electrode potential is scanned back and forth between two predetermined limits, and the resulting current is measured. Figure 12.4 shows a simplified measurement circuit incorporating a source meter, specifically designed for conducting CV experiments. This setup is pivotal for electrochemical analysis, providing precise control and measurement of electrode potential and current. The source meter enables the application of a potential sweep to the working electrode while concurrently measuring the resultant current, facilitating the characterization of redox processes and the determination of electrochemical parameters such as reaction kinetics and diffusion coefficients. This configuration is essential for obtaining accurate and reproducible CV data, which is fundamental in the investigation of electrochemical systems and the development of novel materials and sensors.

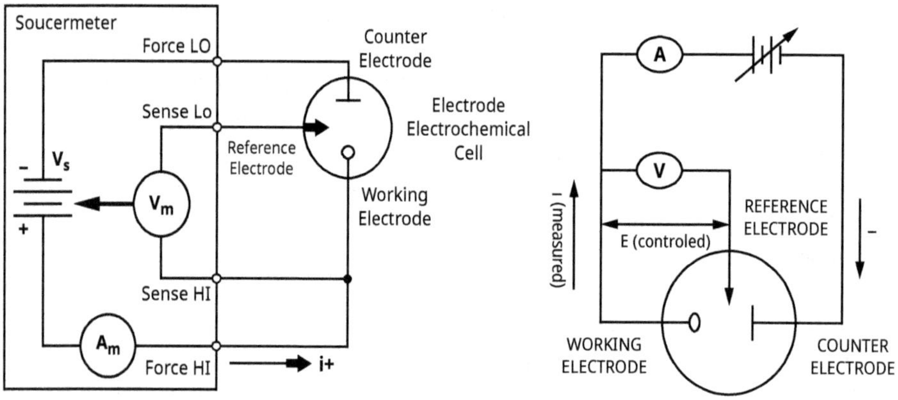

Figure 12.4: Simplified measurement circuit for performing cyclic voltammetry.

This CV technique is particularly useful for studying redox processes, evaluating electrochemical stability, and characterizing surface modifications. CV can reveal the reversibility of electrochemical reactions, the presence of redox-active species, and the kinetics of electron transfer processes. Sandhyarani [54] reviewed CV to investigate surface modifications and electrochemical reactivity, demonstrating its applicability in assessing material performance and understanding corrosion mechanisms.

ENA is a technique that examines spontaneous fluctuations in current and potential to gain insights into the corrosion mechanisms and effectiveness of protective coatings. By analyzing these noise signals, researchers can detect the presence of localized corrosion, identify corrosion mechanisms, and assess the stability of passive films. ENA is valuable for studying materials in real-time under dynamic conditions, providing data on the randomness and magnitude of electrochemical noise. Jaquez-Munoz et al. [55] employed ENA to investigate the corrosion behavior of titanium alloys, illustrating its effectiveness in detecting corrosion events and evaluating the effectiveness of protective coatings.

Anodic and cathodic polarization studies involve applying positive (anodic) or negative (cathodic) potentials to a material to assess its corrosion resistance and passivation properties. Anodic polarization helps in understanding the material's tendency to undergo oxidation and form passive films, while cathodic polarization reveals the material's behavior under reducing conditions. These studies provide valuable data on the material's electrochemical stability and resistance to degradation. Pound [56] reviewed the corrosion behavior of titanium and its alloys using anodic and cathodic polarization techniques, demonstrating their role in assessing the material's performance and suitability for biomedical implants.

Chronoamperometry is a technique that measures the current as a function of time after a sudden change in the electrode potential. By applying a potential step, researchers can study the kinetics of electrochemical reactions including corrosion, film formation, and dissolution. The resulting current-time transients provide information on reaction rates and mechanisms. Amiri et al. [57] reviewed some methodologies of chronoamperometry for detecting and monitoring microbial populations, highlighting its versatility in various electrochemical studies, including those related to material corrosion and bioactivity.

Chronopotentiometry involves applying a constant current to a material and measuring the resulting changes in potential over time. This technique provides insights into the kinetics of electrochemical reactions and the stability of electrochemical systems. By analyzing the potential response, researchers can study electrode kinetics, corrosion behavior, and reaction mechanisms. Liliana et al. [58] studied chronopotentiometry to investigate the electrochemical behavior of stainless steel, offering valuable data on corrosion mechanisms and electrode stability under constant current conditions.

EQCM is a technique that measures the mass changes of a thin film deposited on a quartz crystal in response to electrochemical reactions. By monitoring the frequency changes of the quartz crystal, researchers can study film growth, dissolution, and biomolecule adsorption on metallic surfaces. EQCM is particularly useful for investigating surface modifications and interactions at a high sensitivity level. Akgönüllüet al. [59] investigated EQCM to use in aptasensors for biomedical applications, demonstrating its capability in analyzing surface modifications and material interactions with biological molecules.

ECSTM integrates scanning tunneling microscopy with electrochemical techniques to visualize surface structures at the atomic level while analyzing electrochemical processes. ECSTM provides high-resolution imaging of surface topography and allows for real-time observation of electrochemical reactions. Yagati et al. [60] used ECSTM to study titanium oxide surfaces in SBFs, showcasing its ability to provide detailed insights into surface modifications and interactions at the atomic scale.

EC-AFM combines AFM with electrochemical techniques to investigate surface topography and electrochemical reactions at the nanoscale. This technique provides high-resolution imaging and analysis of surface features including topography and local electrochemical activity. EC-AFM is valuable for studying nanoscale phenomena and understanding material interactions in physiological environments. Chen et al. [61] applied EC-AFM to characterize metallic biomaterials in physiological conditions, demonstrating its utility in detailed surface and interaction analysis.

Electrochemical potentiokinetic reactivation (EPR-DL) is an advanced technique for evaluating corrosion potential and intergranular corrosion susceptibility, crucial for characterizing corrosion resistance and sensitization of metallic biomaterials. It provides insights into electrochemical behaviors such as passivation, depassivation, and localized corrosion kinetics [62–67]. EPR-DL measures the degree of chromium depletion at grain boundaries, predicting material performance in corrosive environments. Its ability to quantify sensitization under various conditions enhances the durability and biocompatibility of metallic implants.

EPR-DL effectively assesses corrosion behavior and sensitization in metallic biomaterials. Studies demonstrate its efficacy in determining sensitization levels in stainless steels, influenced by tempering temperatures [62]. Optimizing EPR-DL conditions, including acid concentration, scan rate, and temperature, enhances accuracy and resolution in evaluating sensitization in austenitic stainless steels [63].

Comparing EPR-DL with techniques like linear sweep voltammetry (LSV) in KOH solution shows LSV-KOH may offer superior accuracy in quantifying deleterious phases in super duplex stainless steels, emphasizing the importance of selecting appropriate techniques for specific applications [64].

Electrophoretic deposition (EPD) in biomaterials offers advantages such as producing a range of microstructures and nanostructures, including bioactive coatings on orthopedic implants and biomedical nanostructures like biopolymer-ceramic nanocomposites [65]. EPD can also deposit proteins and biological entities for sensors and functional coatings, showcasing its potential in advanced biomaterial processing [65, 66]. EPD plays a significant role in bone tissue engineering, enabling the creation of patient-specific bioengineering scaffolds with enhanced mechanical properties and antibacterial capabilities when combined with 3D printing technology [66].

Linear dichroism (LD) spectroscopy provides valuable structural information about biomaterials, aiding in understanding subunit orientations and overall structure [67]. EPD and LD spectroscopy are critical techniques in biomaterials, contributing to advancements in biomedical implants, tissue engineering, and structural characterization [65–67].

EPR-DL enhances biocompatibility by providing insights into material-biological system interactions, aiding in developing safer and more effective biomaterials. EPR-DL studies the toxicity profile of nanostructured biomaterials at the cellular level, investigating the impact of postprocessing methods on cross-linking in 3D scaffolds to improve biocompatibility [68–70]. EPR-DL contributes to developing materials with enhanced physical and mechanical characteristics, crucial for better integration with surrounding bone tissue [71].

EPR-DL is a powerful tool in biomaterials research, driving the development of safer, more biocompatible solutions across toxicity assessment, postprocessing optimization, and material property enhancement [68–71]. EPR spin labels are essential in site-directed spin-labeling protocols, offering increased resistance to chemical reduction in biological environments and compatibility with structural analysis software [72].

These characteristics make them valuable tools in biomolecular research, ensuring biocompatibility in various biological systems.

12.4 Conclusion

This integrative literature review systematically collected data on electrochemical investigations and the corrosion stability of metallic implants to explore their role in promoting safer and more effective medical implants. The primary research question focused on understanding how electrochemical investigations and corrosion stability contribute to enhancing the safety and effectiveness of medical implants.

The findings support the hypothesis that advanced surface modifications, such as the formation of protective titanium dioxide films and the application of mechanical treatments, significantly enhance the corrosion stability of metallic implants. These improvements in stability reduce the release of toxic metal ions and particulate debris, thereby minimizing adverse biological responses and improving the overall performance and safety of the implants. The evidence suggests that these enhancements contribute to the development of more durable and biocompatible implants, leading to better patient outcomes and advancing technological progress in the field of biomedical materials.

The systematic approach adopted in this study underscores the importance of continual research efforts in advancing healthcare technologies. The significance of electrochemical investigations and corrosion stability in addressing evolving healthcare needs and improving patient care outcomes is highlighted. The continuous development and refinement of these techniques are crucial for enhancing the safety and effectiveness of medical implants.

By meticulously collating relevant evidence and critically evaluating studies, this research provides valuable insights into the development of novel materials. The focus on fostering safer medical implants and contributing to scientific advancements in the field is paramount. This comprehensive synthesis of literature not only informs evidence-based clinical practice but also propels scientific innovation in the realm of biomedical materials and implantology.

References

[1] Teixeira, R. L. P., de Lacerda, J. C., Conceição, I. C., da Silva, S. N., Siqueira, G. O., & Moura Filho, F. (2021). The effects of Niobium on the bioactivity of Ni-Ti-Al-Nb shape memory alloys. *Archives of Metallurgy and Materials*, *66*(2), 437–442. https://journals.pan.pl/Content/118798/PDF/AMM-2021-2-12-Teixeira.pdf

[2] Teixeira, R. L. P., Godoy, G. C. D., & Pereira, M. D. M. (2004). Calcium phosphate formation on alkali-treated titanium alloy and stainless steel. *Materials Research*, *7*, 299–303. https://doi.org/10.1590/S1516-14392004000200013

[3] Yarram, A., Ramesh, K., Rao, G. N., Yasmeen, S., Sajjan, S. M., & Ramaraju, A. V. Corrosion in titanium dental implants – A review. doi: 10.37983/IJDM.2019.1301

[4] Bormann, T., Mai, P. T., Gibmeier, J., Sonntag, R., Müller, U., & Kretzer, J. P. (2020). Corrosion behavior of surface-treated metallic implant materials. *Materials*, *13*(9), 2011. https://doi.org/10.3390/ma13092011

[5] Marques, A. V. M., Carmo, K. M. D., Lage, W. C., Teixeira, R. L. P., Lacerda, J. C., Teixeira, C. H. S. B., & Shitsuka, R. (2020). Evaluation of the effect of plastic deformation on the microstructure, hardness and magnetic properties of AISI type 316L stainless steel. *Matéria (Rio de Janeiro)*, *25*, e–12611. https://doi.org/10.1590/S1517-707620200002.1011

[6] Teixeira, R. L. P., de Lacerda, J. C., Florencio, K. C., da Silva, S. N., & Henriques, A. B. (2023). TRIP effect produced by cold rolling of austenitic stainless steel AISI 316L. *Journal of Materials Science*, *58*(7), 3334–3345. https://doi.org/10.1007/s10853-023-08235-7

[7] Dikici, B., Esen, Z., Duygulu, O., & Gungor, S. (2015). Corrosion of metallic biomaterials. *Advances in Metallic Biomaterials: Tissues, Materials and Biological Reactions*, 275–303. doi: 10.1007/978-3-662-46836-4_12

[8] Eliaz, N. (2019). Corrosion of metallic biomaterials: A review. *Materials*, *12*(3), 407. https://doi.org/10.3390/ma12030407

[9] Nagay, B. E., Cordeiro, J. M., & Barão, V. A. R. (2022). Insight into corrosion of dental implants: From biochemical mechanisms to designing corrosion-resistant materials. *Current Oral Health Reports*, *9*(2), 7–21. https://doi.org/10.1007/s40496-022-00306-z

[10] Matusiewicz, H., & Richter, M. (2020). Potentially toxic metallic wear nanoparticles and trace metal ions release from metal-on-metal orthopedic implants in human biological specimens: An overview of in vivo and ex vivo clinical studies. *World Journal of Advanced Research and Reviews*, *08*(03), 242–292. https://doi.org/10.30574/wjarr.2020.8.3.0469

[11] Matusiewicz, H., & Richter, M. (2022). Metal ions release from metallic orthopedic implants exposed to tribocorrosion and electrochemical corrosion conditions in simulated body fluids: Clinical context and in vitro experimental investigations. *World Journal of Advanced Research and Reviews*, *14*(02), 261–283. https://doi.org/10.30574/wjarr.2022.14.2.0438

[12] Mu, Y., Kobayashi, T., Sumita, M., Yamamoto, A., & Hanawa, T. (2000). Metal ion release from titanium with active oxygen species generated by rat macrophages in vitro. *Journal of Biomedical Materials Research*, *49*, 238–243. https://doi.org/10.1002/(SICI)1097-4636(200002)49:2<238::AID-JBM12>3.0.CO;2-J

[13] Richter, M., & Matusiewicz, H. Review of the local tissue reaction to metallic spinal implant debris. *Ions and Nanoparticles World Journal of Advanced Research and Reviews*, *9*(3), 167–187. https://doi.org/10.30574/wjarr.2021.9.3.0085

[14] Nicolli, A., Trevisan, A., Bortoletti, I., Pozzuoli, A., Ruggieri, P., Martinelli, A., Gambalunga, A., & Carrieri, M. (2020). Biological monitoring of metal ions released from hip prostheses. *International Journal of Environmental Research and Public Health*, *17*(9), 3223. https://doi.org/10.3390/ijerph17093223

[15] Doering, M., Kieninger, J., Urban, G., & Weltin, A. (2022). Electrochemical microelectrode degradation monitoring: In situ investigation of platinum corrosion at neutral pH. *Journal of Neural Engineering*, *19*, 016005. https://iopscience.iop.org/article/10.1088/1741-2552/ac47da

[16] Ibrahem Khaled, N., & Santhiya, D. (2023). Zein-bioactive glass nanocomposite coating on magnesium alloy substrate for orthopedic applications. *Advanced Composite Materials*, *33*(3), 324–346. https://doi.org/10.1080/09243046.2023.2253110

[17] Silva, A., & Plaine, A. H. (2023). Electrochemical corrosion study of biomaterials: A bibliometric study based on co-word analysis. *Results in Engineering*, *20*, 101489. https://doi.org/10.1016/j.rineng.2023.101489

[18] Nagay, B. E., Cordeiro, J. M., & Barão, V. A. R. (2022). Insight into corrosion of dental implants: From biochemical mechanisms to designing corrosion-resistant materials. *Current Oral Health Reports*, *9*, 7–21. https://doi.org/10.1007/s40496-022-00306-z

[19] Chen, S. T., Li, Y., Gu, N., Wen, C., & Lu, J. J. (2023). Metallurgical manipulation of surface Volta potential in bimetals and cell response of human mesenchymal stem cells. *Biomaterials Advances*, *153*, 213529. https://doi.org/10.1016/j.bioadv.2023.213529

[20] Ryu, J. J., Cudjoe, E., Patel, M., & Caputo, M. (2022). Sliding corrosion fatigue of metallic joint implants: A comparative study of CoCrMo and Ti_6Al_4V in simulated synovial environments. *Lubricants*, *10*(4), 65. https://doi.org/10.3390/lubricants10040065

[21] Teixeira, R. L. P., & Silva, P. C. D. (2024). Advancing metallic biomaterials for biomedical implants: A comprehensive integrative review. *Revista de Gestao Social e Ambiental*, *18*(5), e05255. https://doi.org/10.24857/rgsa.v18i5.5255

[22] Ehrensberger, M. T., & Gilbert, J. L. (2021). A review of the electrochemical mechanobiology of titanium and titanium alloys for implant applications. *Acta Biomaterialia*, *120*, 277–292. https://doi.org/10.1016/j.actbio.2020.06.027

[23] Li, B., Huang, J., Ruan, J., Peng, Q., Huang, S., Li, Y., & Li, F. (2023). Dosimetric impact of CT metal artifact reduction for spinal implants in stereotactic body radiotherapy planning. *Quantitative Imaging in Medicine and Surgery*, *13*(12), 8290.

[24] Cao, Z., Gao, X., Liu, G., & Pei, Y. (2023). Effect of metal implants and metal artifacts on back-projected two-dimensional entrance fluence determined by EPID dosimetry. *Journal of Applied Clinical Medical Physics*, *24*(11), e14115.

[25] Gao, Y., Qi, D., Li, C., & Zhao, H. (2023). Safety Assessment of Human with Metal Implants Exposed to Magnetic Field Based on Numerical Method. In *2023 4th International Conference on Computer Engineering and Application (ICCEA)* (pp. 56–60). IEEE.

[26] Bakıcıerler, G., Şişman, G., & Akgüngör, K. (2022). Estimating the dose differences nearby the metal implant by means of artificial contouring errors via Monaco and Geant4. *Revista Mexicana de Física*, *68*(5).

[27] Cronin, M. A., & George, E. (2023). The why and how of the integrative review. *Organizational Research Methods*, *26*(1), 168–192.

[28] Yemini, M., Engel, L., & Ben Simon, A. (2023). Place-based education–a systematic review of literature. *Educational Review*, 1–21.

[29] Silva, P. C. D., de Oliveira, L. L. V., Teixeira, R. L. P., Brito, M. L. D. A., & Filippe, A. R. T. M. (2022). Executive functions in Alzheimer's disease: A systematic review. *Journal of Alzheimer's Disease Reports*, *6*(1), 81–99.

[30] Bakkum, L., Schuengel, C., Sterkenburg, P. S., Frielink, N., Embregts, P. J., de Schipper, J. C., & Tharner, A. (2022). People with intellectual disabilities living in care facilities engaging in virtual social contact: A systematic review of the feasibility and effects on well-being. *Journal of Applied Research in Intellectual Disabilities*, *35*(1), 60–74.

[31] Toronto, C. E., & Remington, R. (Eds.). *A Step-by-step Guide to Conducting an Integrative Review* (pp. 1–9). Cham: Springer International Publishing.

[32] Mendes, K. D. S., Silveira, R. C. D. C. P., & Galvão, C. M. (2019). Use of the bibliographic reference manager in the selection of primary studies in integrative reviews. *Texto & Contexto-Enfermagem*, *28*, e20170204.
[33] Moher, D., Liberati, A., Tetzlaff, J., Altman, D. G., & Prisma Group (2010). Preferred reporting items for systematic reviews and meta-analyses: The PRISMA statement. *International Journal of Surgery*, *8*(5), 336–341.
[34] Sarkis-Onofre, R., Catalá-López, F., Aromataris, E., & Lockwood, C. (2021). How to properly use the PRISMA statement. *Systematic Reviews*, *10*(1), 1–3.
[35] Lambermon, F., Vandenbussche, F., Dedding, C., & van Duijnhoven, N. (2020). Maternal self-care in the early postpartum period: An integrative review. *Midwifery*, *90*, 102799.
[36] Souza, M. T. D., Silva, M. D. D., & Carvalho, R. D. (2010). Integrative review: What is it? How to do it? *Einstein (São Paulo)*, *8*, 102–106.
[37] Mantsiou, C., Liakos, A., Mainou, M., Papanas, N., Tsapas, A., & Bekiari, E. (2023). A simple guide to systematic reviews and meta-analyses. *The International Journal of Lower Extremity Wounds*, 15347346231169842.
[38] Livinski, A., Joubert, D., & Terry, N. (2015). *Undertaking a Systematic Review: What You Need to Know*. National Institutes of Health Library.
[39] Wang, J., Zhao, G., Feng, L., & Chen, S. (2022). Metallic nanomaterials with biomedical applications. *Metals*, *12*(12), 2133.
[40] Li, H., Wang, P., & Wen, C. (2022). Recent progress on nanocrystalline metallic materials for biomedical applications. *Nanomaterials*, *12*(12), 2111.
[41] Mehmood, S., Ali, Z., Khan, S. R., Aman, S., Elnaggar, A. Y., Ibrahim, M. M., & Trukhanov, A. V. (2022). Mechanically stable magnetic metallic materials for biomedical applications. *Materials*, *15*(22), 8009.
[42] Nakano, T. (2019). Physical and Mechanical Properties of Metallic Biomaterials. In *Metals for Biomedical Devices* (pp. 97–129). Woodhead Publishing.
[43] Niinomi, M. (2022). Recent metallic materials for biomedical applications. *Metallurgical and Materials Transactions A*, *33*, 477–486.
[44] Bordbar-Khiabani, A., & Gasik, M. (2023). Electrochemical and biological characterization of Ti–Nb–Zr–Si alloy for orthopedic applications. *Scientific Reports*, *13*(1), 2312.
[45] Hulka, I., Uțu, I. D., Brito-Garcia, S., Verdu-Vazquez, A., & Mirza-Rosca, J. C. (2024). Electrochemical study and mechanical properties of Ti-Zr Alloy for biomedical applications. *Crystals*, *14*(6), 493.
[46] Sivaprakash, V., Natrayan, L., Suryanarayanan, R., Narayanan, R., & Paramasivam, P. (2021). Electrochemical anodic synthesis and analysis of TiO2 nanotubes for biomedical applications. *Journal of Nanomaterials*, *2021*(1), 9236530.
[47] Vinogradova, M., Klyatskina, E., Amigó, V., & Segovia, F. (2024). Properties and prospects for biomedical application of new Ti-Nb-Sn alloys obtained by electrical resistance sintering. *Materials Letters*, 136864.
[48] Wang, L. N., Jin, M., Zheng, Y., Guan, Y., Lu, X., & Luo, J. L. (2014). Nanotubular surface modification of metallic implants via electrochemical anodization technique. *International Journal of Nanomedicine*, 4421–4435.
[49] Jambagi, S. C., & Malik, V. R. (2021). A review on surface engineering perspective of metallic implants for orthopaedic applications. *Journal of Management*, *73*(12), 4349–4364.
[50] Utku, F. S. (2017). Electrochemically designed interfaces: Electrical properties and morphology of micro-nanostructured Titania implant surfaces. *Surface and Coatings Technology*, *324*, 546–551.
[51] Furko, M., Lakatos-Varsányi, M., & Balázsi, C. (2016). Complex electrochemical studies on silver-coated metallic implants for orthopaedic application. *Journal of Solid State Electrochemistry*, *20*, 263–271.

[52] Hou, Y., Chen, C., Zhou, S., Li, Y., Wang, D., & Zhang, L. (2016). Fabrication of an integrated cartilage/bone joint prosthesis and its potential application in joint replacement. *Journal of the Mechanical Behavior of Biomedical Materials*, *59*, 265–271.

[53] Eliaz, N., & Metoki, N. (2017). Calcium phosphate bioceramics: A review of their history, structure, properties, coating technologies and biomedical applications. *Materials*, *10*(4), 334. https://doi.org/10.3390/ma10040334

[54] Sandhyarani, N. (2019). Surface Modification Methods for Electrochemical Biosensors. In *Electrochemical Biosensors* (pp. 45–75). Elsevier.

[55] Jaquez-Munoz, J. M., Gaona-Tiburcio, C., Cabral-Miramontes, J., Nieves-Mendoza, D., Maldonado-Bandala, E., Olguin-Coca, J., & Almeraya-Calderon, F. (2021). Electrochemical noise analysis of the corrosion of titanium alloys in NaCl and H2SO4 solutions. *Metals*, *11*(1), 105.

[56] Pound, B. G. (2014). Corrosion behavior of metallic materials in biomedical applications. I. Ti and its alloys. *Corrosion Reviews*, *32*(1–2), 1–20.

[57] Amiri, M., Bezaatpour, A., Jafari, H., Boukherroub, R., & Szunerits, S. (2018). Electrochemical methodologies for the detection of pathogens. *ACS Sensors*, *3*(6), 1069–1086.

[58] Liliana, P., Elena, S. C., Virgil, C. L., Laurențiu, D. M., & Daniel, P. S. (2018). Corrosion behavior of Ni-Cr dental casting alloys. *International Journal of Electrochemical Science*, *13*(1), 410–423.

[59] Akgönüllü, S., Özgür, E., & Denizli, A. (2022). Quartz crystal microbalance-based aptasensors for medical diagnosis. *Micromachines*, *13*(9), 1441.

[60] Yagati, A. K., Lee, J. Y., Nam, E. S., Cho, S., & Choi, J. W. (2017). Electrochemical scanning tunneling microscopy analysis on protein based electronic devices. *Science of Advanced Materials*, *9*(1), 102–121.

[61] Chen, H., Qin, Z., He, M., Liu, Y., & Wu, Z. (2020). Application of electrochemical atomic force microscopy (EC-AFM) in the corrosion study of metallic materials. *Materials*, *13*(3), 668.

[62] Tavares, S. S., de Sampaio, M. T., Perez, G., de Almeida, B. B., & Ponzio, E. A. (2022). DL-EPR and AFM study of sensitization of a 17% Cr multiphase stainless steel. *Materials and Corrosion*, *73*(6), 866–875.

[63] Zhang, M. X., Yang, B., Wang, S. L., Wu, H. C., & Shi, Y. Z. (2016). Parametric optimization of DL-EPR technique for evaluating IGC susceptibility of 316LN type stainless steel using the Taguchi method. *Materials and Corrosion*, *67*(3), 271–276.

[64] de Sampaio, M. T., Furtado, A. B., Ignácio, M. D. C., Tavares, S. S., Pardal, J. M., Pimenta, A. R., & Ponzio, E. A. (2023). DL-EPR vs. LSV-KOH: From simple detection of deleterious phases to analysis of morphology and high accuracy on determination of sigma phase in UNS S32750 stainless steel. *Corrosion Engineering Science and Technology*, *58*(1), 1–11.

[65] Boccaccini, A. R., Keim, S., Ma, R., Li, Y., & Zhitomirsky, I. (2010). Electrophoretic deposition of biomaterials. *Journal of the Royal Society Interface*, *7*(suppl_5), S581–S613.

[66] Tian, L., Zhang, Z., Tian, B., Zhang, X., & Wang, N. (2020). Study on antibacterial properties and cytocompatibility of EPL coated 3D printed PCL/HA composite scaffolds. *RSC Advances*, *10*(8), 4805–4816.

[67] Hicks, M. R., Kowałski, J., & Rodger, A. (2010). LD spectroscopy of natural and synthetic biomaterials. *Chemical Society Reviews*, *39*(9), 3380–3393.

[68] Thomas, N. G., Varghese, N., Kalarikkal, N., Thomas, S., Sreedharan, M., George, S. S., & George, V. T. (2023). Toxicity Evaluation and Biocompatibility of Nanostructured Biomaterials. In *Cytotoxicity-Understanding Cellular Damage and Response*. IntechOpen.

[69] Jeršovaitė, J., Šarachovaitė, U., Matulaitienė, I., Niaura, G., Baltriukienė, D., & Malinauskas, M. (2023). Biocompatibility enhancement via post-processing of microporous scaffolds made by optical 3D printer. *Frontiers in Bioengineering and Biotechnology*, *11*, 1167753.

[70] John, A. A., Subramanian, A. P., Vellayappan, M. V., Balaji, A., Jaganathan, S. K., Mohandas, H., & Yusof, M. (2015). Physico-chemical modification as a versatile strategy for the biocompatibility enhancement of biomaterials. *RSC Advances*, *5*(49), 39232–39244.

[71] Nicolae, L. C. (2016). *The Development of Light Curable Biomaterials with Enhanced Biocompatibility for Orthopaedic Surgery*. Doctoral dissertation, University of Birmingham. https://etheses.bham.ac.uk/id/eprint/6951/

[72] Bleicken, S., Assafa, T. E., Zhang, H., Elsner, C., Ritsch, I., Pink, M., & Bordignon, E. (2019). Gem-diethyl pyrroline nitroxide spin labels: Synthesis, EPR characterization, rotamer libraries and biocompatibility. *ChemistryOpen*, *8*(8), 1057–1065.

Samir Das*, Sachin Latiyan, Arbind Prasad, Bidyanand Mahto, and Sudipto Datta

Chapter 13
In vitro studies of metallic biomaterials

Abstract: Metallic materials and their alloys are extensively used as biomaterials. These have received a lot of attention in biomedical applications because of their good mechanical properties, biocompatibility, and ability to adapt to surface modifications. These play an important role in the healthcare industry, and at the current stage, it is difficult for ceramics and polymers to replace metals and alloys due to their high strength and toughness properties. As safety is among the most important intended characteristics of biomaterials, so use of Co–Cr–Mo alloy, stainless steel, commercially pure titanium, and its alloys is in increased demand. In vitro investigations give critical information about the long-term behavior and safety of metallic biomaterials before in vivo implantation. These investigations are crucial in evaluating these materials for usage in medical devices like implants and prostheses to analyze their interaction with biological systems. The corrosion evaluation methods using cell cultures, the identification of reconstructive features of surface oxide film, fretting fatigue, cytotoxicity, and biocompatibility are discussed in this chapter.

Keywords: Metallic implants, biomaterials, corrosion, alloy, preliminary biocompatibility, biomaterials

Acknowledgment: Sudipto Datta acknowledges SERB for the National Postdoctoral Fellowship (PDF/2022/000990).

***Corresponding author: Samir Das**, School of Medical Science and Technology, Indian Institute of Technology, Kharagpur 721302, West Bengal, India
Sachin Latiyan, Department of Materials Engineering, Indian Institute of Science, Bangalore 560012, India
Arbind Prasad, Mechanical Engineering Department, Katihar Engineering College (Under Department of Science, Technology and Technical Education, Government of Bihar), Katihar 854109, Bihar, India
Bidyanand Mahto, Government Engineering College, Vaishali (Under Department of Science, Technology and Technical Education, Government of Bihar), Vaishali 844118, Bihar, India
Sudipto Datta, Department of Materials Engineering, Indian Institute of Science, Bangalore 560012, Karnataka, India, e-mail: sudiptodatta1990@gmail.com

https://doi.org/10.1515/9783111571423-013

13.1 Introduction

The use of metallic materials in medical implants began in the nineteenth century, coinciding with the rise of the metal industry [1]. The necessity for effective bone repair treatments, particularly for internal fractures in long bones, prompted the creation of metallic implants. However, before the use of Lister's aseptic surgical techniques in the 1860s, attempts to implant metallic devices such as spinal wires and bone pins made of iron, gold, or silver were usually unsuccessful [2].

Metallic biomaterials are widely employed in medical applications, particularly in the design of implants, prosthetics, and tissue engineering scaffolds, due to their high mechanical strength, durability, and biocompatibility. Metals like titanium, magnesium, and cobalt-chromium alloys have distinct advantages, including corrosion resistance, biodegradability (in the case of magnesium), and the ability to tolerate physiological loads [2]. Recently, there has been a growing interest in using non-traditional reconstructive surgery techniques on hard tissues and organs, which has spurred research into metallic biomaterials. Notable developments include new magnesium-based alloys designed for bone tissue engineering and regeneration [2] as well as nickel-titanium shape memory alloys being explored for use as vascular stents.

The interaction of these biomaterials with biological systems is still an important field of research to assure safety and functionality. To emphasize, the Food and Drug Administration (FDA) has yet to approve many of these newly developed alloys for medical implants due to significant concerns regarding their biocompatibility – a critical requirement for any biomedical implant [3]. Consequently, in vitro assessments of metallic implants are essential before they can be considered for clinical applications. In vitro studies are an important first step in determining the appropriateness of metallic biomaterials for therapeutic applications. These controlled laboratory studies enable researchers to recreate the physiological environment while testing for important characteristics such as biocompatibility, corrosion behavior, and mechanical performance. Such studies provide vital insights into how these materials behave in the human body by elucidating cellular reactions, ion release, and surface degradation, allowing designers to improve their designs and forecast long-term performance more accurately. This article discusses the relevance of in vitro research in the advancement of metallic materials for biomedical applications, with a focus on maximizing safety and efficacy before clinical adoption.

13.2 Types of metallic implant

The US FDA has approved a range of medical implants made of metallic materials in the first three classes [4], and these implants are often utilized in orthopedic therapy [2]. Depending on the primary alloying element, the metallic implant can be broadly divided into the following groups: bioinnert metallic implants and bioresorbale metallic implants. Common investigated metallic biomaterials include titanium, magnesium, stainless steel, and cobalt-chromium alloys as discussed further [3].

13.2.1 Stainless steel

Stainless steel-based implants are widely used in orthopedic and dental applications due to their strength, durability, and corrosion resistance, particularly 316L, which is widely used in surgical implants. It exhibits excellent mechanical strength but may suffer from corrosion issues, which can lead to metal ion release and adverse biological responses. It can be used for temporary devices like fracture plates, screws, hip nails, and total hip replacements.

13.2.2 Titanium and its alloys

Titanium is renowned for its biocompatibility and corrosion resistance. Titanium alloys, such as Ti-6Al-4V, are commonly employed in load-bearing applications. In vitro studies often focus on cell adhesion, proliferation, and differentiation on titanium surfaces.

13.2.3 Cobalt-chromium alloys

Cobalt-chromium alloys are characterized by their high wear resistance and strength. These materials are frequently used in dental applications. In vitro studies assess their biocompatibility, particularly the cytotoxic effects of metal ion leaching, total joint replacements (wrought alloy), and dentistry castings [4].

13.2.4 Magnesium alloys

Emerging as biodegradable alternatives, magnesium alloys are being investigated for temporary implants. The researchers' in vitro studies primarily focus on corrosion rates and the effects of degradation products on cell behavior, and biodegradable orthopedic implants' (Class III) animal trial [5, 6].

The effectiveness of the aforesaid melic biomaterials is assessed through various bioassays, including cytotoxicity testing to evaluate cellular toxicity, biocompatibility tests for interaction with biological tissues, corrosion testing for resistance to degradation in physiological conditions, osseointegration studies in animal models to observe bone integration, and inflammatory response evaluations through histological analysis. Based on those assays the performance, safety, and longevity of metallic implants can be significantly enhanced for better patient outcomes in clinical settings.

13.3 Cytocompatibility of different metallic implants

13.3.1 Cell adherence to substances

One of the most crucial characteristics of any implant is its affinity for cells as they are constantly placed next to living tissue. Quantitative assessment of the materials' cell affinity is required for the creation of novel materials with improved cytocompatibility. Measuring the adhering cell's detachment force from a substance is one way to assess a material's affinity for cells quantitatively. A unique technique has been developed to directly assess the shear force necessary to separate a cell from a material [7]. The detachment force in the cell culture medium was evaluated by applying lateral stress to the clinging cell with a microcantilever.

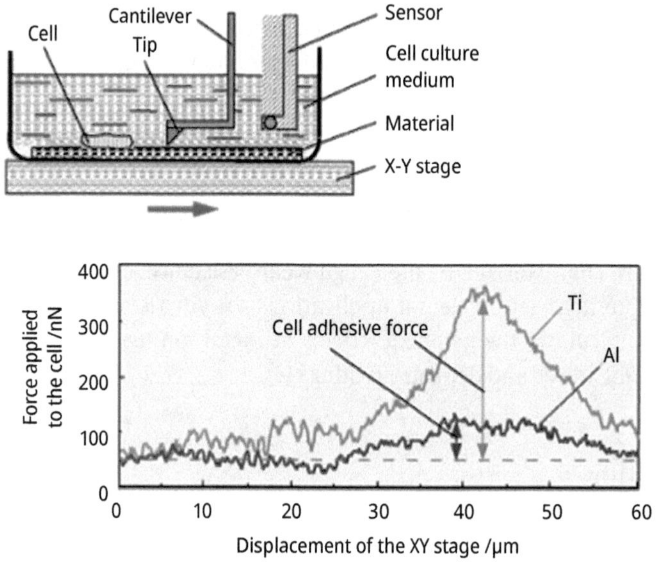

Figure 13.1: The principle of measuring shear for cell detachment from a material with force-displacement curves and a newly created instrument.

During the measurement, an optical microscope was used to view the cell, and a CCD camera recorded the image on a videotape. Before detachment, the sticky area of the cell was assessed by evaluating the video image created. Figure 13.1 shows a schematic illustration of the system. The new technique directly assessed the shear force and total energy required to extract L929 fibroblast from glass, polystyrene, and polystyrene coated with collagen or fibronectin [8]. Cells on collagen-coated polystyrene, fibronectin-coated polystyrene, and polystyrene and glass exhibit the highest cell adhesive shear force, cell adhesive shear strength, cell detachment energy, and cell detachment surface energy, which is consistent with empirical facts. The cell adhesive shear strength and cell detachment surface energy are determined by the number of binds between a cell and a material's surface and not the strength of each binding. In comparison to the cells on glass, the cells on other metals that were dissolved in a cell culture media showed a lower cell adhesive shear strength and a lower cell detachment surface.

A metal that exhibits favorable properties for cells includes a stable oxide layer on the surface, a moderate hydrophilic surface, low cytotoxicity from released metal ions, and strong resistance to medium-level corrosion.

13.4 In vitro cytotoxicity test for metallic implants

In vitro cytotoxicity testing is crucial in the development of metallic implants, as it evaluates the biological safety and compatibility of new biomaterials before they undergo in vivo testing. Various methodologies exist, each with its advantages and limitations. The toxicity of metallic materials is determined by their composition, the toxicity of their elements, and their corrosion and wear resistance. To ensure safety in new alloys, it is essential to exclude toxic elements from their composition. Corrosion resistance is particularly important, as it affects both the longevity of implants and their biocompatibility. Degradation products from corrosion can lead to adverse biological reactions, making thorough testing for both corrosion and biocompatibility crucial before clinical approval of any new metallic biomaterial.

13.4.1 Colony formation assay for cytotoxicity

The evaluation of the cytotoxicity and mutagenicity of metallic materials is crucial for their application in biomedical implants. A systematic assessment of 43 metal salts revealed varying levels of cytotoxicity, with IC_{50} values ranging from 1.36 to 1.026 mol/L for L929 fibroblasts and from 1.42 to 1.022 mol/L for MC3T3-E1 osteogenic cells as shown in Figure 13.2 [9, 10]. The study indicated a strong correlation in cytotoxicity across different cell lines, suggesting a general trend in metal salt toxicity.

Additionally, mutagenicity was assessed using the umu test on 41 metal salts, which included various metals and valences. Notably, $K_2Cr_2O_7$, $IrCl_4$, $RhCl_3$, and $MgCl_2$ showed significant mutagenic potential in the absence of the S-9 mix, while several salts tested positively with it [11].

These findings emphasize the importance of understanding both the cytotoxic and mutagenic properties of metallic biomaterials to ensure their safety and efficacy in medical applications. The development of new alloys must focus on nontoxic elements and improved biocompatibility to enhance patient outcomes.

L929	$Cr^{6+} > Cd^{2+} > V^{3+} > Ag^{+} > Hg^{2+} > Sb^{3+} > Hg^{+} > Tl^{3+} > Ga^{3+} > Cu^{2+} > Mn^{2+} > Co^{2+} > Zn^{2+} > Be^{2+} > Ni^{2+} > Sn^{2+} > In^{3+} > Ir^{4+} > Tl^{+} > Pd^{2+} > Y^{3+} > Cu^{+} > Rh^{3+} > Pb^{2+} > W^{6+} > Cr^{3+} > Bi^{3+} > Ti^{4+} > Cs^{+} > Hf^{4+} > Mo^{5+} > Zr^{4+} > Ta^{5+} > Ba^{2+} > Rb^{+} > Nb^{5+} > Al^{3+} > Fe^{3+} > Ru^{3+} > Fe^{2+} > Sr^{2+} > Li^{+} > Sn^{4+}$
MC3T3-E1	$Cd^{2+} > In^{3+} > V^{3+} > Be^{2+} > Sb^{3+} > Ag^{+} > Hg^{2+} > Cr^{6+} > Co^{2+} > Bi^{3+} > Ir^{4+} > Cr^{3+} > Hg^{+} > Cu^{2+} > Rh^{3+} > Tl^{3+} > Sn^{2+} > Ga^{3+} > Pb^{2+} > Cu^{+} > Mn^{2+} > Tl^{+} > Ni^{2+} > Zn^{2+} > Y^{3+} > W^{6+} > Fe^{3+} > Pd^{2+} > Fe^{2+} > Ti^{4+} > Hf^{4+} > Ru^{3+} > Sr^{2+} > Sn^{4+} > Ba^{2+} > Cs^{+} > Nb^{5+} > Ta^{5+} > Zr^{4+} > Al^{3+} > Mo^{5+} > Rb^{+} > Li^{+}$

Figure 13.2: The cytotoxicity rankings of metal ions in the MC3T3-E1 and L929 cell lines demonstrate a consistent pattern, indicating that the toxicity levels of various metal salts are similar across these two types of cells. (Source: Metallic implant biomaterials).

The analysis of metallic elements in phosphate-buffered saline after conducting fretting fatigue and cytotoxicity tests reveals that metallic biomaterials, including 316L stainless steel, titanium, and the Ti-6Al-4V alloy, can release metallic ions or debris when implanted in regions experiencing fretting fatigue. Additionally, the release of these metallic elements at concentrations as low as 100 ppb could have toxic effects on cells

13.4.2 ISO 10993-5

This standard outlines the guidelines for assessing the cytotoxicity of medical devices. It allows for variability in test conditions, which can lead to inconsistent results across studies. The standard recommends using indirect methods, where extracts from the material are tested on cell cultures [12, 13]. The colony formation assay is a widely used method to assess cytotoxicity and evaluate the effects of various treatments on cell survival and proliferation. Commonly used cell lines include L929 murine fibroblasts and osteoblast-like cells (e.g., U-2 OS) to assess cellular responses [14], depending on the research focus. Once the cell line is chosen, the cells are cultured under optimal conditions, including appropriate media, temperature, and CO_2 levels, to ensure healthy growth. Next, the cells are prepared by trypsinization to create a single-cell suspension, which is then seeded into a multiwell plate, typically at a density of 100–1,000 cells per well. After allowing the cells to adhere for about 24 h, treatments with the compound of interest are applied at various concentrations, with control wells receiving no treatment for baseline comparison. The cells are then incubated for an extended period, usually between 7 and 14 days, to allow sufficient time for colonies to form. Following the incubation period, the colonies are fixed using a suitable fixative like methanol or formaldehyde. Staining with a dye, such as crystal violet, enhances visibility, making it easier to count the colonies. A colony is generally defined as a cluster of more than 50 cells, although this definition can vary depending on the specific protocol. Counting is performed using a microscope or a colony counter, allowing researchers to determine the number of colonies formed in each treatment group. Finally, data analysis involves calculating the percentage of surviving colonies in comparison to the control wells, which can be expressed as a survival fraction or plotted to illustrate the dose-response relationship. It is essential to include multiple replicates for each treatment group and incorporate both positive and negative controls to enhance the statistical validity of the results. This method provides a comprehensive approach to assessing the long-term effects of treatments on cell viability and is a cornerstone technique in cancer research and drug development.

13.4.3 Direct contact assays

This method involves placing the metallic material directly in contact with cultured cells [15, 16]. It can provide insights into how the material interacts with cells but may yield different cytotoxicity results compared to extract-based methods due to direct exposure to metal ions [17]. In this approach, cultured cells are first prepared and allowed to adhere to a culture plate. Once the cells are ready, the test substance – such as a drug, chemical, or biomaterial – is applied directly to the cell monolayer. The cells are then incubated for a specific duration, allowing the substance to interact with them. After the incubation period, various assays can be employed to assess cell viability, such as Trypan blue exclusion, MTT assays, or live/dead staining techniques. These methods differentiate between live and dead cells, providing insight into the cytotoxic effects of the tested substance. The data obtained are analyzed to determine the percentage of viable cells compared to control groups, offering a clear indication of cytotoxicity. Direct contact assays are particularly valuable in toxicology, pharmacology, and materials science, as they provide critical information on the safety and effectiveness of substances intended for medical use.

13.4.4 Tetrazolium salt-based assays (e.g., MTT and WST-8)

Tetrazolium salt-based assays, including MTT (3-(4,5-dimethylthiazol-2-yl)-2,5-diphenyltetrazolium bromide) and WST-8 (water-soluble tetrazolium salt), are widely used methods for assessing cell viability and proliferation in vitro. These assays rely on the reduction of tetrazolium salts by metabolically active cells to produce colored formazan products, which can be quantified spectrophotometrically.

13.4.4.1 MTT assay

The MTT assay was first described by Mossman in 1983 and has since become a standard method for measuring cell viability and proliferation. The assay is based on the ability of viable cells to reduce MTT to purple formazan crystals, which are then solubilized for quantification [18].

First, cells are plated in a 96-well plate at a density appropriate for the specific cell type. Following a suitable incubation period of 1–4 h, MTT solution is added to each well, allowing the cells to metabolize the dye. The plates are then incubated at 37 °C to facilitate the formation of formazan crystals. After this incubation, a solvent, commonly dimethyl sulfoxide, is added to dissolve the formazan crystals. Finally, absorbance is measured at 570 nm, using a reference wavelength of 630 nm, with a microplate reader to quantify cell viability based on the extent of formazan formation. It is measured at 570 nm (with a reference wavelength of 630 nm) using a microplate reader [1]. The MTT assay offers several advantages, including high sensitivity and reproducibility, the ability to assess multiple samples simultaneously, and a quantitative measure of cell viability that correlates with metabolic activity. However, it also has notable disadvantages; potential interference from various compounds, such as glycolysis inhibitors, can affect result accuracy. Additionally, the assay does not differentiate between cytostatic (growth-inhibiting) and cytotoxic (cell-killing) effects, which can limit its interpretative value. Furthermore, the multistep process, particularly the solubilization of formazan crystals, may introduce variability and impact the consistency of results.

13.4.4.2 WST-8 assay

WST-8 is a newer alternative to the MTT assay that offers several advantages. It is reduced by cellular dehydrogenases to produce a soluble formazan dye that does not require solubilization steps. In the WST-8 assay, cells are initially plated in a 96-well format similar to the MTT assay. Following an appropriate incubation period, WST-8 dye is added directly to the wells. The plate is then incubated at 37 °C for a specified duration to allow for the formation of the soluble formazan dye. After incubation, absorbance is measured directly at 450 nm using a microplate reader, eliminating the need for solubilization steps, as the dye remains soluble in the culture medium, which simplifies the process and reduces potential variability in results.

13.4.5 3D cell culture models

Recent advancements in cell culture techniques have led to the development of 3D models that more accurately replicate the tissue environment than traditional monolayer cultures. Unlike flat, 2D cultures that often fail to mimic the complexities of real tissues, 3D models allow cells to grow in a spatial configuration that better resembles their natural surroundings [19]. This three-dimensional architecture facilitates more realistic cell-cell and cell-matrix interactions, enhancing cellular behaviors such as proliferation, differentiation, and response to stimuli. As a result, these models provide valuable insights into how implants or therapeutic agents will behave in vivo, allowing researchers to assess complex biological interactions, predict healing responses, and evaluate the efficacy and safety of new treatments more effectively. Such advancements mark a significant shift toward more physiologically relevant systems in biomedical research, improving the translational potential of findings from the lab to clinical applications [20].

13.4.6 Biochemical assays

Utilizing multiple biochemical endpoints instead of relying solely on cell viability offers a more nuanced understanding of material toxicity. For example, assessing oxidative stress through fluorescence assays can provide critical insights into the toxicological effects that traditional methods might overlook [21]. In such assays, specific fluorescent probes are used to detect reactive oxygen species within cells, indicating oxidative damage caused by a material. This approach allows researchers to evaluate the impact of substances on cellular health beyond mere survival rates, revealing changes in metabolic activity, cell signaling pathways, and overall cellular function. By integrating various endpoints – such as apoptosis markers, inflammatory cytokines, and DNA damage assays – scientists can develop a comprehensive toxicity profile that better informs the safety and biocompatibility of materials in biological systems [22]. This holistic view is particularly valuable in applications like drug development and materials science, where understanding the multifaceted interactions between cells and materials is crucial for predicting in vivo behavior and potential adverse effects.

13.4.7 Inflammatory response assessment

Inflammatory response assessment through the measurement of cytokine release, such as IL-6 and TNF-α, is crucial for evaluating the inflammatory potential and biocompatibility of new implant materials. By quantifying these cytokines, researchers can gain insights into how the immune system responds to an implant, helping to identify any proinflammatory effects that may arise following implantation [23, 24]. This approach is particularly relevant for understanding long-term effects, as chronic inflammation can lead to complications such as tissue damage, implant failure, or rejection. By assessing cytokine profiles, scientists can better predict the interaction between the material and the host tissue, enabling the design of more biocompatible implants that promote healing and minimize adverse inflammatory responses [25]. This comprehensive evaluation is essential for ensuring the safety and efficacy of new materials in clinical applications.

13.5 Challenges and future directions

In vitro studies of metallic biomaterials are essential for evaluating their biocompatibility and cytotoxicity. Yet they face significant challenges including a lack of standardized testing conditions, biological complexity, and limited predictive power regarding in vivo outcomes. Variability in results can arise from differences in cell types, culture conditions, and material-specific behaviors, particularly with emerging biodegradable metals like magnesium and zinc alloys. To enhance the relevance and reliability of these studies, future directions should focus on improving standardization to better mimic physiological conditions, integrating advanced 3D cell culture models or organ-on-a-chip technologies, expanding biochemical assessments to include inflammatory responses, conducting longitudinal studies to simulate chronic exposure, and fostering interdisciplinary collaboration among materials scientists, biologists, and clinicians. Addressing these challenges will ultimately lead to safer and more effective metallic implants for clinical applications.

13.6 Conclusions

In vitro studies play a pivotal role in the development of metallic biomaterials, serving as a foundational step in understanding how these materials interact with biological systems. These investigations give crucial preclinical data that help develop metallic biomaterials before they are tested in vivo and eventually used in clinical settings. Through controlled laboratory experiments, researchers can assess critical factors such as biocompatibility, which determines how well the material integrates with the body without causing adverse reactions. These studies also evaluate the mechanical properties of the materials, including strength and flexibility, to ensure they can withstand the stresses of the human body. Additionally, corrosion behavior is critically analyzed, as the implant's durability in the physiological environment is vital for long-term success. By continuously refining these aspects through rigorous research, scientists aim to enhance the performance of metallic biomaterials, ultimately leading to safer and more effective clinical applications. This ongoing quest seeks to improve patient outcomes and extend the longevity of implants, ensuring that they serve their purpose over many years.

References

[1] Park, J. B., & Lakes, R. S. (2007). *Biomaterials: An Introduction*. New York: Springer.

[2] Black, J. (1984). *Biological Performance of Materials*. New York: Plenum Press.

[3] Kelly, E. J. (1982). Electrochemical behavior of titanium. *Modern Aspects of Electrochemistry, 14*, 319–424.

[4] Hanawa, T., & Ota, M. (1992). Characterization of surface film formed on titanium in electrolyte using XPS. *Applied surface sciences, 55*, 269–276.

[5] Hanawa, T., Asami, K., & Asaoka, K. (1998). Repassivation of titanium and surface oxide film regenerated in simulated bioliquid. *Journal of Biomedical Materials Research, 40*, 530–538.

[6] Hanawa, T., Hiromoto, S., & Asami, K. (2001). Characterization of the surface oxide film of a Co–Cr–Mo alloy after being located in quasi biological environments using XPS. *Applied surface sciences, 183*, 68–75.

[7] Yamamoto, A., Mishima, S., Maruyama, N., & Sumita, M. (1998). A new technique for direct measurement of the shear force necessary to detach a cell from a material. *Biomaterials, 19*, 871–879.

[8] Yamamoto, A., Mishima, S., Maruyama, N., & Sumita, M. (2000). Quantitative evaluation of cell attachment to glass, polystyrene, and fibronectin- or collagen-coated polystyrene by measurement of cell adhesive shear force and cell detachment energy. *Journal of Biomedical Materials Research, 50*, 114–124.

[9] Yamamoto, A., Honma, R., & Sumita, M. (1998). Cytotoxicity evaluation of 43 metal salts using murine fibroblasts and osteoblastic cells. *Journal of Biomedical Materials Research, 39*, 331–340.

[10] Yamamoto, A., Honma, R., Tanaka, A., & Sumita, M. (1999). Generictendencyof metal salt cytotoxicity for six cell lines. *Journal of Biomedical Materials Research, 47*, 396–403.

[11] Yamamoto, A., Kohyama, Y., & Hanawa, T. (2002). Mutagenicity evaluation of forty-one metal salts by the umu test. *Journal of Biomedical Materials Research, 59*, 176–183.

[12] Test conditions can significantly affect the results of in vitro cytotoxicity testing of degradable metallic biomaterials

[13] Preclinical in vitro evaluation of implantable materials: conventional approaches, new models and future directions

[14] In vitro cytocompatibility and antibacterial studies on biodegradable Zn alloys supplemented by a critical assessment of direct contact cytotoxicity assay

[15] Fazal, F., et al. (2019). Evaluation of cytotoxicity of materials using direct contact assays. *Journal of Biomedical Materials Research*. doi: 10.1002/jbm.a.36800

[16] Assessment of cytotoxicity of materials using direct contact and extraction methods. *Journal of Materials Science Materials in Medicine, 26*(10), 1–10. doi: 10.1007/s10856-015-5645-1

[17] In vitro evaluation of cytotoxicity and corrosion behavior of commercially pure titanium and Ti-6Al-4V alloy for dental implants

[18] Limitations of the 3-(4,5-dimethylthiazol-2-yl)-2,5-diphenyl-2H-tetrazolium bromide (MTT) assay when compared to three commonly used cell enumeration assays

[19] Zhang, Y., et al. (2019). 3D cell culture systems for drug development and testing: A review. *Journal of Pharmaceutical Sciences, 108*(6), 1876–1887. doi: 10.1016/j.xphs.2019.02.019

[20] Three-dimensional cell culture systems in regenerative medicine. *Advanced Drug Delivery Reviews, 127*, 140–156. doi: 10.1016/j.addr.2018.02.009
[21] Gonçalves, J., et al. (2020). In vitro assays for assessing cytotoxicity: A review of the current state and future perspectives. *Environmental Toxicology and Pharmacology*, *76*, 103346. doi: 10.1016/j.etap.2020.103346
[22] Integrating multiple toxicity endpoints to assess the safety of engineered nanomaterials. *Nature Nanotechnology*, *12*(6), 501–508. doi: 10.1038/nnano.2017.34
[23] Friedman, A., & Michael, E. (2019). Cytokines as biomarkers of inflammation and biocompatibility in biomaterials. *Journal of Biomedical Materials Research, Part A*, *107*(5), 931–942. doi: 10.1002/jbm.a.36712
[24] Ding, Y., et al. (2020). Cytokine release and macrophage polarization in response to different implant materials. *Biomaterials*, *253*, 120103. doi: 10.1016/j.biomaterials.2020.120103
[25] García-Gareta, E., et al. (2018). Inflammatory response to biomaterials: The role of cytokines in tissue regeneration and remodeling. *Acta Biomaterialia*, *78*, 1–12. doi: 10.1016/j.actbio.2018.09.006

Ranjit Barua, Arbind Prasad*, and Siddharth Mohan Bhasney

Chapter 14
Surface modification and coating of metallic implants

Abstract: Medical science cannot exist without high-performance metallic biomaterials that display superior physical traits, resistance to wear, and the human body reaction. This chapter examines multiple methods to create biomaterials such as casting, powder metallurgy, additive manufacturing, and surface modification. The chapter utilizes multiple advanced ways to test materials like mechanical strength tests plus corrosion monitoring and looks at how the materials are built from small to large scale. Through its titanium alloy examination alongside studies of biodegradable magnesium implants and nitinol technology, the chapter displays material uses and predicts future biomaterial progress. The chapter shows how different research teams combine knowledge to build better metal biomaterials for medical needs.

Keywords: Metallic biomaterials, titanium alloys, magnesium-based implants, nitinol, additive manufacturing, biocompatibility, corrosion resistance, surface modification, medical implants, biomaterial characterization

14.1 Introduction

Modern medical practice depends heavily on high-performance metal materials to make implants and other medical devices [1–3]. Metals prove essential because they have better strength than other materials plus, work well with living tissue, and stop erosion when inside the body [4, 5]. The manufacturing process creates metallic biomaterials that match the body's natural functions and avoid harmful side effects [6, 7]. In order to assist the host, bioimplants are artificial medical devices designed to replace biological structural components of the human body that are defective or nonfunctional. A major factor in deciding how human biology and materials interact is surface modification of biomaterials. The variety of orthopedic and dental implants, such as joint arthrodesis, dental screws, fracture fixation, and complete bone replacement, has expanded due to signifi-

*Corresponding author: **Arbind Prasad**, Mechanical Engineering Department, Katihar Engineering College (Under Department of Science, Technology and Technical Education, Government of Bihar), Katihar 854109, Bihar, India, e-mail: arbind.iitg@gmail.com
Ranjit Barua, Mechanical Engineering Department, Omdayal Group of Institutions, Howrah 711316, West Bengal, India
Siddharth Mohan Bhasney, Machino Polymers Limited, Plot No. 2, Sector 33, Delhi–Jaipur Expressway, Phase 2, Info Technology Park, Sector 33, Gurugram 12200, Haryana, India

https://doi.org/10.1515/9783111571423-014

cant advancements in biomaterials research. Because of their soft consistency (viscoelasticity and stiffness are crucial for guiding the immune response), high water content, porosity, and biocompatibility, hydrogels are appealing soft biomaterials [59]. They are frequently employed as coatings to encourage cell adhesion or in 3D cell cultures to imitate the biological extracellular matrix. Polysaccharides are another type of natural polymer-based hydrogel that is utilized as a material for bone tissue engineering.

Developments in medical solutions drive research transformations of metallic biomaterials. These new methods focus on making implants last longer and fit each patient better while helping them integrate with their biological environment [8]. Metal surgeons use titanium, steel, cobalt, chromium, magnesium, and nitinol extensively in orthopedic cardiovascular and dental medical treatments [9, 10]. Pins, nails, rods, plates, and even joint prostheses are among the medical applications for which 316L steel is well-suited due to its mechanical characteristics. Even though 316L steel has a good corrosion resistance, stress cracking and crevice corrosion can still occur. Access to chlorides, which have been discovered in biological fluids, is the source of the first. The ultrathin, protective oxide layer that passivates the material's outer surface breaks, resulting in the second. Exposure to biological conditions that are high in chloride enhances the vulnerability to stress corrosion cracking. As soon as the stainless steel implants have served their purpose, they must be removed due to this exposure, which limits the material's use in trauma procedures. Crevice corrosion can also happen when the implant gives way from sporadic bending. The oxide's fracture makes it possible for the core material to corrode, which degrades its mechanical qualities and ultimately leads to failure. Two strategies were suggested to prevent crevice corrosion of the implant material: applying a thicker oxide passivation layer to the surface and applying it carefully to prevent oxide fracture. The first strategy depends on the implant's surface being specially prepared. Polishing makes its surface smoother, which lessens interaction with the outside world; nitric acid chemical preparation deepens the oxide layer. However, by removing free iron particles from its surface, electrolytic passivation (anodizing) raises the concentration of chromium and nickel locally, which gives the material its corrosion resistance. Tetrafluoroethylene, sometimes referred to as polytetrafluoroethylene (PTFE), is a synthetic fluoropolymer (C_2F_4) that was originally produced in 1938 and is more commonly known as Teflon or Syncolone. Due to a number of special qualities, it found a wide variety of uses alongside its enlarged version (ePTFE; Gore-Tex). The mechanical characteristics are the most crucial from an orthopedic perspective. The substance greatly reduces friction between working components since it is nonstick and extremely slippery. Furthermore, because of its remarkable nonreactivity, corrosion resistance, and biocompatibility, it is well tolerated in between tissues. Teflon was mostly utilized in medicine to make vascular, biliary, and ophthalmic prostheses, endovascular and urinary catheters, and soft tissue restoration materials. In orthopedics, it was utilized to repair tendinous and ligamentous abnormalities. Teflon is also used to stabilize bone fractures, and in arthroplasty. Because of the material's improper strength, the latter use appears to be dubious. However, it appears to be an unquestionably good substitute for other fluorine-based polymers, such as ethylene tetrafluoroethylene, as a lubricant that lowers the friction between an implant's functioning components. Several facets of surface modification of metallic biomaterials are shown in figure 14.1.

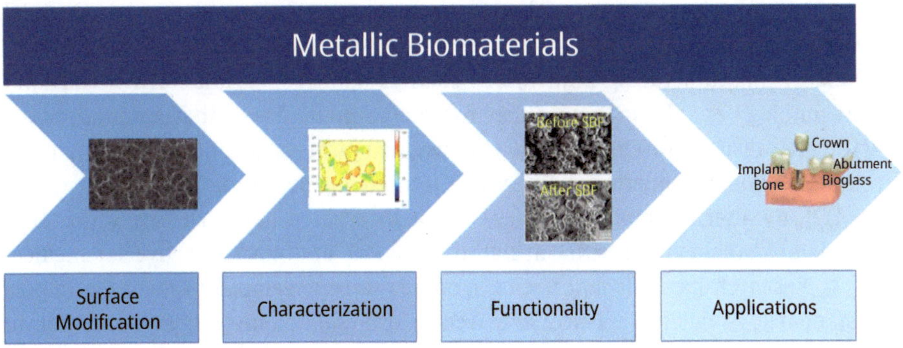

Figure 14.1: The several facets of surface modification of metallic biomaterials that will be covered in this special issue are shown in the following diagram: tribomechanical and/or biofunctional surface characterization, surface modification techniques, and application [11].

Researchers combine materials science knowledge with medical and engineering expertise to create metallic medical devices [12]. The selection of materials depends on the desired application demands which need balanced combinations of mechanical strength and toughness with corrosion resistance and body-friendly features [13]. The specific features of metallic biomaterial surfaces strongly impact their ability to bond with tissue which makes surface engineering essential during development [14, 15]. The titanium nitride (TiN) coating reduces wear and distortion of the implant's surface under load by smoothing, hardening, and improving its resistance to abrasion and corrosion. Because it has characteristics with materials that are located deeper, manufacturing costs can be reduced. Furthermore, it shields the implant from nearby tissue and materials, minimizing metal ion exchange. By using TiN to coat the implant, undesirable side effects caused by chromium release from CoCrMo alloys are successfully reduced. The majority of the materials utilized in orthopedic treatments were discovered in industrial applications from the very beginning of implant manufacturing. However, problems resulting from the lack of biocompatibility were quickly discovered, necessitating either alloy modification or the creation of new materials to address the problems with standardized materials. New materials, including as titanium alloys, CoCrMo, and stainless "surgical" steel, were consequently introduced to enable the production of implants with far more medically appropriate qualities, leading to improved treatment outcomes. However, such materials also showed a number of drawbacks, indicating the need for more modification. A suitable material should ideally have mechanical properties that are similar to those of human bone and whose qualities would allow the use of cutting-edge manufacturing processes like laser-beam sintering to create three-dimensional, customized implants based on the patient's size and unique needs. Furthermore, the area of orthopedics would greatly benefit from materials that could also operate as a carrier of molecular compounds that promote and regulate biological processes at the implant-bone interface, such as fracture healing, implant osseointegration, and regulation of microbial development.

Modern manufacturing innovations such as additive manufacturing (AM) and nanotechnology enable us to produce customized biomaterials that respond to our specific technical requirements [16]. The production of porous bone-like structures through AM boosts implant bonding while minimizing stress on bone during movement [17]. By altering implant surfaces scientists created nanostructure layers that made metal implants both better at fighting bacteria and more effective for medical use [18]. Research methods and testing methods were developed to meet the important standards of medical science. Researchers use updated imaging systems, laboratory testing methods, and biological testing to verify that developed materials satisfy medical requirements [19, 20]. Using digital computer models to test material responses helps doctors find better medical solutions through their research faster [21]. In the early stages of implant technology, form and sufficient mechanical strength were prioritized over chemical content and microstructure. However, as biocompatibility was more understood, the significance of chemical composition grew. When evaluating novel implant materials, the production process, microstructure, and the consequent durability, stiffness, elasticity, toxicity, and irritability become critical considerations. With implants being used in patients who are considerably younger than in the past – between the ages of 30 and 40 – this significance became even more, necessitating even longer implant lifetime. It is assumed that younger patients would utilize the implant for a longer period of time than older ones, which will result in a longer and more severe intoxication from the metal ions that are discharged. Furthermore, the potential for teratogenic issues may accompany the ability to reproduce.

This chapter details all the methods experts use today to produce top-tier metallic biomaterials. This chapter explores how these materials will contribute to healthcare progress while helping to achieve better results for patients.

14.2 Metallic biomaterials: an overview

Around the world, bone fractures are a common form of severe injury, and treating them frequently places a heavy financial strain on society. Even with the availability of sophisticated therapeutic approaches, 10% of instances result in problems including nonunion or delayed fracture healing, which lengthen recovery times and cause patients more suffering. Fracture healing problems are more likely to occur in high-

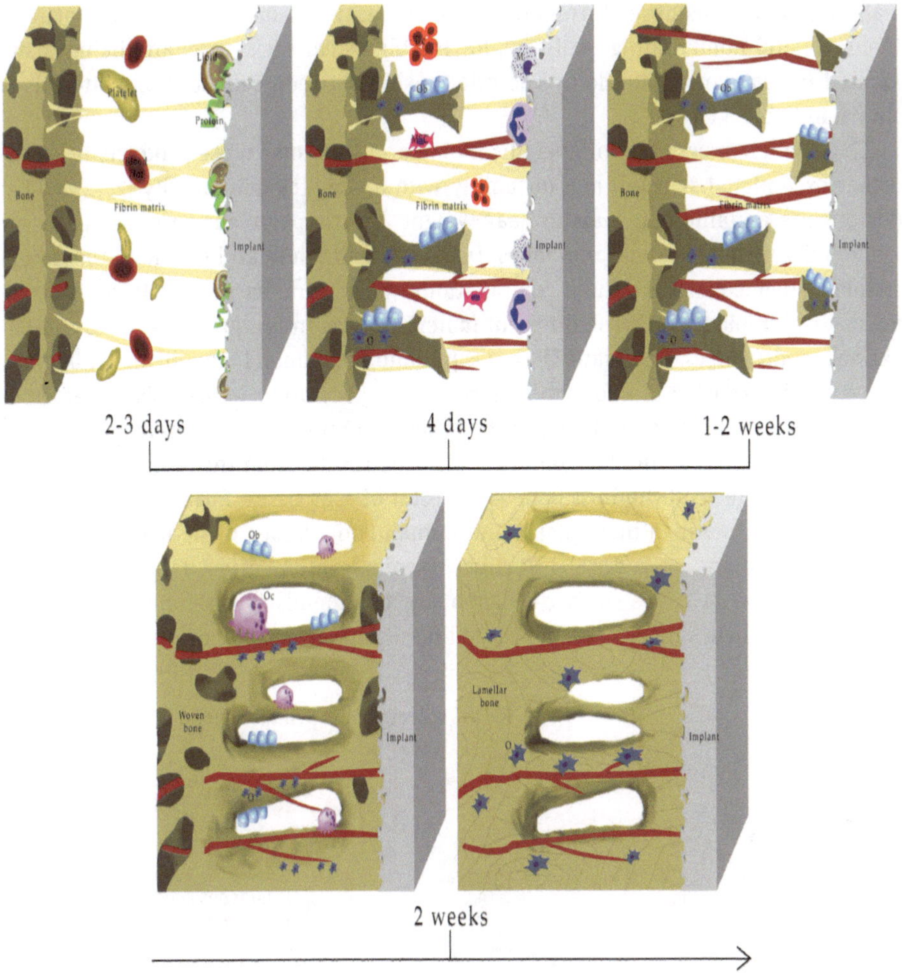

Figure 14.2: The relationship between cells and biomaterials [22].

risk categories, including those with osteoporosis, the elderly, malnourished people, postmenopausal women, and people with poor blood flow. Additionally, surgical procedures like hip and knee replacements may be necessary for people with joint conditions including osteoarthritis and rheumatoid arthritis. Orthopedic implants are employed in these situations, along with temporary fracture fixation equipment and parts such plates, nails, and wires. In orthopedics, biomaterials are utilized to either replace or repair the structural integrity of injured bone. Every biomaterial needs to meet a number of fundamental requirements, including having the right mechanical qualities (e.g., precise weight and elastic modulus), exhibiting good biostability (resistance to oxidation, corrosion, and hydrolysis), guaranteeing biocompatibility, especially for bone prostheses (promoting osseointegration), having high bio-inertness (nontoxic and nonirritating qualities), having high wear resistance, and being easily applied in real-world situations. The relationship between cells and biomaterials is shown in figure 14.2

Medical engineers rely on metallic biomaterials because these products bring together exceptional strength and durability with a body-friendly quality [23]. Medical doctors use metallic materials to bear weight in orthopedic surgery and dentistry while building heart valves and bones. Doctors use titanium and titanium alloys plus stainless steel, cobalt-chromium combinations, magnesium-based metals, and nitinol for medical implants [24]. The different materials excel in unique ways to serve well-defined medical uses. Any material that is put into the human body during an implant procedure is regarded as a foreign material. Layers of fibrous tissues, also referred to as scar tissues, start to form between the tissue and implant if the foreign substances are not biocompatible. Implant failure eventually results from the implant's inability to osteointegrate with the host bone as a result of scar tissue formation. Therefore, full integration between bioimplants and human body tissues is the main need for a successful implant procedure. Effective osseointegration after dental implant is shown in figure 14.3. The surface morphology and chemistry of biomedical devices better govern their biological responses to lifetime and performance.

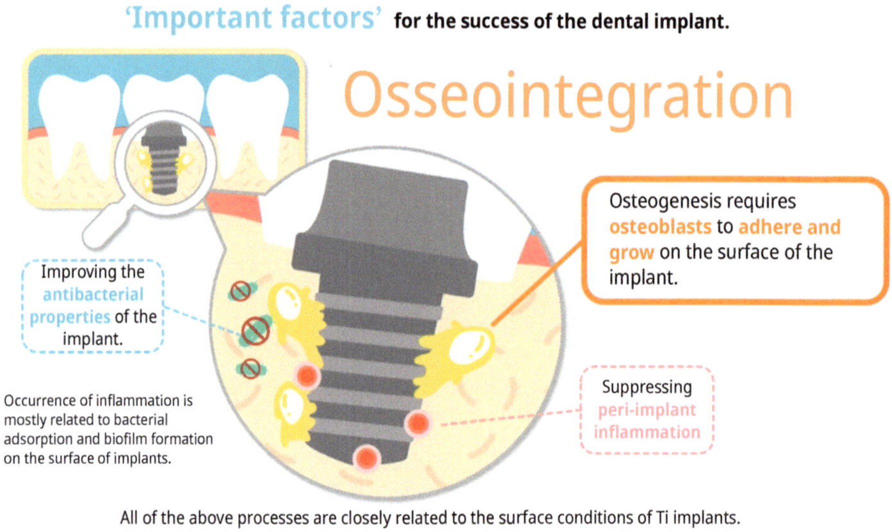

Figure 14.3: Crucial elements for the effective osseointegration after dental implant implantation [25].

14.2.1 Titanium and its alloys

Titanium metal and its alloy types lead the pack in medical device materials because they thrive in biological environments and resist corrosion while staying strong and lightweight [26]. The basic titanium alloy Ti-6Al-4V finds widespread use in both bone joint repairs and dental implantation. Titanium maintains its stability and safety when it produces an oxide layer as a surface protection [27]. Titanium is perfect for implants since it is non-dielectric, does not raise its temperature when subjected to alternating magnetic fields, and does not obstruct magnetic resonance imaging. Due to this important benefit, titanium has almost monopolized the market for implants used in spine surgery and dominated the usage of the material in traumatology and joint replacements. Furthermore, its elasticity is far closer to that of viable bone than it is to steel. Because bone and implants have comparable characteristics, it is possible to prevent unwanted strain components and overload at the bone-implant interface, which lowers the chance of loss or periprosthetic fracture. The surface receives special treatment through anodization and plasma spraying technology to improve its natural bonding with bone tissue and reduce bacterial colonization.

14.2.2 Stainless steel

Stainless steel, especially type 316L, proves to be effective for emergency bone fracture support implants [28]. Manufacturing stainless steel devices remains cheap while the material offers excellent strength and ease of production [29]. The high salt levels inside the human body quickly damage 316L stainless steel making it suitable only for temporary applications. The application of bioactive coatings and passivation methods now helps overcome stainless steel's inherent problems [30].

14.2.3 Cobalt-chromium alloys

The strong resistance to wear and amazing mechanical properties of cobalt-chromium alloys make them suitable for artificial joint surgery and dental product production. Cobalt-chromium alloys work well with the body because they create strong chromium oxide protection [31]. Stress shielding occurs in load-bearing implants because cobalt-chromium alloys show high elastic strength. Conventionally manufactured permanent metals and their alloys, including titanium, stainless steel, and Co-Cr alloys, make up the bulk of commercially used bio-implant materials. Because the elasticity modulus of these bio-implants is substantially different from that of natural bone, they provide a stress-protective effect even though they have good mechanical strength, biocompatibility, and adequate wear resistance properties. Figure 14.4 shows various metal implant usage in clinics. The bone can withstand a significantly reduced strain as a result,

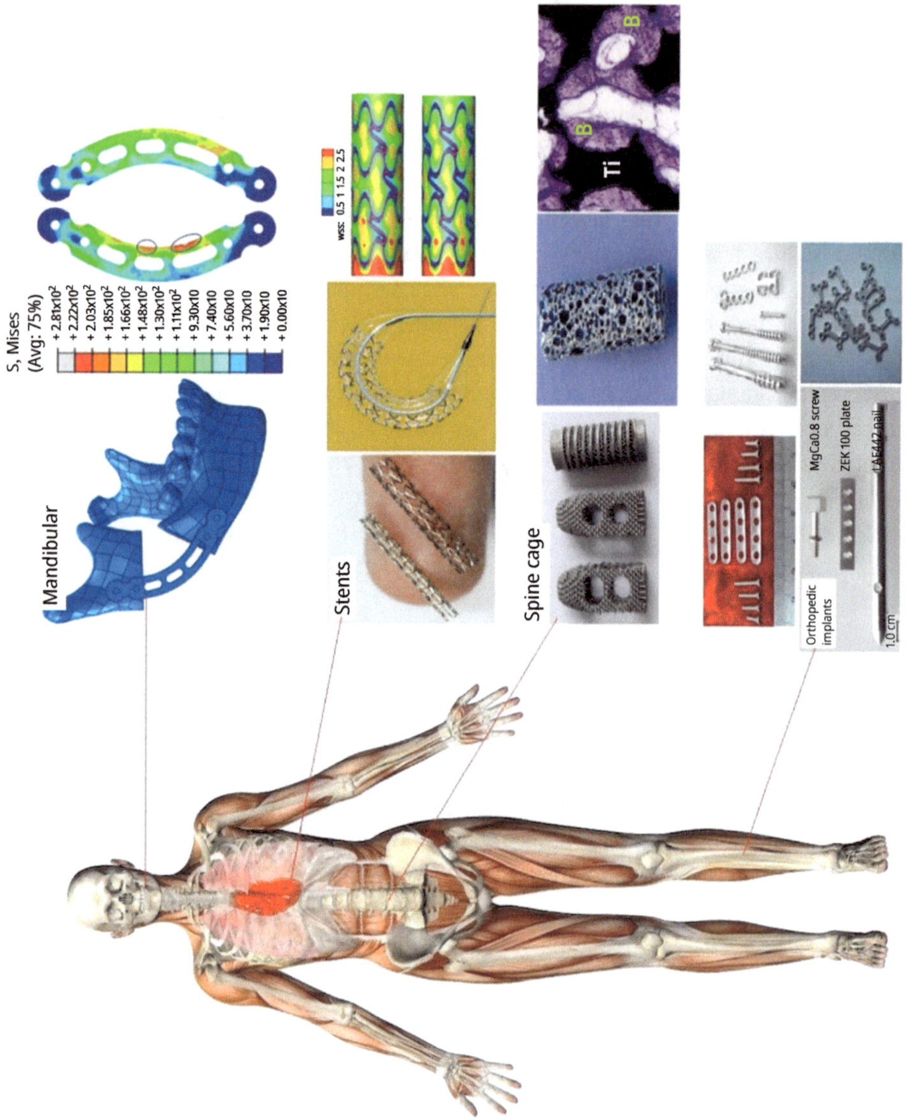

Figure 14.4: Various metal implant usage in clinics. Stents and hard tissue repair, including maxillofacial, spine, and orthopedic fixation implants, are the primary applications for metal implants. B, new bones; WSS, wall shear stress [32].

which gradually leads to implant loosening and refracture as well as a significant decline in bioefficacy. Furthermore, a second operation is required to remove the implant after the fracture has healed, which causes the patient immense anguish.

14.2.4 Magnesium-based alloys

Magnesium-based materials show strong potential as implant materials that break down over time. The implants disappear naturally through their biological processes which cut the need for post-treatment procedures and saves patients from possible harm and medical expenses. Magnesium behaves similarly to natural bone so it creates minimal protective layer stress [33]. Researchers now solve magnesium's quick corrosion problem in body fluids through material blending and protective layer development.

14.2.5 Nitinol

The nickel-titanium alloy nitinol stands out for both shape memory and superelasticity characteristics [34]. This material works well for stent design and orthodontic wires, thanks to its special traits. Minimally invasive medical procedures work better because nitinol recovers its original form after being deformed. Science teams now study different surface solutions to improve nitinol's compatibility with living tissues because nickel ions from this material create worries.

14.2.6 Other metallic biomaterials

Many other metal implants use tantalum along with zirconium and gold. The body accepts tantalum very well and the material resists corrosion better than others for orthopedic devices [36]. People test zirconium alloys because they have a lower elasticity level plus good compatibility with the body. Scientists use dental gold alloys to test new systems that might deliver drugs and fight cancer. Synthetic polymeric gels have more molecular modification options than natural polymeric gels, which makes it easier to customize the candidate qualities to meet particular needs such as adjusting mechanical characteristics and biophysical and biochemical signals. For example, poly(ethylene glycol) hydrogels that have been altered with the adhesion ligand arginine–glycine–aspartic acid provide enhanced cell attachment and differentiation along with mechanical qualities that may be adjusted. Ceramics have a limited utility in load-bearing applications due to their weak fracture toughness and brittle behavior despite their good biocompatible performance. Therefore, when great strength and load-bearing capability are needed for implants, metals and alloys are often utilized.

Metallic implants are used in the majority of medical industry areas. The hip, plates, knee prosthesis, pins, dental materials, screws, and cardiovascular applications are among the load-bearing applications that they are often utilized to replace. Despite their great strength and endurance, metals can lose their qualities in physiological settings, potentially releasing different ions and debris that might cause a biological reaction. The majority of alloys discharge metal ions into the blood's plasma. Excessive blood ion discharge increases the likelihood that it may build up in organs like the liver and spleen and eventually form particulates that impair these organs' ability to function normally. When this condition accumulates over time, it causes cytotoxicity and organ failure. Prosthetics may be made from a variety of biocompatible materials. Metals, particularly titanium alloys and low carbon steels, are used extensively. Zirconium dioxide (ZrO_2), aluminum oxide (alumina, Al_2O_3), alumina-toughened zirconia, hydroxyapatite, tricalcium phosphate (TCP), calcium aluminates (C_3A), titanium oxides (TiO_2), and others are also utilized as ceramic materials. Furthermore, novel material compositions that are specifically produced in accordance with specifications, such as nanomaterials, metal-carbon or metal-nitrogen ceramics, and other intermetallic alloys, are employed, along with polymers like Teflon, nylon, silicones, and a few others [17]. The ceramic family is the one that may improve the best biocompatibility with bones in terms of absorbability and bioactivity when comparing the various qualities of the materials used to interact with the environment. Appropriate spacers have been established using a variety of calcium spacers, including nanoforms in pure or composite form, including calcium sulfate, calcium phosphate, and hydroxyapatite [12]. However, because of their high density and limited elasticity, they offer the lowest mechanical qualities among the many options. In order to prevent postoperative infection, bone spacers are typically utilized as temporary implants during revision surgery [5]. Bone spacers are most commonly made of polymethyl methacrylate (PMMA) cement [17]. PMMA is a polymer that is not biodegradable, which means that a second surgery is required to remove it from the body. Additionally, the polymer exhibits an exothermic reaction during polymerization, which means that only heat-stable antibiotics can be maintained, and the material has an inadequate elution profile [11]. The use of a thin coating layer covering the implant's exterior made of less complex, more easily machined, and less expensive material is a highly promising approach to implant fabrication. One of the most significant developments is the ability to alter and regulate surface characteristics at the micro and even nanoscale, which may lead to a new set of approaches for establishing the intended interaction with the biological environment. The creation of biologically active, absorbable polymers and composites that could act as substrates or scaffolds for biologically active compounds capable of inducing cellular processes, such as adhesion, activations, proliferation, and differentiation into a desired cellular lineage, is another exciting avenue for research. Desired uses would include osteoblasts and bone extracellular matrix, ideally strengthened by angiogenesis assistance to promote skeletal tissue regeneration. The creation of novel materials with more appealing, useful, and desirable qualities might result from a successful collaboration between materials scientists, biologists, and orthopedists. It is impossible to predict today if polymers and composites will soon replace metals and their alloys. Although steels and CoCr alloys appear to have reached the limit of their potential because of an increasing awareness of unwanted side effects, titanium and its alloys are still presumably considered to be a material appropriate for orthopedic applications. Despite the obvious necessity for the creation of more suitable materials, surface alterations may extend their adoption.

The design of metallic biomaterials combines research into their mechanical strength while staying safe for body tissue and underlies these advancements [22]. Researchers today work on different metal compositions and enhance surface quality while using advanced methods to produce new biomaterials that fix current product weaknesses. The human body does not completely absorb metallic materials, and tissue development is hindered due to insufficient implant adhesion, resulting in pain or discomfort in the implant area. In the case of metallic implants, the healing period is longer and the danger of infection is higher than with ceramic materials. Despite certain drawbacks, preference should be given to metallic implants because of their mechanical strength, affordability, and resistance to corrosion. Although they are costly, the chemically inert platinum and gold do not corrode in situ and can be utilized as bioimplants. Because of their superior mechanical strength and biocompatibility in human body circumstances, Ti-based and Mg-based alloys are therefore used in the biomedical sectors today. Stainless steel and alloys based on Ti and Co are the most often utilized metallic materials in biomedical equipment. The range of materials now accessible for orthopedic implant manufacturing is quite limited. Almost all manufacturers provide implants made from at least three different alloys: titanium, cobalt-chromium, and steel; there are also at least two different kinds of ceramic, alumina and zirconia, and polyethylene. In orthopedics, PMMA bone cement essentially eliminates the limited list of materials that are frequently utilized. Although there are not many material options, implant makers may alter the chemical makeup, use certain production techniques, and add new coatings or surface treatments to the materials to enhance performance even further.

14.3 Processing techniques

To meet medical needs and work well with body tissue high-performance metallic biomaterials need proper manufacturing steps that make them functional enough and long-lasting. Special production methods exist to create these materials based on their unique medical uses. Below we examine every main method used for processing high-performance metal biomaterials. The antibacterial activity of medicinal compounds released from the surface-modified MSS is shown in figure 14.5

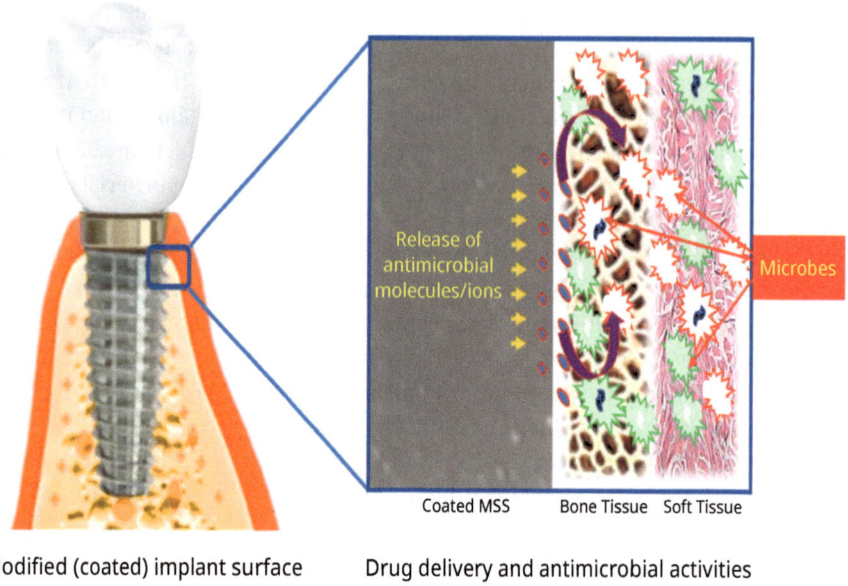

Figure 14.5: The antibacterial activity of medicinal compounds released from the surface-modified MSS: the coated implant surfaces release antimicrobial drugs; the released antibiotic is then delivered to the tissue; and the medication has antibacterial properties [36].

14.3.1 Casting

For many years casting remains the main technique to make metallic medical parts from raw materials. When material is melted and poured into shaped molds it creates the required design [36]. The molds used in casting must handle extreme heat so they are prepared from ceramic or graphite materials. Using casting lets you create difficult geometric shapes and big portions while keeping production expenses affordable [12]. The casting method produces defects like porosity and inclusions that weaken both the mechanical strength and biological compatibility of the finished product. Vacuum casting and investment casting improvements now help create medical parts with superior purity for medical use.

14.3.2 Powder metallurgy

During powder metallurgy, metal powders form the desired shape before being heated to combine at high temperatures. Powder metallurgy gives users total control over material structure and element makeup for making exact parts with complex shapes [23]. During hot isostatic pressing, metal powder transforms through simultaneous heat and pressure application to achieve higher density and strength. In metal production, powder metallurgy works best for making implants that help new bone grow while lowering implant pressure on the surrounding bone [25]. The figure 14.6 shows the dental implants schematics showing the coatings over the implants.

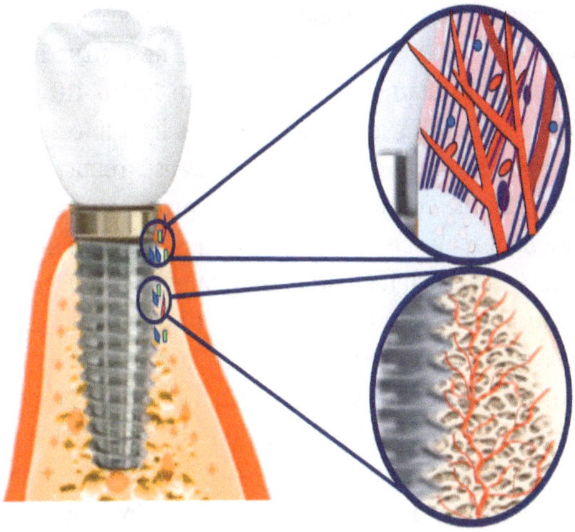

Figure 14.6: To preserve the integrity of sound hard and soft tissues, drug-eluting modified MSS is delivered to bone and soft tissue: release the molecules of the medication; create an atmosphere that is non-inflammatory and biocompatible; and preserve the integrity of both soft and hard tissues [36].

14.3.3 Additive manufacturing

Layer-by-layer material synthesis from a computerized 3D model is the foundation of AM, a fast evolving range of manufacturing techniques [21]. Typical materials include metals, alloys (steel, titanium, and aluminum), photocurable polymers, thermoplastics (polyamides, polycarbonates, elastomers, acrylonitrile butadiene styrene, etc.), and photocurable polymers. Additionally, there are some ceramic materials accessible, although they are an emerging group that needs more research and examination. In order to derive analytical expressions of the pore sizes acquired as a function of factors like infill %, nozzle diameter, and layer height, it is important to control the grid structure and examine the models [28]. The process of making metallic biomaterials has changed through the rise of AM, commonly known as 3D printing technology [37]. By building elements one layer at a time, technology offers better control over design customization and fabrication options. The manufacturing methods SLM and EBM generate titanium and cobalt-chromium implants through additive processes [38]. AM enables the design and production of metal parts that match bone characteristics including mechanical strength and tissue compatibility [39–41]. Current studies focus on understanding and improving the challenges of residual stress formation, surface quality control, and production costs using AM techniques. Currently, there are robocasting-suitable inks made of hydroxyapatite, β-TCP, and bioactive glasses that have average particle sizes of 1–3 μm. In this sense, controlling the materials' morphology, particular surface, and reactivity are crucial factors to maximize process performance. Some writers in the literature have successfully deposited filaments with a diameter of 250 μm using these kinds of inks. It was showed that it is possible to create tall green structures with controlled porosity using automatic 3D printing robocasting. These structures are made up of tenths of horizontal layers with diameters ranging from 150 to 410 μm. Based on lithography ceramic manufacturing (LCM) is a method that uses light to consolidate photopolymer in a suspension to create intricate ceramic pieces. Dense, robust, and accurate complicated three-dimensional ceramic components may be produced quickly using this processing technique [37]. Since light may activate polymerization of all the material included in a layer simultaneously, LCM can reach high processing rates in broad structures. Ceramic pieces are made in green, just like in robocasting, and require sintering, drying, and debinding processes to achieve a solid interior structure. Depending on how many pieces can fit onto the building platform, the procedure is carried out in batches. Porous PEEK for medical purposes is produced using a variety of methods. Injection molding, extrusion, compression molding, machining, and other techniques are examples of traditional manufacturing process processes [6]. Furthermore, developments in AM continue to open up new avenues for biomedical applications by making it possible to create increasingly intricate constructions, such as scaffolding for tissue engineering and implants that are customized for each patient. Ten years ago, PEEK 3D printing – a kind of powder-based AM technology – used the selective laser sintering (SLS) approach. More design freedom is made possible by its ability to create porous PEEK-based composites with intricate structures [12]. PEEK has been commercialized for use in SLS machines by both EOS and 3D systems.

14.3.4 Surface modification

Special treatments improve how well metallic biomaterials work with living tissues while making them longer-lasting [42]. Plasma spraying techniques along with anodization and CVD methods help deposit bioactive coatings onto the material's exterior. These surface coatings help metals resist corrosion better while demonstrating bone-fusing qualities and lowering bacteria attachment. Biological surface technology breakthroughs now produce nanoscale coatings that improve the material's ability to interact with living tissue [28].

14.3.5 Cold spray processing

Cold spray technology uses high-speed particle acceleration to deposit metal powders at low heat levels onto any surface [32]. At mild temperature levels, this technique safeguards the material structure and creates few thermal issues. Cold spray technology helps fix damaged implants while producing durable metallic coatings that bond strongly to surfaces [27].

14.3.6 Rapid solidification processing

When metal cools quickly from molten form it produces tiny grain patterns and stronger product materials. The manufacturing process produces biocompatible materials that provide superior strength as high-performance alloys [13]. Companies use fast cooling methods to produce stents and medical equipment that work well under movement.

14.3.7 Electroforming

Electroforming allows us to create slim-walled complex items using highly precise methods. Metal ions are deposited across a conductive mold during electroplating. After removing the mold we get a detailed metallic part ready for use [38]. The figure 14.7 shows

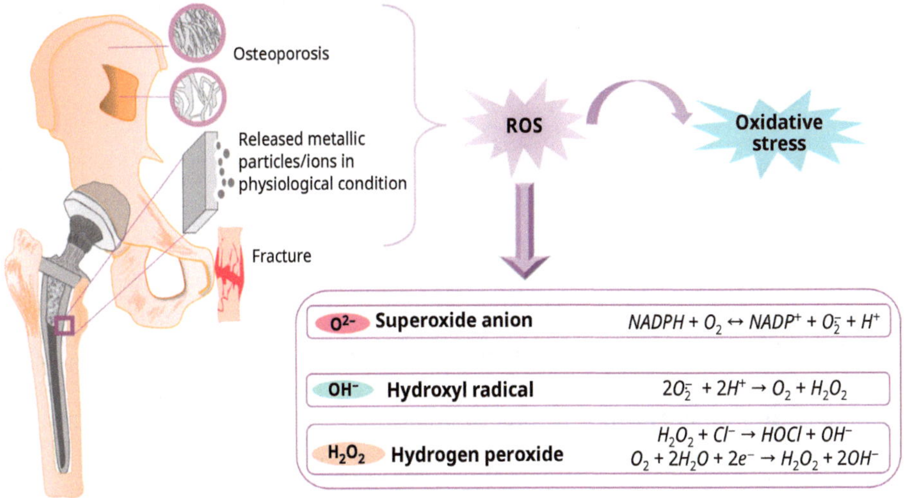

Figure 14.7: The production of ROS linked to titanium implants and the corresponding metabolic processes [43].

production of ROS linked to titanium implants and corresponding metabolic processes. Electrically deposited metal components from a mold work perfectly for producing tiny components needed in medical surgical equipment.

Based on specific use needs and material properties, the selections of processing technologies bring clear pros and cons. Researchers and manufacturers use advanced processes to expand the capabilities of metallic biomaterial products.

14.4 Characterization techniques

14.4.1 Mechanical testing

Metallic biomaterials depend on their mechanical strength and toughness to perform as needed. Standard tests include:

- **Tensile testing**: This determines the material's ultimate tensile strength (UTS), yield strength, and elongation.
- **Hardness testing**: This measures resistance to localized deformation using methods like Vickers, Brinell, or Rockwell hardness tests.
- **Fatigue testing**: This assesses the material's ability to withstand cyclic loading.

14.4.2 Corrosion testing

Metals used for biological materials require protection from corrosion to operate appropriately within human fluids [16]. Researchers check material resistance to corrosion using specialized testing machines named EIS and potentiodynamic polarization to measure performance in human body fluids.

14.4.3 Microstructural analysis

Understanding the microstructure of metallic biomaterials is essential for optimizing their properties. Techniques include:

- **Optical microscopy**: This provides an overview of the grain structure.
- **Scanning electron microscopy (SEM)**: SEM offers high-resolution imaging of surface features.
- **Transmission electron microscopy (TEM)**: TEM reveals detailed information about dislocations, precipitates, and grain boundaries.

14.4.4 Surface characterization

Surface properties influence the biocompatibility and osseointegration of metallic biomaterials. Common techniques include:
- **Atomic force microscopy (AFM)**: AFM measures surface roughness and topography.
- **X-ray photoelectron spectroscopy (XPS)**: APS analyzes surface chemistry and elemental composition.
- **Contact angle measurements**: They evaluate surface wettability, a key factor for cell adhesion.

14.4.5 Biocompatibility testing

Biocompatibility is assessed through in vitro and in vivo studies to ensure that the material does not elicit adverse biological responses. Tests include:
- **Cytotoxicity assays**: They evaluate the material's effect on cell viability.
- **Hemocompatibility tests**: They measure the interaction of the material with blood components.
- **Animal studies**: They provide insights into the material's integration with living tissue.

14.5 Case studies

14.5.1 Titanium alloy implants

Ti-6Al-4V is a widely used titanium alloy for orthopedic and dental implants due to its excellent combination of strength, corrosion resistance, and biocompatibility [24]. AM of Ti-6Al-4V has enabled the production of patient-specific implants with porous structures, improving osseointegration and reducing stress shielding.

14.5.2 Magnesium-based biodegradable implants

Magnesium alloys are gaining attention for their biodegradability and mechanical properties similar to natural bone. Surface treatments and alloying with elements like zinc and calcium have been employed to enhance their corrosion resistance and biocompatibility.

14.5.3 Nitinol stents

Nitinol's shape memory and superelastic properties make it an ideal material for cardiovascular stents. Surface modifications such as electropolishing and coating with drug-eluting layers have further improved their performance and longevity [17]. The figure 14.8 shows selective laser sintering technology to create microporous structures in orthopedic implants

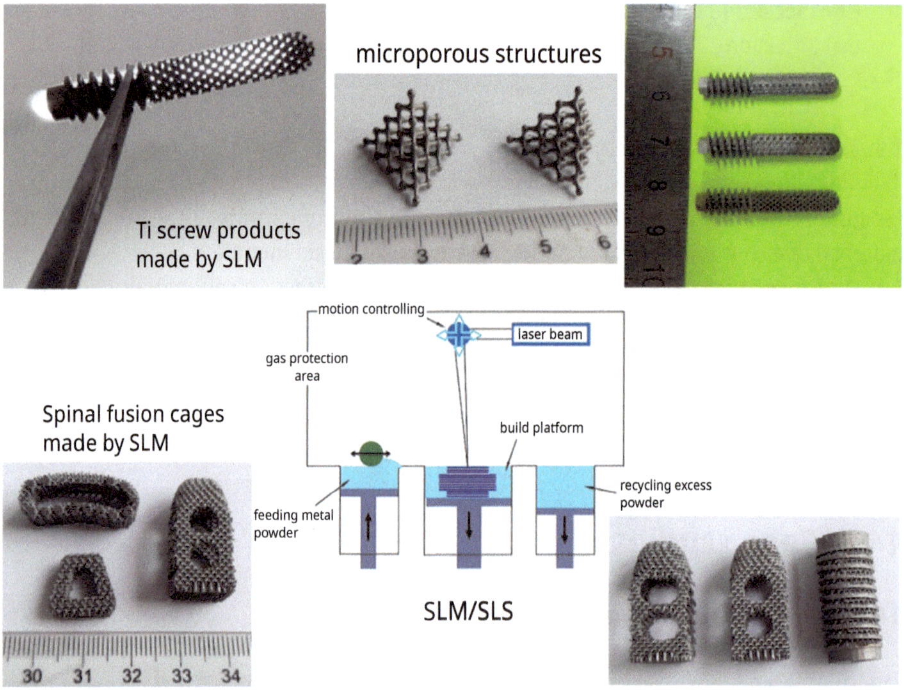

Figure 14.8: Selective laser melting or selective laser sintering (SLM/SLS) technology is used to create some porous orthopedic implants. In SLM/SLS, an upward arrow indicates that the printer's platform is pushed up to provide print powders, while a downward arrow indicates that the platform drops to recycle the print powders. A bidirectional arrow shows that the print particles are reciprocated [32].

14.6 Future directions

The field of metallic biomaterials is poised for transformative advancements, driven by the integration of emerging technologies and interdisciplinary research. Key future directions are given in the following sections.

14.6.1 Advanced manufacturing technologies

The technology of AM, also known as 3D printing, will keep changing how medical materials are made. Biomaterial processing through AM technology will help implants fit each patient better and create designs that bond with bones stronger while protecting that connection [8]. The production of advanced implants now relies on combined material manufacturing and hybrid production strategies to match desired mechanical and biocompatible properties.

14.6.2 Nanotechnology and surface engineering

Metallic biomaterials will gain major surface improvements through the use of nanotechnology. Nanostructured coatings made from hydroxyapatite and bioactive glass add both improved body compatibility and antibacterial actions [23]. Laser texturing and nanopatterning help cells stick and multiply better within the surface of tissues.

14.6.3 Smart and functional biomaterials

In the future metallic biomaterials will gain new features including self-repair functions along with the ability to remember shapes and deliver medicine. Researchers will develop nitinol shape memory alloys to create adaptive implants that react to body functions [41]. Medicines release implants that automatically supply medication will fight infections and reduce swelling to help us heal better.

14.6.4 Biodegradable metallic biomaterials

Magnesium-based alloy research will concentrate on improving both structural strength and corrosion resistance as a bioidgradable medical material. Research in metal treatment and protected surfaces will increase the medical use of these items while decreasing the number of additional procedures patients require.

14.6.5 Computational modeling and simulation

The combination of computational tools including finite element analysis and machine learning will speed up the process of designing enhanced metallic biomaterials [19]. These resources help us forecast how materials behave within living tissues while assisting metal alloy creation and tuning implant design for distinct usage areas.

14.6.6 Regulatory and standardization advances

The more advanced metallic biomaterials require regulatory systems to modify their evaluation methods to maintain safety and effectiveness [22]. A unified system to test new biomaterials will help research turn into real-world medical use. A combined effort from researchers' medical product makers and testing offices will help speed up the product approval steps.

14.6.7 Interdisciplinary collaboration

Research teams containing medical physicians' scientific engineers and materials specialists will create advances in metallic biomaterials [18]. When medical scientists work together with biologists and doctors, they can create better treatments for medical needs. The best solution for medical problems appears when medical researchers work together with healthcare providers and industrial businesses.

Our progress in metallic biomaterial technology will help achieve better medical results and make life better for patients [27]. Through advanced technology integration and team effort, researchers can make medical applications of these materials more effective.

14.7 Conclusion

Medical science has reached new heights through the work done in high-performance metallic biomaterial development. These materials meet medical implant standards best because they have strong mechanical properties and they resist both corrosion and harmful tissue interactions. Different biomaterial production methods allow us create special medical materials through casting powder metallurgy 3D printing and surface engineering. Specialized tests of mechanical strength corrosion resistance and compatibility help make sure these materials can safely perform medical duties [47–49]. The medical field depends heavily on titanium alloy and magnesium-based biomaterials in addition to stainless steel and cobalt-chromium alloys, thanks to nitinol material properties. Because it is bio-inert, well-tolerated in living tissues, and resistant to wear and corrosion, polyethylene is a promising material. Furthermore, there are many opportunities to synthesize novel polymer compounds, and the manufacture is reasonably priced. Therefore, it is likely that in the near future, polymers will be used for orthopedic purposes instead of metals and their alloys. Healthcare applications need different materials since each material handles medical problems through its special features from supporting orthopedic parts to disappearing devices that cut down future surgical requirements. Manufacturing improvements and modeling techniques with surface engineering methods are producing next-level biomaterials with better results for patients and performance benefits. Metallic biomaterials will evolve further through ongoing development of emerging technologies, including AM nanotechnology and artificial intelligence systems. These developments will help doctors make unique implants while designing biodegradable medical tools that function better than current options [45–47]. Bringing industry professionals and medical teams together with research experts will speed up the move from experimental discoveries to medical technology tools. Development of better processing methods and testing tools for high-performance metallic biomaterials will improve how doctors use them and open new medical treatment possibilities. Patient lives globally will benefit more as research and scientific advancement develop these materials further. In conclusion, the features of the most widely used materials and implant techniques have been discussed, and the focus of future research is suggested to be on new materials to create customized and very resilient bio-inert implants. Due to their excellent durability, accessibility, and standardized production methods and know-how, metals today make up the great majority of implants. However, there are practical restrictions to additional enhancements such as surface or structural alteration or reinforcement using more resilient and elastic materials. Since there are several options that outperform metals in terms of elasticity and durability without the risk of sensitization and intoxication, polymers and composites may prove to be more appropriate in the future.

References

[1] Lorenzetti, M., Dogša, I., Stošicki, T., Stopar, D., Kalin, M., Kobe, S., & Novak, S. (2015). The influence of surface modification on bacterial adhesion to titanium-based substrates. *ACS Applied Materials & Interfaces*, *7*(3), 1644–1651. https://doi.org/10.1021/am507148n

[2] Echeverry-Rendón, M., Galvis, O., Quintero Giraldo, D., Pavón, J., López-Lacomba, J. L., Jiménez-Piqué, E., Anglada, M., Robledo, S. M., Castaño, J. G., & Echeverría, F. (2015). Osseointegration improvement by plasma electrolytic oxidation of modified titanium alloys surfaces. *Journal of Materials Science. Materials in Medicine*, *26*(2), 72. https://doi.org/10.1007/s10856-015-5408-4

[3] Ma, Y., Zhang, Z., Liu, Y., Li, H., Wang, N., Liu, W., Li, W., Jin, L., Wang, J., & Chen, S. (2015). Nanotubes functionalized with BMP2 knuckle peptide improve the osseointegration of titanium implants in rabbits. *Journal of Biomedical Nanotechnology*, *11*(2), 236–244. https://doi.org/10.1166/jbn.2015.2006

[4] Ghaith, E.-S., Hodgson, S., & Sharp, M. (2015). Laser surface alloying of 316L stainless steel coated with a bioactive hydroxyapatite-titanium oxide composite. *Journal of Materials Science. Materials in Medicine*, *26*(2), 83. https://doi.org/10.1007/s10856-015-5399-1

[5] Razavi, M., Fathi, M., Savabi, O., Vashaee, D., & Tayebi, L. (2015). In vivo assessments of bioabsorbable AZ91 magnesium implants coated with nanostructured fluoridated hydroxyapatite by MAO/EPD technique for biomedical applications. *Materials Science & Engineering. C, Materials for Biological Applications*, *48*, 21–27. https://doi.org/10.1016/j.msec.2014.11.020

[6] Cochis, A., Azzimonti, B., Della Valle, C., Chiesa, R., Arciola, C. R., & Rimondini, L. (2015). Biofilm formation on titanium implants counteracted by grafting gallium and silver ions. *Journal of Biomedical Materials Research. Part A*, *103*(3), 1176–1187. https://doi.org/10.1002/jbm.a.35270

[7] Waidi, Y. O., Jain, N., Barua, R., Vannaladsaysy, V., Das, S., Prasad, A., & Datta, S. (2024). Smart and Self-healing Hydrogels for Biomedical Applications. In *Nanomanufacturing Techniques in Sustainable Healthcare Applications* (pp. 110–130). CRC Press.

[8] Chowdhury, S., Jain, N., Waidi, Y. O., Barua, R., Das, S., Prasad, A., & Datta, S. (2024). Biological Smart Biomaterials: Materials for Biomedical Applications. In *Applications of Biotribology in Biomedical Systems* (pp. 313–325). Cham: Springer Nature Switzerland.

[9] Jain, N., Waidi, Y. O., Chowdhury, S., Barua, R., Das, S., Prasad, A., & Datta, S. (2024). Medical Devices Tribology. In *Applications of Biotribology in Biomedical Systems* (pp. 235–250). Cham: Springer Nature Switzerland.

[10] Barua, R., & Datta, S. (2024). Beyond Boundaries: Navigating Future Horizons with Augmented and Virtual Reality in Healthcare and Biomedical Engg. In *Navigating the Augmented and Virtual Frontiers in Engineering* (pp. 199–211). IGI Global.

[11] Beltrán, A. M., & Begines, B. (2022). Surface modification, functionalization and characterization of metallic biomaterials. *Metals*, *12*(4), 667. https://doi.org/10.3390/met12040667

[12] El-Hadad, S., Khalifa, W., & Nofal, A. (2015). Surface modification of investment cast-316L implants: Microstructure effects. *Materials Science & Engineering. C, Materials for Biological Applications*, *48*, 320–327. https://doi.org/10.1016/j.msec.2014.12.038

[13] Liu, X. H., Wu, L., Ai, H. J., Han, Y., & Hu, Y. (2015). Cytocompatibility and early osseointegration of nanoTiO2-modified Ti-24 Nb-4 Zr-7.9 Sn surfaces. *Materials Science & Engineering. C, Materials for Biological Applications*, *48*, 256–262. https://doi.org/10.1016/j.msec.2014.12.011

[14] Zhao, J. M., Park, W. U., Hwang, K. H., Lee, J. K., & Yoon, S. Y. (2015). Biomimetic deposition of hydroxyapatite by mixed acid treatment of titanium surfaces. *Journal of Nanoscience and Nanotechnology*, *15*(3), 2552–2555. https://doi.org/10.1166/jnn.2015.10266

[15] Datta, S., & Barua, R. (2023). Advanced materials for surgical tools and biomedical implants. *Advanced Materials and Manufacturing Techniques for Biomedical Applications*, 25–35.

[16] Waidi, Y. O., Barua, R., & Datta, S. (2023). Metals, Polymers, Ceramics, Composites Biomaterials Used in Additive Manufacturing for Biomedical Applications. In *Modeling, Characterization, and Processing of Smart Materials* (pp. 165–184). IGI Global.

[17] Cho, Y., Hong, J., Ryoo, H., Kim, D., Park, J., & Han, J. (2015). Osteogenic responses to zirconia with hydroxyapatite coating by aerosol deposition. *Journal of Dental Research*, *94*(3), 491–499. https://doi.org/10.1177/0022034514566432

[18] Mao, L., Shen, L., Chen, J., Wu, Y., Kwak, M., Lu, Y., Xue, Q., Pei, J., Zhang, L., Yuan, G., Fan, R., Ge, J., & Ding, W. (2015). Enhanced bioactivity of Mg-Nd-Zn-Zr alloy achieved with nanoscale MgF_2 surface for vascular stent application. *ACS Applied Materials & Interfaces*, *7*(9), 5320–5330. https://doi.org/10.1021/am5086885

[19] Kunz, F., Rebl, H., Quade, A., Matschegewski, C., Finke, B., & Nebe, J. B. (2015). Osteoblasts with impaired spreading capacity benefit from the positive charges of plasma polymerised allylamine. *European Cells & Materials*, *29*, 177–189. https://doi.org/10.22203/ecm.v029a13

[20] Milleret, V., Ziogas, A., Buzzi, S., Heuberger, R., Zucker, A., & Ehrbar, M. (2015). Effect of oxide layer modification of CoCr stent alloys on blood activation and endothelial behavior. *Journal of Biomedical Materials Research. Part B, Applied Biomaterials*, *103*(3), 629–640. https://doi.org/10.1002/jbm.b.33232

[21] Li, Y., Yang, W., Li, X., Zhang, X., Wang, C., Meng, X., Pei, Y., Fan, X., Lan, P., Wang, C., Li, X., & Guo, Z. (2015). Improving osteointegration and osteogenesis of three-dimensional porous Ti_6Al_4V scaffolds by polydopamine-assisted biomimetic hydroxyapatite coating. *ACS Applied Materials & Interfaces*, *7*(10), 5715–5724. https://doi.org/10.1021/acsami.5b00331

[22] Omarov, S., Nauryz, N., Talamona, D., & Perveen, A. (2022). Surface modification techniques for metallic biomedical alloys: A concise review. *Metals*, *13*(1), 82. https://doi.org/10.3390/met13010082
[23] Sowa, M., Piotrowska, M., Widziołek, M., Dercz, G., Tylko, G., Gorewoda, T., Osyczka, A. M., & Simka, W. (2015). Bioactivity of coatings formed on Ti-13Nb-13Zr alloy using plasma electrolytic oxidation. *Materials Science & Engineering. C, Materials for Biological Applications*, *49*, 159–173. https://doi.org/10.1016/j.msec.2014.12.073
[24] Shi, X., Xu, L., Munar, M. L., & Ishikawa, K. (2015). Hydrothermal treatment for TiN as abrasion resistant dental implant coating and its fibroblast response. *Materials Science & Engineering. C, Materials for Biological Applications*, *49*, 1–6. https://doi.org/10.1016/j.msec.2014.12.059
[25] Tuikampee, S., Chaijareenont, P., Rungsiyakull, P., & Yavirach, A. (2024). Titanium surface modification techniques to enhance osteoblasts and bone formation for dental implants: A narrative review on current advances. *Metals*, *14*(5), 515. https://doi.org/10.3390/met14050515
[26] Kim, H., Choi, S. H., Ryu, J. J., Koh, S. Y., Park, J. H., & Lee, I. S. (2008). The biocompatibility of SLA-treated titanium implants. *Biomedical Materials (Bristol, England)*, *3*(2), 025011. https://doi.org/10.1088/1748-6041/3/2/025011
[27] Minagar, S., Wang, J., Berndt, C. C., Ivanova, E. P., & Wen, C. (2013). Cell response of anodized nanotubes on titanium and titanium alloys. *Journal of Biomedical Materials Research. Part A*, *101*(9), 2726–2739. https://doi.org/10.1002/jbm.a.34575
[28] Stio, M., Martinesi, M., Treves, C., & Borgioli, F. (2015). In vitro response of human peripheral blood mononuclear cells to AISI 316L austenitic stainless steel subjected to nitriding and collagen coating treatments. *Journal of Materials Science. Materials in Medicine*, *26*(2), 100. https://doi.org/10.1007/s10856-015-5446-y
[29] Ghaith, E.-S., Hodgson, S., & Sharp, M. (2015). Laser surface alloying of 316L stainless steel coated with a bioactive hydroxyapatite-titanium oxide composite. *Journal of Materials Science. Materials in Medicine*, *26*(2), 83. https://doi.org/10.1007/s10856-015-5399-1
[30] El-Hadad, S., Khalifa, W., & Nofal, A. (2015). Surface modification of investment cast-316L implants: Microstructure effects. *Materials Science & Engineering. C, Materials for Biological Applications*, *48*, 320–327. https://doi.org/10.1016/j.msec.2014.12.038
[31] Batal, A., Sammons, R., & Dimov, S. (2019). Response of Saos-2 osteoblast-like cells to laser surface texturing, sandblasting and hydroxyapatite coating on CoCrMo alloy surfaces. *Materials Science & Engineering. C, Materials for Biological Applications*, *98*, 1005–1013. https://doi.org/10.1016/j.msec.2019.01.067
[32] Yang, K., Zhou, C., Fan, H., Fan, Y., Jiang, Q., Song, P., Fan, H., Chen, Y., & Zhang, X. (2017). Bio-functional design, application and trends in metallic biomaterials. *International Journal of Molecular Sciences*, *19*(1), 24. https://doi.org/10.3390/ijms19010024
[33] Shangguan, Y., Sun, L., Wan, P., Tan, L., Wang, C., Fan, X., Qin, L., & Yang, K. (2016). Comparison study of different coatings on degradation performance and cell response of Mg-Sr alloy. *Materials Science & Engineering. C, Materials for Biological Applications*, *69*, 95–107. https://doi.org/10.1016/j.msec.2016.06.073
[34] Etminanfar, M. R., Khalil-Allafi, J., Montaseri, A., & Vatankhah-Barenji, R. (2016). Endothelialization and the bioactivity of Ca-P coatings of different Ca/P stoichiometry electrodeposited on the nitinol superelastic alloy. *Materials Science & Engineering. C, Materials for Biological Applications*, *62*, 28–35. https://doi.org/10.1016/j.msec.2016.01.036
[35] Sultana, N., Nishina, Y., & Nizami, M. Z. (2024). Surface modifications of medical grade stainless steel. *Coatings*, *14*(3), 248. https://doi.org/10.3390/coatings14030248
[36] Barua, R., Das, S., RoyChowdhury, A., & Datta, P. (2023). Simulation and experimental investigation of the surgical needle deflection model during the rotational and steady insertion process. *The International Journal of Artificial Organs*, *46*(1), 40–51.

[37] Datta, S., Barua, R., & Prasad, A. (2023). Additive manufacturing for the development of artificial organs. *Advanced Materials and Manufacturing Techniques for Biomedical Applications*, 411–427.
[38] Barua, R., & Datta, S. (2024). The Ongoing Advancements in Surgical Robotics: Transforming the Landscape of Surgery. In *AI-Driven Innovations in Digital Healthcare: Emerging Trends, Challenges, and Applications* (pp. 288–303). IGI Global.
[39] Kato, E., Sakurai, K., & Yamada, M. (2015). Periodontal-like gingival connective tissue attachment on titanium surface with nano-ordered spikes and pores created by alkali-heat treatment. *Dental Materials: Official Publication of the Academy of Dental Materials*, *31*(5), e116–e130. https://doi.org/10.1016/j.dental.2015.01.014
[40] Prasad, A., Bhasney, S. M., Prasannavenkadesan, V., Sankar, M. R., & Katiyar, V. (2023). Polylactic acid reinforced with nano-hydroxyapatite bioabsorbable cortical screws for bone fracture treatment. *Journal of Polymer Research*, *30*(5), 177.
[41] Prasad, A., Bhasney, S. M., Prasannavenkadesan, V., Sankar, M. R., & Katiyar, V. (2023). Nano-hydroxyapatite reinforced polylactic acid bioabsorbable cancellous screws for bone fracture fixations. *Journal of Applied Polymer Science*, *140*(43), e54577.
[42] Prasad, A., Datta, S., Kumar, A., & Gupta, M. (2023). Introduction to next-generation materials for biomedical applications. *Advanced Materials and Manufacturing Techniques for Biomedical Applications*, 1–24.
[43] Prasad, A., Kumar, A., & Gajrani, K. K. (Eds.). (2022). *Biodegradable Composites for Packaging Applications*. CRC Press.
[44] Chakraborty, G., Prasad, A., & Kumar, A. (2022). Processing of Biodegradable Composites. In *Biodegradable Composites for Packaging Applications* (pp. 33–48). CRC Press, USA.
[45] Chakraborty, G., Pandey, V., Prasad, A., & Kumar, A. (2023). Introduction to Sustainable Manufacturing for Industries 4.0. In *Sustainable Smart Manufacturing Processes in Industry 4.0* (pp. 1–17). CRC Press.
[46] Liang, Y., Xu, J., Chen, J., Qi, M., Xie, X., & Hu, M. (2015). Zinc ion implantation-deposition technique improves the osteoblast biocompatibility of titanium surfaces. *Molecular Medicine Reports*, *11*(6), 4225–4231. https://doi.org/10.3892/mmr.2015.3311
[47] Vishnu, J., Kesavan, P., Shankar, B., Dembińska, K., & Swiontek Brzezinska, M. (2023). Engineering antioxidant surfaces for titanium-based metallic biomaterials. *Journal of Functional Biomaterials*, *14*(7), 344. https://doi.org/10.3390/jfb14070344

Nisha Shankhwar*, Satyendra Singh, and Arbind Prasad

Chapter 15
Challenges and prospects of metallic biomaterials: innovations, advances, and future directions

Abstract: Metallic biomaterials have transformed modern medicine, playing a vital role in orthopedic implants, dental prosthetics, and cardiovascular devices. Their strength, durability, and biocompatibility make them essential for life-saving medical applications. However, despite their advantages, these materials come with challenges; corrosion, wear, ion release, and immune responses can all impact the long-term success of implants. This chapter dives into the world of metallic biomaterials, exploring the different types, like titanium, magnesium-based alloys, nitinol, stainless steel, and cobalt-chromium (Co-Cr) based alloys, along with how they're used in medicine. We'll also look at cutting-edge manufacturing techniques, from traditional methods to game-changing innovations like 3D printing and nanotechnology-based coatings. But it's not just about the materials; regulations and ethical considerations are just as important. With safety and patient well-being at the forefront, we'll discuss global standards and the future of personalized implants driven by AI and advanced engineering. Looking ahead, the next generation of metallic biomaterials is set to push boundaries with biodegradable implants, smart coatings, and sustainable materials. By the conclusion of this chapter, readers, practitioners, doctors, and researchers will gain a comprehensive understanding of the current state of the metallic biomaterials, the challenges it encounters, and the promising advancements shaping the future of biomedical engineering.

Keywords: Metallic biomaterials, biomedical implants, biocompatibility, corrosion resistance, additive manufacturing, biodegradable metals, smart coatings

Acknowledgments: Dr. Nisha sincerely appreciates the Department of Biotechnology (DBT), Ministry of Science and Technology of India, for their generous financial assistance provided under project no. DBT/2022-23/July/N/2240 through the DBT-RA Program. She specially thanks Dr. Mohit Kumar for his invaluable contributions and unwavering support in the successful completion of this work. The author is also deeply thankful to her daughter, Vanshika, whose constant support and understanding gave her the flexibility to dedicate herself to this project.

***Corresponding author: Nisha Shankhwar**, Special Centre for Nanoscience, Jawaharlal Nehru University, New Delhi 110067, India, e-mail: shankhwarnisha3july@gmail.com, nisha2023@jnu.ac.in
Satyendra Singh, Special Centre for Nanoscience, Jawaharlal Nehru University, New Delhi 110067, India
Arbind Prasad, Mechanical Engineering Department, Katihar Engineering College (Under Department of Science, Technology and Technical Education, Government of Bihar), Katihar 854109, Bihar, India

https://doi.org/10.1515/9783111571423-015

15.1 Introduction

15.1.1 Overview of metallic biomaterials

Metallic biomaterials have revolutionized modern medicine by providing durable, biocompatible solutions for implants and medical devices. Their exceptional mechanical strength, corrosion resistance, and ability to integrate with biological tissues make them indispensable in fields such as orthopedics, dentistry, and cardiovascular medicine [1, 2]. Unlike polymers and ceramics, metals offer superior load-bearing capacity, making them ideal for applications like hip replacements, bone plates, dental implants, and stents [1].

Metallic biomaterials play a crucial role in medical applications, offering high mechanical strength, durability, and biocompatibility. This figure highlights five essential aspects: surface modification, characterization, applications, target, and functionality. Understanding these aspects is fundamental for optimizing biomaterial performance in biomedical applications. Figure 15.1 presents an overview of the critical properties of metallic biomaterials in biomedical engineering.

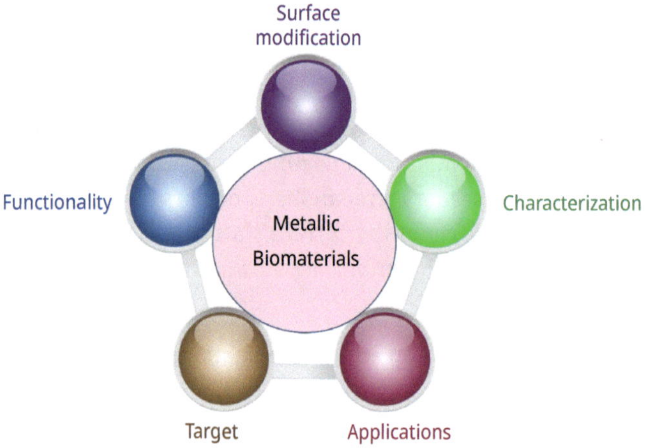

Figure 15.1: Key aspects governing the design and use of metallic biomaterials.

Commonly used metallic biomaterials include titanium (Ti) and its alloys, stainless steel, Co-Cr alloys, and magnesium-based materials [3]. Each of these materials has unique properties tailored for specific medical applications. Ti alloys are highly corrosion-resistant and promote bone integration, while biodegradable Mg alloys are designed for temporary implants that gradually dissolve in the body, reducing the need for secondary removal surgeries [4].

15.1.2 Importance in modern medicine and engineering

The continuous development of metallic biomaterials has significantly improved patient outcomes, increasing the durability and functionality of medical implants. As the global population ages and the demand for surgical implants rises, advancements in biomaterials engineering have become more crucial than ever [5]. Emerging technologies such as additive manufacturing (AM, 3D printing) [6] nanostructured coatings [7] and bioactive surface modifications [8, 9] are enabling the creation of next-generation implants with enhanced performance, including better osseointegration, wear resistance, and antibacterial properties.

Metal AM offers numerous benefits, making it a revolutionary approach in biomedical engineering. As illustrated in Figure 15.2, metal AM offers significant advantages such as improved material utilization and complex design capabilities. This figure presents key advantages, including flexibility, zero wastage, complex structures, enhanced mechanical properties, fast production, scalability, and effective design. These attributes make metal AM a preferred choice for manufacturing customized medical implants [10].

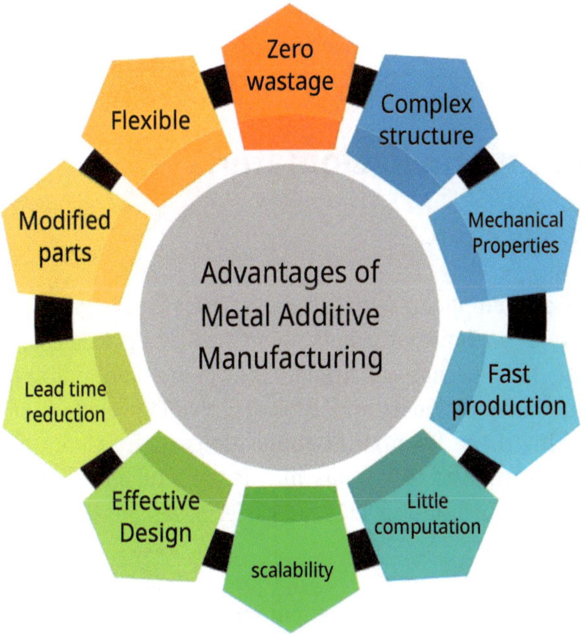

Figure 15.2: Key benefits of metal additive manufacturing, including design flexibility, material efficiency, reduced waste, and enhanced mechanical properties.

Beyond traditional medical applications, metallic biomaterials also play a crucial role in bioengineering, robotics, and regenerative medicine [11, 12]. Shape-memory alloys like nitinol are widely used in self-expanding stents and orthodontic wires due to their unique ability to return to their original shape after deformation [13, 14]. The integration of computational modeling and AI-driven biomaterials design is further pushing the boundaries of personalized medicine, enabling the development of patient-specific implants and predictive failure analysis for long-term reliability [15, 16].

This chapter follows a structured approach to provide a comprehensive understanding of metallic biomaterials, their applications, challenges, and future prospects. It begins with a discussion of the fundamental properties and classifications of metallic biomaterials, followed by an in-depth look at different metal types used in biomedical applications. The discussion then shifts to modern fabrication techniques, surface modification strategies, and the challenges associated with corrosion, wear, and biocompatibility. Regulatory and ethical considerations are explored to highlight safety and compliance aspects in medical applications. Finally, the chapter concludes with insights into emerging trends and future directions in metallic biomaterials research.

By examining both the technical aspects and real-world applications, this chapter aims to provide readers with a clear perspective on how metallic biomaterials are shaping the future of medicine and biomedical engineering.

15.2 Fundamentals of metallic biomaterials

15.2.1 Definition and classification

Metallic biomaterials are a class of engineered materials specifically designed for medical applications, particularly in implants, prosthetics, and medical devices. These materials are used primarily for their mechanical strength, durability, and ability to integrate with biological tissues. Unlike polymers and ceramics, metals provide superior load-bearing capacity, making them essential in applications such as joint replacements, bone fixation devices, cardiovascular stents, and dental implants [17].

Metallic biomaterials can be classified into several categories based on their composition and functional properties:

Titanium and titanium alloys: Known for their excellent strength-to-weight ratio, corrosion resistance, and biocompatibility, Ti-based materials are widely used in orthopedic and dental implants [18].

Stainless steel: Offers high mechanical strength and cost-effectiveness, making it common in temporary implants and surgical tools. However, its susceptibility to corrosion limits long-term applications [19].

Cobalt-chromium (Co-Cr) alloys: Characterized by high wear resistance and superior mechanical properties, these alloys are commonly used in joint replacements and cardiovascular stents [20].

Magnesium-based alloys: Biodegradable metals that gradually dissolve in the body, reducing the need for removal surgeries. These are particularly promising for temporary implants such as bone scaffolds [21].

Nickel-titanium (nitinol) alloys: Shape-memory alloys with superplastic properties, widely used in self-expanding stents and orthodontic wires [13].

The selection of metallic materials for biomedical applications depends on their mechanical properties and biocompatibility. This figure categorizes various metallic materials used in different body organ parts, such as cranium, dental, joint replacement, bone fracture, spine, and bone scaffolds. Notable materials include stainless steel (316L), titanium alloys (Ti-6Al-4V), and Co-Cr alloys (Co-Cr-Mo), which are widely used in medical implants [22]. Figure 15.3 illustrates how metallic materials are utilized in different body organ parts, emphasizing their importance in medical implants and prosthetics.

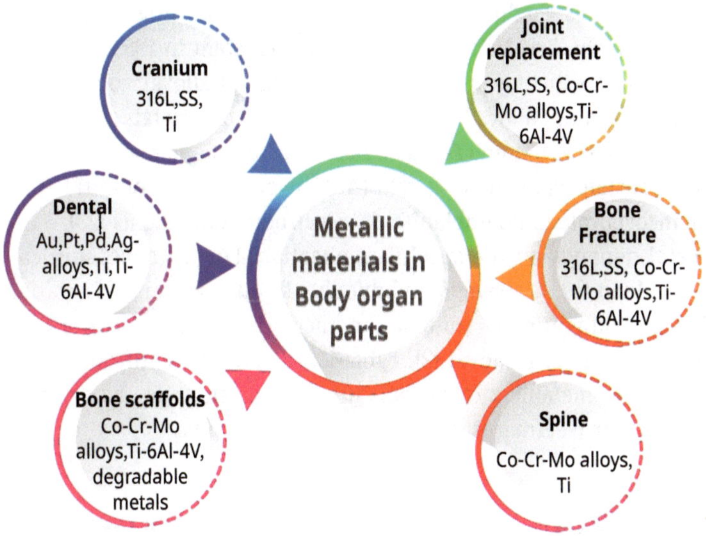

Figure 15.3: Application of metallic materials in various body organ implants, such as orthopedic joints, dental implants, and cardiovascular stents.

15.2.2 Historical evolution and milestones

The use of metals in medicine dates back thousands of years, with ancient civilizations utilizing gold and silver for dental restorations due to their corrosion resistance and ease of shaping [23, 24]. Early implantable medical devices were limited in type and application due to material constraints. It was only in the past century that these devices became a common part of healthcare. The first significant use of metal implants, beyond dental applications, began in the late nineteenth century with the introduction of the "Lane Plate," an internal fixation plate for bone fractures [24]. A major breakthrough came with the development of aseptic surgical techniques, which reduced bacterial exposure and made implantation safer. However, early metal implants, made from available alloys, were prone to corrosion. The introduction of stainless steel and vitallium (a Co-Cr-Mo alloy) in the early 1900s marked the beginning of modern metal implants [20, 25]. Throughout the mid-twentieth century, metal implant technology expanded into orthopedic, dental, and cardiac applications. Hip implants, introduced in the 1950s and 1960s, and featured metal-on-metal designs using Co-Cr alloys [20]. Around the same time, advances in miniaturized circuitry enabled the first pacemaker implantation in 1958 [26]. By the late twentieth century, metal implants had become integral across medical specialties, leading to innovations such as the first implantable neurostimulator in 1967 and the first balloon-mounted arterial stent in 1985. The most widely used metals in implants include stainless steel (iron-based alloys), cobalt-based alloys, pure titanium, and titanium alloys (Ti-6Al-4V) [27]. Refractory metals like molybdenum (Mo), tungsten (W), and tantalum (Ta) have been used as alloying elements due to their resistance to corrosion. Noble metals; gold (Au), silver (Ag), platinum (Pt), and iridium (Ir), are often incorporated into implants with electronic components [28].

The choice of metal depends on the specific application. For instance, surgical-grade stainless steel is valued for its strength and ductility but lacks strong integration with bone or soft tissue, making it more suitable for fracture fixation devices or temporary implants designed for later removal [2]. Table 15.1 illustrates significant developments and applications of metallic biomaterials in various fields such as orthopedics, dentistry, cardiovascular treatments, and personalized implants.

15.2.3 Properties and requirements for biomedical applications

For a metal to be considered suitable for biomedical applications, it must meet several critical requirements, ensuring both functionality and biocompatibility [29]. These include:

Mechanical properties: The material must have sufficient strength, toughness, and fatigue resistance to withstand the mechanical loads encountered in the human body. Orthopedic implants need high tensile strength and resistance to cyclic loading to prevent fracture over time [18].

Table 15.1: The evolution of metallic biomaterials in medical applications, highlighting key breakthroughs from the twentieth century onward.

Time period	Advancement description	Applications
1930s	Introduction of stainless steel in orthopedic implants, improving durability and affordability	Stainless steel used in early orthopedic implants such as joint replacements
1940s to 1950s	Development of Co-Cr alloys, offering superior wear resistance for joint replacements	Co-Cr used in hip and knee prosthetics
1950s to 1960s	Introduction of Ti and its alloys, becoming the gold standard in orthopedic and dental applications due to excellent biocompatibility and corrosion resistance	Ti used in dental implants, bone plates, and joint replacements
1970s to 1980s	Emergence of Ni-Ti leading to the development of self-expanding stents for cardiovascular treatments	Nitinol used in cardiovascular stents, guidewires
2000s to present	Shift towards biodegradable metallic implants (e.g., Mg-based alloys) and advanced fabrication techniques like 3D printing for patient-specific implants	Mg-based alloys for bone fixation, 3D-printed implants for custom surgical solutions

Corrosion resistance: Given the body's aqueous and ionic environment, biomaterials must resist degradation and ion release that could cause adverse biological reactions. Ti and Co-Cr alloys exhibit excellent corrosion resistance, ideal for long-term implants [3, 27].

Biocompatibility: The material should not trigger immune responses, toxicity, or inflammation. The formation of a stable oxide layer on Ti-based materials enhances their compatibility with human tissues [10, 30].

Osseointegration: Certain implants, particularly orthopedic and dental, must bond effectively with surrounding bone tissue [23, 30]. Ti's surface properties promote direct bone attachment, improving implant stability and longevity [31, 32].

Wear resistance: Joint replacements, such as hip and knee prosthetics, require high resistance to friction and wear. Co-Cr alloys are commonly used in such applications due to their exceptional wear resistance [4].

Advancements in surface engineering, such as hydroxyapatite (HA) coatings and nanostructured modifications, further enhance these properties by improving biocompatibility and implant longevity [7, 9, 33, 34].

15.2.4 Biocompatibility and corrosion resistance considerations

The long-term success of metallic implants depends heavily on their biocompatibility and resistance to degradation in the human body.

Biocompatibility: A biomaterial must not provoke harmful immune responses or toxicity. Some metals, like nickel in stainless steel or Co-Cr alloys, can cause allergic reactions or inflammatory responses in certain patients. Therefore, surface modifications, coatings, or alternative alloys are often used to minimize adverse reactions [35].

Corrosion resistance: Metallic implants are constantly exposed to bodily fluids, which can accelerate corrosion and ion release. Corrosion can lead to implant failure, inflammation, and metal ion toxicity (e.g., chromium and cobalt ions from Co-Cr implants). The use of passivation layers, such as the TiO_2 layer on Ti implants, significantly enhances corrosion resistance [36].

Strategies to improve biocompatibility and corrosion resistance: To improve biocompatibility and corrosion resistance, several strategies have been employed. Surface coatings such as HA, diamond-like carbon, and ceramic layers are commonly used to enhance biocompatibility and provide a protective barrier against corrosion. Alloying with elements like molybdenum and niobium further improves corrosion resistance while also enhancing the mechanical properties of the materials. Nanostructured surfaces have been developed to promote better cell adhesion, which helps facilitate improved integration with surrounding tissues [37]. These strategies collectively aim to enhance the performance and longevity of metallic biomaterials in medical applications.

15.3 Types of metallic biomaterials and their applications

Metallic biomaterials are widely used in the medical field due to their exceptional mechanical properties, durability, and biocompatibility. Each metal or alloy has distinct advantages and limitations that make it suitable for specific biomedical applications. This section explores the most commonly used metallic biomaterials, their properties, and how they contribute to improving patient outcomes.

15.3.1 Titanium and titanium alloys

15.3.1.1 Properties and biocompatibility

Titanium and its alloys, such as Ti-6Al-4V, are among the most widely used materials for medical implants due to their exceptional strength-to-weight ratio, corrosion resistance, and biocompatibility. One of the key advantages of titanium as a biomaterial is its high corrosion resistance, which is attributed to the natural formation of a stable oxide layer (TiO_2) that prevents ion release and degradation in the body's physiological environment [27]. Ti has a low elastic modulus compared to stainless steel and Co-Cr alloys, making it more similar to natural bone. This characteristic helps reduce stress shielding and ensures better load distribution, which is crucial for long-term implant success. Furthermore, titanium exhibits excellent osseointegration, meaning its surface promotes direct bone attachment, enhancing implant stability and longevity. These properties make titanium an ideal choice for permanent medical implants.

15.3.1.2 Applications in orthopedics and dentistry

Ti and its alloys are used extensively in:

Orthopedic implants: Hip and knee replacements, bone plates, screws, and spinal fusion devices [18, 38].

Dental implants: Ti dental screws integrate well with bone, providing long-term stability for artificial teeth [24].

Craniofacial and maxillofacial Reconstruction: Custom Ti implants are used for skull and jawbone repair following trauma or disease [39, 40].

15.3.2 Stainless steel alloys

15.3.2.1 Mechanical properties and limitations

Stainless steel, particularly 316L stainless steel, is a cost-effective material known for its high mechanical strength and resistance to mechanical wear, making it a widely used option for medical implants. It offers good strength and toughness, making it suitable for load-bearing applications, although it is not as strong as Co-Cr alloys. However, stainless steel has certain limitations, including higher susceptibility to corrosion. It is more prone to pitting and crevice corrosion in the body's physiological environment, which can lead to metal ion release and implant degradation over time [19, 25]. Another drawback is its limited biocompatibility, primarily due to its nickel content, which may trigger allergic reactions in some patients, restricting its long-term use in implants. Despite these challenges, stainless steel remains a commonly used material for temporary and cost-sensitive implant applications.

15.3.2.2 Applications in surgical implants

Due to its affordability and adequate performance, stainless steel is commonly used in:

Temporary implants: Bone plates, screws, and intramedullary rods that are later removed after healing [41].

Surgical instruments: Scalpels, forceps, retractors, and other medical tools [42].

Cardiovascular applications: Stainless steel is used in some stents and heart valve frames but is increasingly being replaced by more advanced materials like Co-Cr alloys [43].

15.3.3 Cobalt-chromium (Co-Cr) alloys

15.3.3.1 Strength and wear resistance

Co-Cr alloys are highly valued for their exceptional mechanical strength, excellent wear resistance, and superior corrosion resistance, making them an ideal choice for long-term medical implants. One of their key advantages is their exceptional hardness, which makes them significantly more wear-resistant than stainless steel and titanium, particularly in applications such as joint replacements. Co-Cr alloys exhibit superior corrosion resistance due to the formation of a protective chromium oxide layer that prevents metal ion release, enhancing their durability within the body. However, despite their general biocompatibility, some patients may develop hypersensitivity reactions to cobalt or chromium ions, which can limit their use in certain individuals. The outstanding mechanical and chemical properties of Co-Cr alloys make them a preferred material for high-performance implants [20].

15.3.3.2 Applications in joint replacements

Co-Cr alloys are widely used in:

Hip and knee prosthetics: The high wear resistance of Co-Cr alloys makes them ideal for artificial joint surfaces [44].

Dental crowns and bridges: Due to their strength and resistance to degradation in the oral environment [45].

Cardiovascular stents: Some self-expanding stents are made from Co-Cr alloys due to their durability and biocompatibility [46].

15.3.4 Magnesium (Mg)-based alloys

15.3.4.1 Biodegradability and challenges

Mg-based alloys are gaining significant attention for their unique ability to degrade naturally within the body, eliminating the need for secondary removal surgeries. A key advantage of Mg is its biodegradability, as it dissolves over time in physiological conditions, making it an excellent candidate for temporary implants. Mg is lightweight and has mechanical properties that are closer to natural bone compared to other metals, which helps in reducing stress shielding and improving implant performance. However, one of the major challenges with Mg alloys is their rapid degradation in bodily fluids, which can lead to premature implant failure and the accumulation of hydrogen gas, potentially causing local tissue damage [21]. To address this issue, strategies such as surface coatings and alloying with elements like calcium and zinc are employed to control the degradation rate and enhance implant longevity. Despite these challenges, Mg alloys hold great promise for next-generation biodegradable implants.

15.3.4.2 Applications in temporary implants

Mg alloys are primarily used in:

Bone fixation devices: Screws, plates, and pins that degrade as the bone heals, reducing the need for surgical removal [47].

Cardiovascular stents: Biodegradable Mg stents are being developed to provide temporary support to arteries before dissolving, reducing long-term complications [43].

15.3.5 Nitinol (nickel-titanium) alloys

15.3.5.1 Shape memory and superelasticity

Nitinol, a nickel-titanium (Ni-Ti) alloy, stands out among metallic biomaterials due to its unique shape memory effect and superelasticity, making it highly valuable for specialized medical implants. The shape memory effect allows nitinol to return to a predetermined shape when heated, making it particularly useful for self-expanding implants such as stents and orthopedic devices [48]. Similarly, shape memory polymers (SMPs) are gaining attention as an alternative to Nitinol in certain applications. These polymers can be programmed to change shape in response to external stimuli; temperature, light, pH, offering additional advantages like tunable degradation and lower stiffness compared to metals [34]. In addition to the shape memory effect, nitinol exhibits superelasticity, enabling it to undergo significant deformation and recover its original shape without permanent damage. This property provides flexibility in dynamic environments such as blood vessels and joints, where implants must withstand continuous mechanical stress. However, the presence of nickel raises potential biocompatibility concerns, as it may cause adverse reactions in some patients. To mitigate this, surface coatings and alloy modifications are employed to enhance its corrosion resistance and biocompatibility [49]. Despite these challenges, nitinol, along with emerging SMPs, remains a critical material for advanced medical applications requiring high flexibility and adaptability.

15.3.5.2 Applications in stents and orthodontics

Nitinol is widely used in:

Cardiovascular stents: Self-expanding stents that adjust to the shape of blood vessels, improving blood flow while minimizing surgical complexity [50].

Orthodontic wires: Used in braces due to their ability to exert continuous, gentle force, reducing the need for frequent adjustments [51].

Minimally invasive implants: Used in medical devices such as filters for preventing blood clots and flexible surgical instruments [52].

Table 15.2 provides a comparison of various metallic biomaterials used in biomedical applications, highlighting their properties, advantages, and specific uses.

15.4 Manufacturing and processing techniques

The manufacturing and processing of metallic biomaterials play a critical role in determining their mechanical properties, biocompatibility, and long-term performance within the human body. Various fabrication methods, including both traditional and advanced techniques, have been developed to tailor metallic implants to meet clinical demands. Surface modifications and coating technologies have emerged as essential strategies to enhance implant functionality and longevity.

Table 15.2: Comparison of metallic biomaterials in biomedical applications.

Material	Key properties	Advantages	Limitations	Medical applications
Titanium and titanium alloys	High strength-to-weight ratio Excellent corrosion resistance Low elastic modulus (closer to bone) Superior osseointegration	High biocompatibility Long-term implant stability Lightweight	Lower wear resistance than Co-Cr Can cause stress shielding in bone due to modulus mismatch Expensive compared to stainless steel	Orthopedic implants (hip/knee replacements, bone plates) Dental implants Craniofacial reconstruction
Stainless steel (316L)	High mechanical strength Good toughness Moderate corrosion resistance	Cost-effective Readily available Easy to process and manufacture	Corrosion susceptibility (pitting/crevice) Higher elastic modulus (causes stress shielding) Nickel content may cause allergies	Temporary bone plates and screws Surgical tools Some cardiovascular stents
Cobalt-chromium (Co-Cr) Alloys	High wear resistance Superior mechanical strength Good corrosion resistance	Excellent long-term durability Ideal for load-bearing implants High fatigue resistance	Higher modulus than bone (stress shielding) Risk of metal ion release (Co, Cr, and Mo) Expensive	Joint replacements (hip, knee, and shoulder) Dental prosthetics Cardiovascular stents
Magnesium-based alloys	Biodegradable in vivo Low density (close to natural bone) Good mechanical strength	Eliminates need for second surgery (temporary implants) Promotes bone healing	Rapid corrosion (can lead to implant failure) Hydrogen gas release can cause local tissue damage Limited mechanical strength for high-load applications	Biodegradable orthopedic screws and plates Temporary cardiovascular stents
Nickel-titanium (nitinol)	Shape memory effect Superelasticity Moderate corrosion resistance	Self-expanding properties (useful for minimally invasive procedures) High flexibility	Nickel content may cause allergic reactions Requires precise manufacturing control Lower wear resistance than Co-Cr	Cardiovascular stents Orthodontic wires Surgical clips and filters

15.4.1 Traditional fabrication methods

Conventional manufacturing techniques, such as casting, forging, and machining, have long been used in the production of metallic biomaterials. Casting involves melting metal alloys and pouring them into molds to achieve the desired shape, followed by controlled cooling to refine the microstructure. Forging enhances the mechanical properties of metals through controlled deformation and recrystallization, making it particularly useful for load-bearing implants like hip and knee prostheses. Machining, including milling, drilling, and turning, provides precise shaping and surface finishing for metallic implants, ensuring dimensional accuracy and compatibility with the surrounding biological environment [53]. Despite their effectiveness, these methods often result in material waste and limited design flexibility, prompting the need for advanced fabrication technologies.

15.4.2 Additive manufacturing (3D printing)

AM, commonly known as 3D printing, has revolutionized the production of metallic biomaterials by enabling the fabrication of complex geometries with high precision and minimal material wastage. Techniques such as selective laser melting (SLM), electron beam melting (EBM), and direct metal laser sintering (DMLS) allow for the layer-by-layer construction of implants with intricate lattice structures that closely mimic natural bone architecture, promoting osseointegration and reducing stress shielding effects. One of the key advantages of AM is its ability to customize implants based on patient-specific anatomical data, improving fit and functionality while minimizing surgical complications. Furthermore, AM facilitates the incorporation of controlled porosity, which enhances bone ingrowth and vascularization, leading to better long-term integration with host tissue [54]. However, challenges such as residual stress accumulation, surface roughness, and potential defects in printed structures necessitate further refinement of process parameters and post-processing treatments.

For AM, metallic biomaterials must possess specific properties to ensure functionality and adaptability. This figure illustrates essential properties such as chemical stability, high tunability, flexible design, large specific surface area, unusual shape, and massive porosity. These characteristics contribute to the successful application of metallic biomaterials in biomedical AM. Figure 15.4 presents an overview of the essential characteristics of metallic biomaterials that enhance their suitability for AM processes.

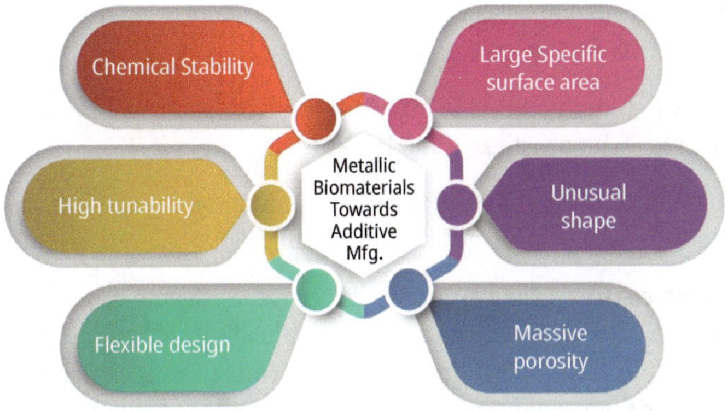

Figure 15.4: Properties of metallic biomaterials for additive manufacturing.

15.4.3 Surface modifications for enhanced performance

To improve the biological performance of metallic implants, surface modification techniques have been extensively explored. These methods aim to optimize surface roughness, wettability, and chemical composition to enhance cell adhesion, reduce bacterial colonization, and minimize immune responses [37]. Common surface modification strategies include:

Mechanical treatments: Processes like grit blasting and shot peening create microroughness, which enhances osseointegration by increasing surface area for bone attachment [1].

Chemical treatments: Techniques such as acid etching and anodization modify surface chemistry to improve corrosion resistance and bioactivity [1].

Physical treatments: Plasma spraying and ion implantation alter surface properties at the micro and nanoscale, promoting improved wear resistance and biocompatibility [50].

Biological functionalization: Surface coatings incorporating bioactive molecules, growth factors, or antibacterial agents further enhance the performance of metallic implants by stimulating tissue regeneration and reducing infection risks [29].

15.4.4 Coating technologies for improved biocompatibility

Coatings play a vital role in enhancing the biocompatibility and functionality of metallic biomaterials by providing a protective interface between the implant and surrounding tissues. Various coating technologies have been developed to improve wear resistance, prevent corrosion, and promote biological interactions [35]. Key coating techniques include:

HA coatings: Mimic the mineral composition of natural bone, enhancing osteoconductivity and facilitating faster bone integration [8, 9, 30].

Diamond-like carbon (DLC) coatings: Provide excellent wear resistance and reduce friction, making them ideal for joint prostheses and load-bearing implants [55].

Plasma-sprayed coatings: Used to apply bioactive or antimicrobial layers that improve implant longevity and reduce post-surgical complications [56].

Polymeric and drug-eluting coatings: Deliver localized therapeutic agents, antibiotics or anti-inflammatory drugs, to prevent infections and improve healing outcomes [57].

By integrating advanced fabrication techniques with innovative surface modifications and coatings, researchers continue to push the boundaries of metallic biomaterial performance. Future developments in nanotechnology, bioactive coatings, and smart materials will further enhance the safety, durability, and efficacy of metallic implants, ultimately improving patient outcomes.

15.5 Challenges in metallic biomaterials

15.5.1 Corrosion and degradation issues

One of the most significant limitations of metallic biomaterials is their susceptibility to corrosion in the physiological environment. Corrosion can degrade implant integrity, leading to mechanical failure and the release of harmful metal ions. Although titanium and its alloys exhibit excellent corrosion resistance due to their stable oxide layer (TiO_2), materials such as stainless steel and cobalt-chromium alloys are prone to localized corrosion (e.g., pitting and crevice corrosion) in chloride-rich body fluids. Strategies such as surface coatings, alloy modifications, and bioactive ceramic layers have been developed to improve corrosion resistance and prolong implant lifespan [2].

15.5.2 Wear and fatigue failure

Metallic implants, especially those in load-bearing applications like hip and knee replacements, are subjected to continuous mechanical stress, leading to wear and fatigue failure. Wear debris generated from metal-on-metal articulations can induce an inflammatory response, resulting in osteolysis and implant loosening. fatigue Failure occurs due to the propagation of microcracks under repeated loading cycles, ultimately leading to implant fracture. Research on novel alloy compositions, surface hardening techniques, and advanced manufacturing methods (e.g., AM) aims to enhance the wear and fatigue resistance of metallic biomaterials [58].

15.5.3 Ion release and toxicity concerns

Corrosion and wear can contribute to the release of metal ions, which may cause adverse biological reactions. Elements like Ni, Co, and Cr; commonly found in some metallic biomaterials – are associated with cytotoxicity, genotoxicity, and hypersensitivity reactions. Nickel, in particular, is a known allergen that can trigger immune responses in sensitive individuals. To mitigate these effects, researchers are exploring nickel-free stainless-steel alloys, biocompatible coatings, and surface modification techniques to minimize ion release while maintaining mechanical strength [59].

15.5.4 Immune response and inflammatory reactions

The body's immune system may recognize metallic implants as foreign objects, leading to inflammation and adverse tissue responses. Wear particles and metal ions can further stimulate macrophage activation, resulting in chronic inflammation, fibrosis, and eventual implant failure. In severe cases, excessive inflammatory responses contribute to aseptic loosening and periprosthetic osteolysis. Solutions such as bioactive coatings, nanotopographical modifications, and anti-inflammatory drug-eluting implants are being developed to reduce immune reactions and enhance implant biocompatibility [60].

15.5.5 Cost and accessibility of advanced alloys

Advanced metallic biomaterials, including titanium and cobalt-chromium alloys, offer superior mechanical properties and biocompatibility. However, their high production costs and complex manufacturing processes limit accessibility, particularly in low-resource healthcare settings. The cost of raw materials, precision machining, and regulatory compliance further adds to the financial burden. To address this issue, research is focused on developing cost-effective alternatives, such as magnesium-based biodegradable metals, hybrid composite materials, and improved fabrication techniques that balance performance with affordability [61].

Developing next-generation metallic biomaterials requires a multidisciplinary approach involving material scientists, engineers, and clinicians to address these challenges and enhance the safety, durability, and accessibility of medical implants. Table 15.3 summarizing the challenges in metallic biomaterials, their causes, effects, and potential solutions.

Table 15.3: Key challenges in metallic biomaterials and their mitigation strategies.

Challenges	Causes	Impact on implants and patients	Potential solutions
Corrosion and degradation	Exposure to body fluids, pitting, crevice corrosion, and galvanic corrosion	Structural weakening, ion release, and implant failure	Passivation layers (TiO_2), ceramic coatings, corrosion-resistant alloys (Ti-6Al-4V and CoCr), and surface treatments (anodization and plasma spraying)
Wear and fatigue failure	Friction in load-bearing joints, cyclic loading, and microcrack formation	Wear debris-induced osteolysis, implant loosening, and mechanical failure	Surface hardening (DLC coatings and nitriding), advanced alloys (TiNb and CoCrMo), and optimized implant designs
Ion release and toxicity	Corrosion, wear, and passive dissolution of metallic ions	Local tissue irritation, systemic toxicity (Co, Ni, and Cr), and allergic reactions	Ion barrier coatings (hydroxyapatite and zirconia), nickel-free stainless steel, and biocompatible alloys (TiNb and Ta-based)
Immune response and inflammation	Wear debris, metal ions, and hypersensitivity reactions	Chronic inflammation, macrophage activation, and implant rejection	Bioactive coatings (drug-eluting layers and growth factors), immunomodulatory surfaces, and controlled porosity for better osseointegration
Cost and accessibility of advanced alloys	High material costs, complex manufacturing, and limited global supply	Expensive implants and reduced accessibility in developing regions	Additive manufacturing (3D printing), alternative biomaterials (Mg-based biodegradable implants), and material recycling techniques

15.6 Innovations and advances in metallic biomaterials

The continuous evolution of metallic biomaterials has led to significant advancements aimed at improving implant performance, biocompatibility, and longevity. Researchers are now integrating smart coatings, hybrid materials, nanotechnology, and AI-driven design approaches to develop next-generation implants that actively interact with biological environments, minimize complications, and enhance patient outcomes. These innovations are shaping the future of metallic biomaterials, moving beyond passive structural support toward dynamic, functional implants.

15.6.1 Smart and bioactive coatings

Traditional metallic implants rely on their bulk material properties for performance, but modern advancements have introduced smart and bioactive coatings that actively influence the biological response of surrounding tissues. These coatings serve multiple functions, including promoting osseointegration, preventing bacterial infections, and delivering therapeutic agents directly at the implant site [57].

15.6.1.1 Types of smart and bioactive coatings

Drug-eluting coatings: These are designed to release antibiotics, anti-inflammatory agents, or growth factors over time, reducing infection risks and enhancing healing. Titanium implants coated with vancomycin-releasing hydrogels to prevent post-surgical infections [18].

Stimuli-responsive coatings: These are materials that respond to environmental changes such as pH, temperature, or mechanical stress. pH-sensitive coatings that release anti-inflammatory drugs in response to localized inflammation around the implant [34, 48].

Bioactive ceramic coatings: HA, bioglasses/ceramics, calcium phosphate, and zirconia-based coatings mimic the mineral structure of bone, enhancing bone-implant integration [9, 62].

Antibacterial and anti-biofilm coatings: Silver nanoparticles, chitosan, and graphene-based coatings actively reduce bacterial adhesion and biofilm formation, addressing implant-associated infections [59].

15.6.2 Hybrid and composite metallic materials

Traditional metallic biomaterials such as Ti, CoCr, and stainless steel are widely used in orthopedic and dental implants. However, their mechanical properties often differ from natural bone, leading to issues like stress shielding (bone resorption due to mismatched stiffness). Hybrid and composite metallic materials aim to bridge this gap by combining different materials to achieve enhanced mechanical properties, biocompatibility, and corrosion resistance [18].

15.6.2.1 Hybrid and composite metallic biomaterials

Titanium-tantalum (Ti-Ta) alloys: They offer excellent corrosion resistance and improved elasticity closer to natural bone, reducing stress shielding [63].

Metal-ceramic composites: Combining metals with bioactive ceramics (e.g., titanium-HA composites) enhances osseointegration while maintaining strength.

Porous metal scaffolds: 3D-printed metal scaffolds with engineered porosity mimic bone structure, facilitating cell infiltration and vascularization [18].

Graphene-reinforced metal composites: These improve mechanical strength, wear resistance, and antibacterial properties when incorporated into titanium and Mg implants [55].

15.6.3 Nanotechnology and surface engineering

At the nanoscale, material properties differ significantly, allowing for precise control over surface characteristics that influence cell adhesion, protein interactions, and immune responses. Nanotechnology-driven surface engineering is a major breakthrough in metallic biomaterials, enhancing biocompatibility, osseointegration, and antibacterial properties [51].

15.6.3.1 Nanotechnology innovations in biomaterials

Nanostructured surfaces: Modifying implant surfaces at the nanolevel (e.g., nanoporous titanium and nanoroughened coatings) improves cell attachment and bone integration.

Nanoparticle-embedded coatings: Incorporating silver, zinc oxide, or copper nanoparticles imparts antimicrobial properties, preventing implant-associated infections.

Nanotextured titanium implants: They mimic natural bone microarchitecture, promoting osteoblast (bone-forming cell) activity while reducing bacterial adhesion.

Self-assembled nanolayers: They are ultra-thin coatings that control protein adsorption, reducing immune rejection and enhancing implant performance.

By leveraging nanotechnology, metallic implants can be engineered at the molecular level, improving biomechanical and biochemical interactions with the human body [64]. These advancements promise faster healing, reduced infection risks, and extended implant longevity.

15.6.4 AI and computational approaches in biomaterials design

Artificial intelligence (AI) and computational modeling are transforming biomaterials research, enabling the development of optimized implant designs, material selection, and predictive simulations before clinical use [65]. These technologies reduce experimental costs, accelerate material discovery, and improve implant performance through data-driven insights.

15.6.4.1 Applications of AI in biomaterials design

Machine learning for material discovery: AI algorithms analyze vast datasets to identify new alloy compositions with superior mechanical and biological properties.

Computational fluid dynamics (CFD): This simulates **corrosion, ion release, and fluid-implant interactions** to optimize material durability and biocompatibility.

Finite element analysis (FEA): This predicts **mechanical performance under physiological conditions**, ensuring implants withstand cyclic loading and stress distribution.

AI-driven personalized implant design: It uses patient imaging data (MRI and CT scans) to create **customized implants** that fit perfectly, reducing surgical complications and improving patient outcomes.

AI-driven approaches enhance precision, efficiency, and predictive capabilities, revolutionizing how metallic biomaterials are designed and tested, ultimately leading to safer and more effective medical implants [66].

15.7 Regulatory and ethical considerations

The development and use of metallic biomaterials in medical applications require rigorous regulatory oversight and ethical considerations to ensure patient safety, efficacy, and long-term reliability. Regulatory agencies set stringent guidelines for material selection, manufacturing processes, biocompatibility testing, and post-market surveillance to minimize risks associated with metallic implants. Additionally, ethical concerns regarding patient consent, long-term safety, and equitable access to advanced biomaterials must be addressed to promote responsible medical innovation.

15.7.1 Global standards and approvals (FDA, ISO, CE)

The regulatory approval process for metallic biomaterials varies across countries but follows a common goal: ensuring that **implants meet high standards of safety, quality, and performance** before they are used in clinical settings.

Key Regulatory bodies and their guidelines:

U.S. Food and drug administration (FDA): Categorizes metallic implants under Class II (moderate risk) or Class III (high risk) medical devices, depending on their function and invasiveness. Requires preclinical testing (ISO 10993 biocompatibility tests), mechanical assessments, and clinical trials before market approval [67]. Post-market surveillance under the Medical Device Reporting (MDR) system ensures ongoing safety monitoring.

International organization for standardization (ISO): Establishes universal material and testing standards, such as ISO 5832 for metallic biomaterials and ISO 10993 for biocompatibility assessment [68]. ISO standards help manufacturers align with international best practices, facilitating global market entry.

European CE marking (Conformité Européenne): Requires compliance with the Medical Device Regulation (MDR 2017/745), focusing on risk assessment, clinical evaluation, and traceability. Ensures that metallic biomaterials meet safety and performance benchmarks before being sold within the European market [67, 68].

In addition to these, countries such as **Japan (PMDA), China (NMPA), and India (CDSCO)** have their own regulatory frameworks, often aligning with ISO and FDA guidelines to ensure global safety standards for biomaterials.

One major challenge in the approval process is the long timeline and high cost associated with regulatory compliance [67, 68]. The development of new metallic biomaterials often requires years of clinical trials, in-depth toxicity studies, and continuous monitoring before full approval, delaying market entry but ensuring safety.

15.7.2 Ethical concerns in biomedical implantation

Beyond regulatory requirements, ethical considerations play a crucial role in the development and use of metallic biomaterials. Ethical concerns arise in areas such as informed consent, implant longevity, and the equitable distribution of advanced implantable technologies [69].

Key ethical challenges in metallic biomaterials:

Informed consent and patient autonomy: Patients must be fully aware of the risks, benefits, and alternative treatments before receiving a metallic implant. Issues arise when patients are not provided with long-term performance data, leading to dissatisfaction or unexpected complications post-surgery.

Implant longevity and reoperation risks: Many metallic implants, such as hip and knee replacements, have a lifespan of 10–20 years, after which revision surgeries may be needed.

Ethical concerns arise when patients, especially younger individuals, are not fully informed about the need for future replacements, potentially leading to multiple surgeries over a lifetime.

Equitable access to advanced biomaterials: While next-generation metallic implants (e.g., customized 3D-printed implants, bioactive coatings) offer superior outcomes, they are often expensive, limiting accessibility for lower-income patients and healthcare systems in developing countries. The ethical dilemma of healthcare inequality emerges when life-improving technologies remain available only to privileged populations. Efforts to develop cost-effective alternatives and government policies promoting affordability are needed to bridge this gap.

Animal and human testing in biomaterial research: Preclinical testing of metallic biomaterials often involves animal studies to evaluate toxicity, osseointegration, and immune responses before human trials. Ethical frameworks such as the **3Rs** (Replacement, Reduction, Refinement) principle aim to minimize animal use by promoting advanced computational modeling, organ-on-a-chip technologies, and *in vitro* testing methods.

By addressing these ethical concerns, clinicians, researchers, and policymakers can ensure that metallic biomaterials are developed and implemented responsibly, balancing innovation with patient welfare and global healthcare equity.

15.7.3 Long-term safety and post-market surveillance

Even after regulatory approval and clinical introduction, metallic biomaterials must undergo continuous monitoring to detect long-term risks and failures. Post-market surveillance (PMS) ensures that implants perform safely under real-world conditions, helping to identify potential issues that were not apparent during initial clinical trials [70].

Key aspects of post-market surveillance:

Implant registries and failure tracking: Many countries maintain national joint registries (e.g., UK National Joint Registry, Australian Orthopedic Association NJR) to track implant performance and detect trends in failure rates, wear-related complications, and adverse reactions. Data from these registries help refine future implant designs and improve patient safety.

Metal ion monitoring and biocompatibility assessment: Patients with CoCr or stainless-steel implants are monitored for elevated metal ion levels (e.g., cobalt, chromium, and nickel), which may cause systemic toxicity or hypersensitivity reactions. Regular screening protocols, such as blood tests and imaging studies, help detect early signs of metal-related complications.

15.8 Future prospects and emerging trends

Different biomaterials exhibit varying advantages and limitations. As shown in Figure 15.5, different biomaterial classes exhibit distinct advantages and limitations, influencing their suitability for specific medical applications. This figure compares metals and alloys, polymeric materials, ceramic materials, and composite materials in terms of their benefits and disadvantages. While metals offer high mechanical strength and biocompatibility, they are susceptible to corrosion. Polymers are lightweight but degrade over time, ceramics are fragile, and composites can be challenging to obtain. Understanding these factors aids in selecting the appropriate biomaterial for medical applications [71].

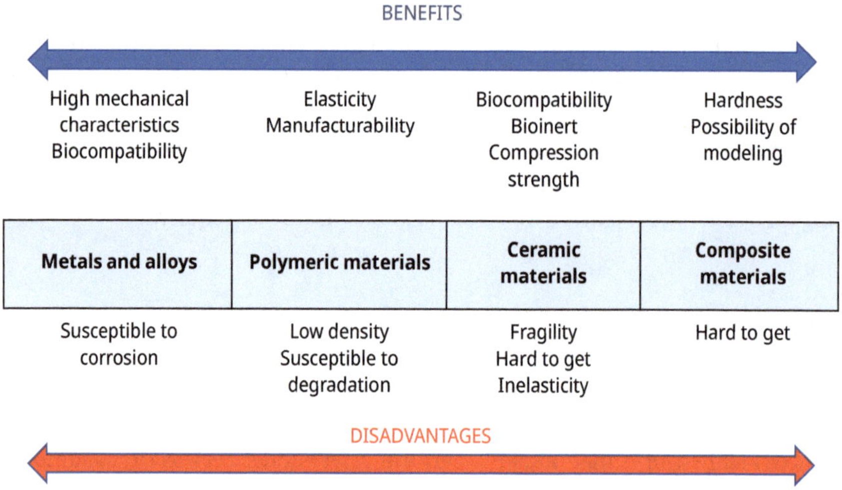

Figure 15.5: A comparative analysis of the advantages, including strength, biocompatibility, and degradability, and disadvantages, brittleness, wear resistance, and potential toxicity, in biomedical applications.

The field of metallic biomaterials is rapidly evolving, driven by advances in materials science, AI, AM, and sustainability. Future innovations are expected to focus on biodegradable metals, personalized implants, eco-friendly biomaterials, and cross-disciplinary research collaborations, all aimed at improving patient outcomes, reducing complications, and enhancing the longevity of medical implants. These emerging trends will shape the next generation of metallic biomaterials, making them more adaptable, intelligent, and environmentally sustainable.

Next-generation biodegradable metallic implants are revolutionizing temporary medical applications such as bone fixation plates, stents, and pediatric implants by naturally degrading within the body. This eliminates the need for secondary removal surgeries and minimizes long-term complications. Mg alloys, known for their lightweight and osteoconductive properties, are ideal for orthopedic and cardiovascular applications but require surface modifications to control rapid corrosion. Zn alloys offer moderate degradation rates, making them suitable for bioresorbable vascular stents, though their lower mechanical strength necessitates reinforcement. Iron (Fe) alloys provide high strength for load-bearing implants but degrade too slowly, prompting research into porous structures and alloying with biodegradable elements. The shift toward biodegradable implants marks a paradigm change in biomaterials, offering temporary support while reducing risks associated with permanent implants.

The personalization of implants through AI and 3D printing is transforming implant design, ensuring better fit, functionality, and biocompatibility. AI-driven models analyze patient imaging data to optimize implant geometry, predict longevity, and aid in surgical planning. Meanwhile, 3D printing technologies like SLM and EBM enable the creation of patient-specific titanium implants with controlled porosity, enhancing osseointegration and vascularization. Smart implants embedded with AI-driven sensors can monitor implant health, detect early signs of failure, and provide real-time data to physicians. These advancements promise more precise and adaptive implant solutions tailored to individual patient needs.

Sustainability in biomaterials is gaining traction as the industry seeks eco-friendly alternatives to traditional metallic implants. Recycling medical-grade metals like titanium and cobalt alloys reduces reliance on virgin metal extraction, although maintaining biocompatibility standards remains a challenge. Green manufacturing techniques, including powder metallurgy and AM, minimize material waste, while low-energy sintering processes reduce the environmental footprint. Magnesium-based biodegradable implants further contribute to sustainability by eliminating long-term disposal concerns. Additionally, efforts are underway to replace scarce and toxic elements, such as vanadium in Ti-6Al-4V, with safer alternatives like niobium and tantalum. Aligning implant technology with environmental responsibility ensures a more ethical and sustainable future for healthcare.

Cross-disciplinary research is driving the next era of metallic biomaterials by integrating nanotechnology, AI, biotechnology, and regenerative medicine. Biohybrid metallic implants, combining biomaterials with living cells, enhance biofunctionality and tissue regeneration. Smart coatings infused with drug-delivery capabilities allow implants to release therapeutic agents in response to infection or mechanical stress. AI simulations are optimizing implant performance, predicting wear patterns, and enhancing surgical outcomes. In neuroprosthetics, bioresorbable Mg-based brain implants are enabling temporary neural stimulation without requiring surgical removal. These innovations are shaping the future of biomaterials into smart, interactive, and regenerative medical solutions, paving the way for the next generation of personalized medicine.

15.9 Conclusion

This chapter examined the dynamic field of metallic biomaterials, highlighting their crucial role in advancing medical science through enhanced implants, tissue engineering, and regenerative medicine. These materials are integral to reshaping healthcare, offering strength, biocompatibility, and degradability. However, their full potential is constrained by persistent challenges such as corrosion resistance, material degradation, and the need for improved biological interactions. The interplay between surface modifications and manufacturing techniques remains critical, yet standardized solutions for consistent clinical outcomes are still lacking.

Despite these hurdles, the impact of metallic biomaterials on healthcare is profound, promising personalized solutions that cater to individual patient needs. Moving forward, interdisciplinary collaboration between materials science, engineering, and healthcare will be essential to overcoming existing barriers. Future research must focus on developing advanced alloys, exploring hybrid materials, and refining testing and manufacturing processes to ensure long-term performance and reliability. With these efforts, metallic biomaterials will continue to evolve, driving the next generation of medical technologies and revolutionizing patient care for years to come.

References

[1] Thanigaivel, S., Priya, A. K., Balakrishnan, D., Dutta, K., Rajendran, S., & Soto-Moscoso, M. (2022, Nov). Insight on recent development in metallic biomaterials: Strategies involving synthesis, types and surface modification for advanced therapeutic and biomedical applications. *Biochemical Engineering Journal*, *187*, 108522. doi: 10.1016/j.bej.2022.108522.

[2] Prasad, K., et al. (2017, Jul). Metallic Biomaterials: Current Challenges and Opportunities. *Materials*, *10*(8), 884. doi: 10.3390/ma10080884.

[3] Goharian, A., & Abdullah, M. R. (2017). Bioinert Metals (Stainless Steel, Titanium, Cobalt Chromium). In *Trauma Plating Systems* (pp. 115–142). Elsevier. doi: 10.1016/B978-0-12-804634-0.00007-0

[4] Hussein, M., Mohammed, A., & Al-Aqeeli, N. (2015, May). Wear Characteristics of Metallic Biomaterials: A Review. *Materials*, *8*(5), 2749–2768. doi: 10.3390/ma8052749.

[5] Gautam, S., Bhatnagar, D., Bansal, D., Batra, H., & Goyal, N. (2022, Jun). Recent advancements in nanomaterials for biomedical implants. *Biomedical Engineering Advances*, *3*, 100029. doi: 10.1016/j.bea.2022.100029.

[6] Shankhwar, N., Saha, P., Zurek-Mortka, M., Thomas, S., & Unal, A. (2022). Tissue Engineered Products – Translational Avenues. In *Tissue Engineering* (pp. 575–625). Elsevier. doi: 10.1016/B978-0-12-824064-9.00007-1

[7] Shankhwar, N., Kothiyal, G. P., & Srinivasan, A. (2013). Sol-gel derived porous bioactive nanocomposites: Synthesis and in vitro bioactivity. *AIP Conference Proceedings, 1536*, 209–210. doi: 10.1063/1.4810174.

[8] Shankhwar, N., & Srinivasan, A. (2016). Evaluation of sol-gel based magnetic 45S5 bioglass and bioglass-ceramics containing iron oxide. *Materials Science and Engineering C, 62*, 190–196. doi: 10.1016/j.msec.2016.01.054.

[9] Shankhwar, N., Kothiyal, G. P., & Srinivasan, A. (2015). Influence of phosphate precursors on the structure, crystallization behaviour and bioactivity of sol-gel derived 45S5 bioglass. *RSC Advances, 5*(122), 100762–100768. doi: 10.1039/c5ra19184j.

[10] Davoodi, E., *et al.* (2022, Sep). Additively manufactured metallic biomaterials. *Bioactive Materials, 15*, 214–249. doi: 10.1016/j.bioactmat.2021.12.027.

[11] Shankhwar, N., Kumar, M., Mandal, B. B., Robi, P. S., & Srinivasan, A. (2016). Electrospun polyvinyl alcohol-polyvinyl pyrrolidone nanofibrous membranes for interactive wound dressing application. *Journal of Biomaterials Science. Polymer Edition, 27*(3), 247–262. doi: 10.1080/09205063.2015.1120474.

[12] Shankhwar, N., Kumar, M., Mandal, B. B., & Srinivasan, A. (2016). Novel polyvinyl alcohol-bioglass 45S5 based composite nanofibrous membranes as bone scaffolds. *Materials Science and Engineering C, 69*, 1167–1174. doi: 10.1016/j.msec.2016.08.018.

[13] Sathishkumar, M., *et al.* (2023, Sep). Possibilities, performance and challenges of nitinol alloy fabricated by directed energy deposition and powder bed fusion for biomedical implants. *Journal of Manufacturing Processes, 102*, 885–909. doi: 10.1016/j.jmapro.2023.08.024.

[14] O'Brien, B., Weafer, F. M., & Bruzzi, M. S. (2017). 1.3 Shape Memory Alloys for Use in Medicine ☆. In *Comprehensive Biomaterials II* (pp. 50–78). Elsevier. doi: 10.1016/B978-0-12-803581-8.10084-0

[15] Suh, J. S., Suh, B.-C., Bae, J. H., & Kim, Y. M. (2023, Jan). Machine learning-based design of biodegradable Mg alloys for load-bearing implants. *Master of Design, 225*, 111442. doi: 10.1016/j.matdes.2022.111442.

[16] Moscol-Albañil, I., Solórzano-Requejo, W., Rodriguez, C., Ojeda, C., & Díaz Lantada, A. (2024, Apr). Innovative AI-driven design of patient-specific short femoral stems in primary hip arthroplasty. *Master of Design, 240*, 112868. doi: 10.1016/j.matdes.2024.112868.

[17] *Fundamental Biomaterials: Metals*. (2018). Elsevier. doi: 10.1016/C2016-0-03502-7.

[18] Abd-Elaziem, W., Darwish, M. A., Hamada, A., & Daoush, W. M. (2024, May). Titanium-Based alloys and composites for orthopedic implants Applications: A comprehensive review. *Master of Design, 241*, 112850. doi: 10.1016/j.matdes.2024.112850.

[19] Patnaik, L., Ranjan Maity, S., & Kumar, S. (2020). Status of nickel free stainless steel in biomedical field: A review of last 10 years and what else can be done. *Materials Today: Proceedings, 26*, 638–643. doi: 10.1016/j.matpr.2019.12.205.

[20] Narushima, T., Ueda, K., & Alfirano,. (2015). *Co-Cr Alloys as Effective Metallic Biomaterials* (pp. 157–178). doi: 10.1007/978-3-662-46836-4_7.

[21] Sezer, N., Evis, Z., Kayhan, S. M., Tahmasebifar, A., & Koç, M. (2018, Mar). Review of magnesium-based biomaterials and their applications. *Journal of Magnesium and Alloys, 6*(1), 23–43. doi: 10.1016/j.jma.2018.02.003.

[22] Davis, R., *et al.* (2022, May). A comprehensive review on metallic implant biomaterials and their subtractive manufacturing. *The International Journal of Advanced Manufacturing Technology, 120*(3–4), 1473–1530. doi: 10.1007/s00170-022-08770-8.

[23] Walter, N., Stich, T., Docheva, D., Alt, V., & Rupp, M. (2022, Nov). Evolution of implants and advancements for osseointegration: A narrative review. *Injury, 53*, S69–S73. doi: 10.1016/j.injury.2022.05.057.

[24] Kandavalli, S. R., *et al.* (2021, May). A brief review on the evolution of metallic dental implants: history, design, and application. *Front Mater, 8*. doi: 10.3389/fmats.2021.646383.

[25] Linder, L., & Lundskog, J. (1975, May). Incorporation of stainless steel, titanium and Vitallium in bone. *Injury, 6*(4), 277–285. doi: 10.1016/0020-1383(75)90174-6.

[26] Aquilina, O. (2006, Apr). A brief history of cardiac pacing. *Images in Paediatric Cardiology, 8*(2), 17–81. [Online]. Available: http://www.ncbi.nlm.nih.gov/pubmed/22368662.

[27] Semenova, I., Polyakov, A., Gareev, A., Makarov, V., Kazakov, I., & Pesin, M. (2023, Oct). Machinability features of Ti-6Al-4V alloy with ultrafine-grained structure. *Metals (Basel), 13*(10), 1721. doi: 10.3390/met13101721.

[28] Tu, H.-L., Zhao, H.-B., Fan, Y.-Y., & Zhang, Q.-Z. (2022, May). Recent developments in nonferrous metals and related materials for biomedical applications in China: A review. *Rare Metals, 41*(5), 1410–1433. doi: 10.1007/s12598-021-01905-y.

[29] Yang, K., *et al.* (2017, Dec). Bio-functional design, application and trends in metallic biomaterials. *International Journal of Molecular Sciences, 19*(1), 24. doi: 10.3390/ijms19010024.

[30] Shankhwar, N., Sharma, K., Kothiyal, G. P., & Srinivasan, A. (2016). *Bioactive Glass and Glass-ceramics Containing Iron Oxide: Preparation and Properties* vol. 9. doi: 10.4032/9789814613996.

[31] Shankhwar, N., Kothiyal, G. P., & Srinivasan, A. (2015). Influence of phosphate precursors on the structure, crystallization behaviour and bioactivity of sol–gel derived 45S5 bioglass. *RSC Advances, 5*(122), 100762–100768. doi: 10.1039/C5RA19184J.

[32] Shankhwar, N., Kothiyal, G. P., & Srinivasan, A. (2014). Understanding the magnetic behavior of heat treated CaO-P2O 5-Na2O-Fe2O3-SiO2 bioactive glass using electron paramagnetic resonance studies. *Physica B: Condensed Matter, 448*, 132–135. doi: 10.1016/j.physb.2014.03.070.

[33] Shankhwar, N., Singh, R. K., & Srinivasan, A. (2017). Evolution of magnetic and bone mineral phases in heat-treated bioactive glass containing zinc and iron oxides. *International Journal of Applied Glass Science, 8*(1), 105–115. doi: 10.1111/ijag.12193.

[34] Shankhwar, N., Unal, A., & Singh, S. 2025). Nanomedicine in Patient-specific Clinical Trend: Shape Memory Polymers for Emerging Biomedical Applications and Their Future Prospects. In *Nanomedicine in Translational Research* (pp. 309–338). Elsevier. doi: 10.1016/B978-0-443-22257-3.00014-3

[35] Bandyopadhyay, A., Mitra, I., Goodman, S. B., Kumar, M., & Bose, S. (2023, Mar). Improving biocompatibility for next generation of metallic implants. *Progress in Materials Science, 133*, 101053. doi: 10.1016/j.pmatsci.2022.101053.

[36] Eliaz, N. (2019, Jan). Corrosion of metallic biomaterials: A review. *Materials (Basel) 12*(3). doi: 10.3390/ma12030407.

[37] Bin A. Fadzil, A. F., Pramanik, A., Basak, A. K., Prakash, C., & Shankar, S. (2022, Oct). Role of surface quality on biocompatibility of implants – A review. *Annals of 3D Printed Medicine, 8*, 100082. doi: 10.1016/j.stlm.2022.100082.

[38] Kaur, M., & Singh, K. (2019, Sep). Review on titanium and titanium based alloys as biomaterials for orthopaedic applications. *Materials Science and Engineering: C, 102*, 844–862. doi: 10.1016/j.msec.2019.04.064.

[39] Lim, H.-K., Choi, Y.-J., Choi, W.-C., Song, I.-S., & Lee, U.-L. (2022, May). Reconstruction of maxillofacial bone defects using patient-specific long-lasting titanium implants. *Scientific Reports, 12*(1), 7538. doi: 10.1038/s41598-022-11200-0.

[40] Riviș, M., Roi, C., Roi, A., Nica, D., Văleanu, A., & Rusu, L.-C. (2020, May). The implications of titanium alloys applied in maxillofacial osteosynthesis. *Applied Sciences, 10*(9), 3203. doi: 10.3390/app10093203.

[41] Barber, C. C., Burnham, M., Ojameruaye, O., & McKee, M. D. (2021, Sep). A systematic review of the use of titanium versus stainless steel implants for fracture fixation. *OTA International, 4*(3), e138. doi: 10.1097/OI9.0000000000000138.

[42] Azeem, N., & Duarte, A. J. (2024). Surgical Instruments. In *Neuromodulation Techniques for the Spine* (pp. 38–45). Elsevier. doi: 10.1016/B978-0-323-87584-4.00014-0

[43] Fu, J., Su, Y., Qin, Y.-X., Zheng, Y., Wang, Y., & Zhu, D. (2020, Feb). Evolution of metallic cardiovascular stent materials: A comparative study among stainless steel, magnesium and zinc. *Biomaterials*, *230*, 119641. doi: 10.1016/j.biomaterials.2019.119641.

[44] Michel, R., Nolte, M., Reich, M., & Löer3, F. (1991, Feb). Systemic effects of implanted prostheses made of cobalt-chromium alloys. *Archives of Orthopaedic and Trauma Surgery*, *110*(2), 61–74. doi: 10.1007/BF00393876.

[45] Majerič, D., Lazić, V., Majerič, P., Marković, A., & Rudolf, R. (2021, Jul). Investigation of CoCr dental alloy: Example from a casting workflow standpoint. *Crystals (Basel)*, *11*(8), 849. doi: 10.3390/cryst11080849.

[46] Youssef, A. A., *et al.* (2010). Cobalt chromium coronary stents and drug-eluting stents in real practice. *International Heart Journal*, *51*(4), 231–237. doi: 10.1536/ihj.51.231.

[47] He, M., Chen, L., Yin, M., Xu, S., & Liang, Z. (2023, Mar). Review on magnesium and magnesium-based alloys as biomaterials for bone immobilization. *Journal of Materials Research and Technology*, *23*, 4396–4419. doi: 10.1016/j.jmrt.2023.02.037.

[48] Oncel, L., Bugdayci, M., & Turan, A. (2025). *NiTi-Based Shape Memory Alloys* (pp. 191–213). doi: 10.1007/978-3-031-57499-3_7.

[49] Castleman, L. S., Motzkin, S. M., Alicandri, F. P., Bonawit, V. L., & Johnson, A. A. (1976, Sep). Biocompatibility of nitinol alloy as an implant material. *Journal of Biomedical Materials Research*, *10*(5), 695–731. doi: 10.1002/jbm.820100505.

[50] Zhang, W., Li, Z., Chai, M., Dong, P., & Wang, Y. (2025, Feb). Selective laser melting of NiTi cardiovascular stents: Processibility, surface quality and mechanical properties. *Optics & Laser Technology*, *181*, 111842. doi: 10.1016/j.optlastec.2024.111842.

[51] Nevárez-Rascón, A., et al. (2025, Jan). Nanomechanical and bending properties of nickel-titanium orthodontics archwires by depth-sensing indentation. *American Journal of Orthodontics and Dentofacial Orthopedics*, *167*(1), 29–38. doi: 10.1016/j.ajodo.2024.08.009.

[52] Akif Dündar, M., Gunoz, A., Kepir, Y., Kara, M., Avcı, A., & Arbağ, H. (2025, May). Comparative biomechanical analysis of titanium fixation plates and NiTi shape memory staples for mandibular fracture treatment using finite element method. *Biomed Signal Process Control*, *103*, 107501. doi: 10.1016/j.bspc.2025.107501.

[53] Pandey, A., Awasthi, A., & Saxena, K. K. (2020, Apr). Metallic implants with properties and latest production techniques: A review. *Advances in Materials and Processing Technologies*, *6*(2), 405–440. doi: 10.1080/2374068X.2020.1731236.

[54] Davoodi, E., *et al.* (2022, Sep). Additively manufactured metallic biomaterials. *Bioactive Materials*, *15*, 214–249. doi: 10.1016/j.bioactmat.2021.12.027.

[55] Zia, A. W., Anestopoulos, I., Panayiotidis, M. I., & Birkett, M. (2023, Jun). Soft diamond-like carbon coatings with superior biocompatibility for medical applications. *Ceramics International*, *49*(11), 17203–17211. doi: 10.1016/j.ceramint.2023.02.085.

[56] Kravanja, K. A., & Finšgar, M. (2022, May). A review of techniques for the application of bioactive coatings on metal-based implants to achieve controlled release of active ingredients. *Mater Design*, *217*, 110653. doi: 10.1016/j.matdes.2022.110653.

[57] Tran, P. A., Sarker, A., Tran, N., Jeffery, C., Rifai, A., & Fox, K. (2020). Coatings on Metallic Implants for Biomedical Applications. In *Metallic Biomaterials Processing and Medical Device Manufacturing* (pp. 359–385). Elsevier. doi: 10.1016/B978-0-08-102965-7.00011-4

[58] Niinomi, M. (2019). Fatigue Failure of Metallic Biomaterials. In *Metals for Biomedical Devices* (pp. 153–188). Elsevier. doi: 10.1016/B978-0-08-102666-3.00005-5

[59] Ali, S., et al. (2020, Oct). Biocompatibility and corrosion resistance of metallic biomaterials. *Corrosion Reviews*, *38*(5), 381–402. doi: 10.1515/corrrev-2020-0001.

[60] Mariani, E., Lisignoli, G., Borzì, R. M., & Pulsatelli, L. (2019, Feb). Biomaterials: Foreign bodies or tuners for the immune response? *International Journal of Molecular Sciences*, *20*(3), 636. doi: 10.3390/ijms20030636.

[61] Wang, L., Liu, C., & Xie, L. (2021, Nov). Editorial: metallic biomaterials for medical applications. *Frontiers in Materials*, *8*. doi: 10.3389/fmats.2021.795642.

[62] Shankhwar, N., & Srinivasan, A. (2016, May). Evaluation of sol–gel based magnetic 45S5 bioglass and bioglass–ceramics containing iron oxide. *Materials Science and Engineering: C*, *62*, 190–196. doi: 10.1016/j.msec.2016.01.054.

[63] Soro, N., Brodie, E. G., Abdal-hay, A., Alali, A. Q., Kent, D., & Dargusch, M. S. (2022, Jun). Additive manufacturing of biomimetic Titanium-Tantalum lattices for biomedical implant applications. *Mater of Design*, *218*, 110688. doi: 10.1016/j.matdes.2022.110688.

[64] Yaqoob, A. A., *et al.* (2020, May). Recent advances in metal decorated nanomaterials and their various biological applications: a review. *Frontiers in Chemistry*, *8*. doi: 10.3389/fchem.2020.00341.

[65] Gokcekuyu, Y., Ekinci, F., Guzel, M. S., Acici, K., Aydin, S., & Asuroglu, T. (2024, Jul). Artificial intelligence in biomaterials: a comprehensive review. *Applied Sciences*, *14*(15), 6590. doi: 10.3390/app14156590.

[66] Wu, C., Xu, Y., Fang, J., & Li, Q. (2024, May). Machine learning in biomaterials, biomechanics/mechanobiology, and biofabrication: State of the art and perspective. *Archives of Computational Methods in Engineering*. doi: 10.1007/s11831-024-10100-y.

[67] Schuh, J. C. L., & Funk, K. A. (2019, Apr). Compilation of international standards and regulatory guidance documents for evaluation of biomaterials, medical devices, and 3-D printed and regenerative medicine products. *Toxicol Pathol*, *47*(3), 344–357. doi: 10.1177/0192623318804121.

[68] Ladani, L., & Palmieri, M. (2024, Sep). Review of the use of metals in biomedical applications: Biocompatibility, additive manufacturing technologies, and standards and regulations. *Metals (Basel)*, *14*(9), 1039. doi: 10.3390/met14091039.

[69] Davis, R., *et al.* (2022, May). A comprehensive review on metallic implant biomaterials and their subtractive manufacturing. *The International Journal of Advanced Manufacturing Technology*, *120*(3–4), 1473–1530. doi: 10.1007/s00170-022-08770-8.

[70] Banerjee, S., Campbell, B., Rising, J., Coukell, A., & Sedrakyan, A. (2019, Jul). Long-term active surveillance of implantable medical devices: An analysis of factors determining whether current registries are adequate to expose safety and efficacy problems. *BMJ Surgery, Interventions, & Health Technologies*, *1*(1), e000011. doi: 10.1136/bmjsit-2019-000011.

[71] Claudia Spataru, M., Simona Baltatu, M., Victor Sandu, A., & Vizureanu, P. (2024). General trends on biomaterials applications: Advantages and limitations. *IntechOpen*. doi: 10.5772/intechopen.114838.

Index

316L 42–46, 48, 50–51, 56–57, 212, 232, 241, 286, 291
316L stainless steel 22
3D printing 205, 208, 213, 215, 217, 221, 297, 303–304
3D printing technology 262, 266
3D scaffolds 266

abrasion 287
abrasive polishing 139
adaptability 73
additive manufacturing 63, 71, 134, 143, 205–208, 210–213, 215–216, 218–220, 222, 297, 303
adhesion 139
adolescence 187
adsorbate 104
adsorption 103, 106, 108, 122, 124
aerospace 4
$AgNO_3$ 168
aluminum 83, 88
anatase 101, 103–108, 110, 113, 116, 118–120, 125
anatomy 205, 207–208, 213–214, 218
anodization 87
anodization 185, 195
antibacterial 229, 233, 240–243
antibiotic 187, 189, 191, 197
anti-inflammatory 65, 73, 324–325, 327
antimicrobial 149, 153
arthroplasty 56
aseptic 274
aseptic loosening 230
Au NPs 110
augmentations 23

bacterial 150, 155–156, 162, 164, 166
binder jetting 209, 214, 217, 219–220
biocompatibility 5, 41–42, 44, 50–51, 53–54, 56, 59, 64, 251–253, 255–257, 259–260, 266–267, 273–275, 277–278, 281–282, 285–289, 291, 294–295, 301, 309–310, 312–321, 323, 325–330, 332–334
biocompatible 3, 5, 7, 9–12
biocompatible materials 143
biodegradability 274
biodegradable 309–310, 315, 319, 325–326, 332–333
bioengineering, 312
biofilm 10, 189, 191
biofunctional 81, 84, 94
bioglass 212
bio-HEAs 13
biohybrid 333
biological 51–53, 55, 150, 155–156, 160–161, 163, 168–173, 310, 312, 315, 322–323, 325–327, 329, 334
biological fluids 286
biological functions 134, 136
biological responses 252–253, 255, 267
biomaterials 21–23, 27, 33, 229–231, 233–234, 236–237, 239–240, 243–244
biomedical implants 133–134, 136, 139, 141, 143, 146
biomimetic 69, 186, 197
bioresorbable 218
blood 49
bolts 46
bone fracture fixation 22
bone healing 46–48, 51, 55
bone regeneration 62, 64–66
bone screws 185
bone tissue 69
brookite 102, 104–106, 108
buccal mucosa 191

calcium 194, 196
calcium fluoride 94
calcium phosphate 294
cancer 105
carbide 43, 45
cardiac 207
cardiac valve 232
cardiovascular 22, 28, 31, 61–62, 66, 73, 83, 188, 215, 286, 294, 302, 309–310, 312–315, 321, 333
casting 141
cell adherence 276
cell adhesion 134, 214, 316, 323, 328
cellular interactions 10
ceramic 294–297
chemical stability 101, 103, 121, 126
chemical surface 87
chemical vapor deposition 87, 112
chitosan 153, 156
chitosan coating 153

https://doi.org/10.1515/9783111571423-016

chromium 45, 188, 286–287, 290–291, 295, 297, 304
chronoamperometry 251–252, 255–256, 265
clarity 104, 116
CNC 151
coatings 63, 65, 68–69, 71
cobalt 3, 286, 290–291, 295, 297, 304
cobalt alloys 231, 233
cobalt-based alloys 22
cobalt-chromium 44, 207, 210–212, 215, 218, 318, 321
cobalt-chromium alloys 313
Co-Cr 309–310, 313–319, 321
CoCrMo 22, 25
cold spray technology 299
collagen 156
computational 303
computational modeling 312, 329, 331
computed tomography 209
controlled drug-release systems 195
corrosion 41–46, 48–59, 239, 251–257, 259–267, 273–275, 277, 282, 285–287, 289, 291, 293, 295, 298, 300–301, 303–304
corrosion rate 64–66, 68, 73
corrosion resistance 315–316
Cp-Ti 82–83
cranial plates 215
craniofacial 22, 72
cranium 313
crevice 45, 48–49, 51, 53, 57–58, 286
cryogenic 23
crystallinity 112, 116
CT 154
cyclic voltammetry 261, 263
cytotoxic 275, 278–280
cytotoxicity 273, 275, 277–279, 282

debinding 298
debris 79–80, 84, 95
decisive 44
deformity 42–43
degradation 61–66, 68–69, 72–73, 102, 120, 123
dental 108, 124
dental implants 2
dental structures 154
dentistry 22, 290
devastating 49
dextran 193
diagnostic 156, 161–162, 168
diamond-like carbon 92
dip coating 114
direct current 110
dissatisfaction 331
diverse manufacturing 22
DLC 324, 326
DMLS 211, 215
dosimetric 252, 257
dosimetry 252–254, 256
durability 309–312, 315–316, 318–319, 321, 324–325, 329

EBM 207, 209–210, 213
ECM 186, 195
effective design 311
effectiveness 251–254, 263–264, 267
elastic modulus 81, 83, 231
elasticity 5
electrochemical 251–252, 255–257, 259, 261–266
electrochemical deposition 69, 71
electrodeposition 113, 115
electroforming 299
electron beam melting 322, 333
electrons 106, 110, 117, 119, 122
electropolished 53
electropolishing 53, 59
electrospinning 159–160
embryonic 187, 191
endochondral 194
endovascular 23
environmental 102, 104–105, 113–114, 121–123
etching 194
extendibility 155
extracellular matrix 186, 199, 286, 294

fatigue 44, 47, 49, 57–58
fatigue testing 300
FDA 169
FDM 207, 210, 212, 220
Fe_3O_4 163, 165
femur 47
Fermi level 106, 108
fibrinogen 108
finishing 143
fishery products 125
fixation 41–42, 44, 47, 50, 54, 56–58
flattening 145
flexibility 80, 309, 311, 320–322
flowability 72

forging 134, 141, 143, 145
formazan 280
Formlabs 206
fractures 274
fragile 332
friction stir processing 236
fused deposition modeling 220

genotoxicity 170, 325
gold nanoparticles 154
gram-negative 164, 166
gram-positive 164, 166
grit particles 94

HAp 164, 166
hardness testing 300
healthcare 205
high tensile 57
high-entropy 234
high-entropy alloys 89
high-performance alloys 299
high-voltage 159
hip and knee 317
hips and knees 232
histocompatibility 24
hollandite 104–105
human 285, 287, 290–291, 295, 300
hydrogels 286, 293
hydrophilicity 161, 194
hydrophobicity 139
hydroxyapatite 6, 10, 12–13, 69, 153, 185, 193–194, 197, 199
hydroxylation 106
hypersensitivity 318, 325–326, 332

iliac 188
immobilizations 44
immune 309, 315–316, 323, 325, 328, 331
immunomodulation 192
implant 286–288, 290–297, 303–304
implant fracture 230
implant loosening 63
implant production 133
implants 41–49, 51–59
in vitro 273–274, 277, 282
infection risks 323, 327–328
inflammation 41–42, 51–52, 54–55, 155, 170–172
inflammatory 51–52, 55, 58
instability 230
insulators 252
integration 4
integrative 251, 253–254, 267
intelligent 218
interdisciplinary 302
interstitial 49
intramedullary 42, 47–49
intricate 71–72
invasive 196
invent 160
ion implantation technique 87
ion release 51–52, 59
ionizes 110
iterative design 73
IVDR 169

knee joint 80

large-scale 126, 150, 160
laser beam processing 235, 237
laser cladding 14, 211, 236
laser engraving 240
laser power 237–238
lasers 25
layer 205, 208, 210–211, 214, 218
layer-by-layer 219
ligamentous 286
lithography 298
live tissues 3
liver 294
load-bearing 8, 42, 46–49, 51, 56–57
localized therapy 154
longevity 302
loosening 51–52, 55, 58

Maggiolo 196
magnesium 11, 61, 213, 274–275, 282, 285–286, 290, 293, 301, 303–304, 309–310, 313, 319, 321, 325, 333
magnesium alloys 233
magnesium hydroxide 64
magnesium (Mg) 62, 65
magnetron 110–111
malnourished 289
manufacturing 195, 198
MAPLE 163
maxillofacial 209, 213, 215, 317
MDPA 168

mechanical properties 311, 313, 316, 319–320, 322, 325, 327
mechanical strength 22, 28, 30, 65–66, 73, 285, 287–288, 291, 295–297, 300, 304
medical 205–206, 208, 213, 216–219, 285–288, 290–291, 293–296, 298–300, 303–304
medical aids 206, 208
medical devices 8, 149–152, 155, 158–161, 164, 166, 168–174
medical implants 311, 313, 317–319, 325, 329, 333
medical technology 251, 253
metal implants 22, 27–28
metallic 285–288, 290, 295–304
metallic biomaterials 2, 309–310, 312, 316
metallic implants 42–43, 48–49, 251–256, 259–260, 266–267, 294
metallic stents 62
methanol 103
meticulous 254
Mg alloys 62–73
Mg-based implants 65, 71–73
Mg-Nd-Ag-Zr 62
microbial 288
microblasting 252, 254–255
microplate 280
microstructural 11, 15, 257, 260, 262
microstructure 62, 64, 66–67, 70, 72–73, 288, 300
microwave 111
miniscrew 188
molecular 288, 293
molecules 86, 94–95
molybdenum 5, 7–8, 44–45, 53
MRI 72, 154
MSS 188, 191
MTT 279–280
mutagenicity 277

nanocoating 149
nanocoatings 71, 149–150, 157, 161, 163, 168
nanofibers 160
nanometer 162
nanostructured 303
nanotechnology 198
nanotopographical 325
nanotoxicity 252, 255–256
natural bone 317, 319, 321–322, 324, 327–328
neuromuscular 23
neuroprosthetics 333
neurovascular 22
nickel 8, 44–45, 52, 58, 188
nickel-titanium 313, 319, 321
niosomes 153, 166
nitinol 258–259, 285–286, 290, 293, 302–304, 313, 315, 319–321
nitrogen ion 70
nonreactivity 286
nontoxic 4

opacifying 125
optical microscopy 300
optimization 158–159
organometallic 158
organs 294
orthodontic 215
orthodontic pressures 187
orthopedic 2, 22, 41–50, 52–59, 62–66, 68–69, 72–73, 185, 188, 197–198, 230–232, 237–238, 290, 293, 301, 304
orthopedics 22, 44, 56, 205, 213, 216
orthophosphate 94
osseointegration 3, 153, 161, 230, 239, 288, 290, 301, 311, 315, 317, 321–323, 326–328, 331, 333
osteoarthritis 289
osteoblast 328
osteoblasts 187, 199
osteoconduction 2
osteoconductive 66, 69
osteogenic 194
osteoinduction 2
osteointegrate 290
osteoporosis 289
otorhinology 22
oxidative stress 171
oxide 286, 291, 294
oxide layer 5, 43, 45, 49, 53–54, 83–85, 93
oxidizing 53

passivates 286
passivation 52–53, 59
pathological 23
patient 205–208, 210, 213–219, 221
pediatric orthopedic 64
PEEK 298
PEG 293
peptides 194
permeability 191
pervasiveness 195
pH 150, 154–155, 157

pharmacokinetics 171
phosphate 63–65, 68–69
phosphate-buffered saline 278
photocatalysis 102–103, 106, 112–113, 116, 123
photocatalyst 103, 107, 121
photocatalytic 102–103, 110–115, 117–123
photopolymerization 220
photothermal therapy 105
physical vapor deposition 6, 16, 235
physical vapor deposition 86
physiological 62–64, 66, 68, 73, 252, 255, 258, 260, 263, 265, 274–275, 282, 317, 319, 324, 329
physiology 186–187
pitting 49
plasma 185, 189, 191, 194, 197–198
plasma spraying 185, 194, 197
plasmonic 110, 122
plastic deformation 145
PLGA 193
PMMA 294–295
polyglycolic acid 191
polylactic acid 191
polymorph 105
polymorphic 105
polysaccharide 189
porosity 106, 112, 155, 157, 159
postnatal 191
postoperative 208
postprocessing 143
potentiodynamic 251–252, 255–256, 263, 300
precursors 157–159
prevent 286, 291, 294
PRISMA 254
profilometry 142
proliferation 275, 278–281
prosthesis 80
protective films 255
Prussian blue 103
PTFE 286
pulsed laser deposition 112
PVD coatings 92, 95

quick prototyping 218

radiotherapy 252, 256
rare earth elements 62
redox-active 108
refractory 141
regeneration 23, 28, 30, 33, 186, 189, 192, 194–195, 197
regenerative medicine 2
regulatory 220, 312, 329
Regulatory 330
reinnervation 188
residual stress 297
rivets 46
rutile 101, 103–108, 116, 119–120, 125

sandblasting 235, 243
scalability, 311
scapular 188
scar tissues 191
second-generation alloys 5
selective laser 322
selective laser melting 333
silk fibroin 153
silver nanoparticles 152
sintering 139
sinusoidal 193
size distribution 106, 108, 125
SLA 210–212
SLM 207, 209–210, 213
SLS 163, 165
sol-gel coating 87
solvent 109–110
spin coater 109–110
spin coating 109–110, 114
spinal 49, 57–58
sputtering 101, 110–111
stability 5
stainless steel 22, 24, 35, 41–59, 312, 317, 321
stent 10, 24
stents 215, 310, 312–313, 315, 318–321, 333
sterilization 162, 168
sterilize 209
stimulate 194
STM 103
strategies 221
strength 207, 209–212, 218
strength-to-weight 312, 317, 321
stress shielding 44, 47–48, 58–59, 291, 317, 319, 321–322, 327–328
stringent 253
structural integrity 251
structural stability 187
substrate 109–110, 112, 114, 118–119, 123
surface area 322–323

surface breaks 286
surface coating 134
surface coatings 316, 323
surface grinding 145
surface irregularities 136
surface modification 63, 68, 229, 234, 236
surface modifications 85, 90, 94, 96, 251, 253, 264–265, 267
surface morphologies 139
surface roughness 133–134, 136, 139–143, 145–146
surface texture 133, 143, 145–146
surface topography 133–134, 139, 141, 143, 145–146 *see*
surface treatments 10, 61, 64–65, 69, 71, 73
surgical 287, 289, 300, 304, 311–312, 314–315, 319–320, 322, 324, 327, 329, 333
surmount 243
sustainable 102, 104
Syncolone 286
synovial joints 230

Teflon 286, 294
tensile strength 5
tensile testing 300
teratogenic 288
tetrafluoroethylene 287
texturing, 186, 194
therapeutic 3, 188–189
therapeutics 174
thermal 156
thermal oxidization 236
thermal spraying 236
thrombosis 62
Ti alloys 81–83, 87–88, 90–91
Ti-6Al-4V 5, 291, 301
Ti6Al4V 82–83, 91–93, 252, 255–256
TiO 101, 103–106, 108, 110–126, 238, 317, 324, 326
TiO_2 101, 103–106, 108, 110–126, 317, 324, 326
titanium 3, 62–64, 70, 134, 139–140, 185, 194, 196–197, 210–212, 216, 229–232, 235–236, 238–239, 241, 285–287, 290–291, 293–295, 297, 299, 301, 304
titanium alloys 5, 22, 30
titanium and titanium 317
titanium and titanium alloys 312
titanium dioxide 5, 101–102, 104–105, 112–113, 116–117, 120–121, 123, 125–126, 153
titanium implants 327
TNTs 124
tomography 207
topography 146
toxic metal 188, 251, 253, 267
traditional 207
transition metal nitrides 90
translational 281
transplantations 219
trapped 108, 121
trauma 80
traumatic 191
tribological 133
trypsinization 278

UHMWPE 56

vascularized 188
Volta potential 252, 255–256
voltammetry 251–252, 255–256, 263–264, 266

wear 49, 51, 53–56, 58, 69–70, 106, 229–232, 236–239, 241, 243
wear and fatigue 324
wear debris 252
wear resistance 1, 5–7, 15, 315, 318
wettability 229, 232, 240–243
wound healing 191
wounds 189

yield strength 46, 57

zinc alloys 234
zinc oxide 153
zirconium 83, 233

α-phase 6
α-stabilizers 5
α + β-phase alloys 6

β-phase alloys 6
β-stabilizers 5
β-TCP 298

De Gruyter series in advanced mechanical engineering

Already published in the series

Volume 14
Ranjith Rajamanickam, Amit Sharma, Dhivya Ranjith, J. Paulo Davim (Eds.)
Machine Learning for Medical Applications:
Volume I: Computational Drug Discovery, Bioimaging, Smart Biomaterials, 2025
ISBN 978-3-11-150318-9, e-ISBN (PDF) 978-3-11-150320-2,
e-ISBN (EPUB) 978-3-11-150374-5
Volume II: Computer Vision, Image Processing, Disease Detection, 2025
ISBN 978-3-11-914782-8, e-ISBN (PDF) 978-3-11-220519-8,
e-ISBN (EPUB) 978-3-11-220529-7

Volume 13
Raj Kumar Arya, George Verros, J. Paulo Davim (Eds.)
Compressors and Blowers: Maintenance, Practical Guidance, Energy-Efficient Technologies, 2025
ISBN 978-3-11-136335-6, e-ISBN (PDF) 978-3-11-136395-0,
e-ISBN (EPUB) 978-3-11-136417-9

Volume 12
Nishant Kumar Singh, Rajesh Kumar Verma, Virendra Kumar,
J. Paulo Davim (Eds.)
Machining and Tribology of Advanced Materials: From Coatings, Lubrications,
Surface Treatments to Modeling and Simulation, 2025
ISBN 978-3-11-137642-4, e-ISBN (PDF) 978-3-11-137729-2,
e-ISBN (EPUB) 978-3-11-137772-8

Volume 11
Kanak Kalita, Ranjan Kumar Ghadai, J. Paulo Davim (Eds.)
Microfabrication and Nanofabrication: Precision Manufacturing, 2024
ISBN 978-3-11-133543-8, e-ISBN (PDF) 978-3-11-133553-7,
e-ISBN (EPUB) 978-3-11-133559-9

Volume 10
Kanak Singh, Ranvijay Kumar, Vinay Kumar and J. Paulo Davim (Eds.)
3D Printing: Polymer, Metal and Gel Based Additive Manufacturing, 2024
ISBN 978-3-11-108061-1, e-ISBN (PDF) 978-3-11-108163-2
e-ISBN (EPUB) 978-3-11-108206-6

Volume 9
Raj Kumar Arya, Anurag Kumar Tiwari, George Verros, Prashant Malik and J. Paulo Davim (Eds.)
Sustainable Hydrogen Energy: Production, Storage & Transportation, 2024
ISBN 978-3-11-124641-3, e-ISBN (PDF) 978-3-11-124647-5
e-ISBN (EPUB) 978-3-11-124728-1

www.degruyter.com

Volume 8
Tatjana Sibalija and J. Paulo Davim (Eds.)
Soft Computing in Smart Manufacturing: Solutions toward Industry 5.0, 2022
ISBN 978-3-11-069317-1, e-ISBN (PDF) 978-3-11-069322-5
e-ISBN (EPUB) 978-3-11-069325-6

Volume 7
J. Paulo Davim (Ed.)
Nonconventional Machining, 2023
ISBN 978-3-11-058410-3, e-ISBN (PDF) 978-3-11-058447-9
e-ISBN (EPUB) 978-3-11-058415-8

Volume 6
J. Paulo Davim (Ed.)
Manufacturing in Digital Industries, 2020
ISBN 978-3-11-057507-1, e-ISBN (PDF) 978-3-11-057542-2,
e-ISBN (EPUB) 978-3-11-057526-2

Volume 5
Kaushik Kumar, Divya Zindani, J. Paulo Davim
Rapid Prototyping, Rapid Tooling and Reverse Engineering, 2020
ISBN 978-3-11-066324-2, e-ISBN (PDF) 978-3-11-066490-4,
e-ISBN (EPUB) 978-3-11-066345-7

Volume 4
J. Paulo Davim (Ed.)
Additive and Subtractive Manufacturing, 2019
ISBN 978-3-11-054816-7, e-ISBN (PDF) 978-3-11-054977-5,
e-ISBN (EPUB) 978-3-11-054833-4

Volume 3
J. Paulo Davim (Ed.)
Drilling Technology, 2017
ISBN 978-3-11-047863-1, e-ISBN 978-3-11-048120-4,
e-ISBN (EPUB) 978-3-11-047871-6

Volume 2
J. Paulo Davim (Ed.)
Progress in Green Tribology, 2017
ISBN 978-3-11-037272-4, e-ISBN 978-3-11-036705-8,
e-ISBN (EPUB) 978-3-11-039252-4

Volume 1
J. Paulo Davim (Ed.)
Metal Cutting Technology, 2016
ISBN 978-3-11-044942-6, e-ISBN 978-3-11-045174-0,
e-ISBN (EPUB) 978-3-11-044947-1, Set-ISBN 978-3-11-045175-7